AF572564

Diagnostic Biosensor Polymers

ACS SYMPOSIUM SERIES **556**

Diagnostic Biosensor Polymers

Arthur M. Usmani, EDITOR
Firestone Building Products Company

Naim Akmal, EDITOR
Teledyne Brown Engineering

Developed from a symposium sponsored
by the Division of Industrial and Engineering Chemistry, Inc.,
at the 205th National Meeting
of the American Chemical Society,
Denver, Colorado,
March 28–April 2, 1993

American Chemical Society, Washington, DC 1994

Library of Congress Cataloging-in-Publication Data

Diagnostic biosensor polymers.

Developed from a symposium sponsored by the Division of Industrial and Engineering Chemistry, at the 205th National Meeting of the American Chemical Society, Denver, Colorado, March 28–April 2, 1993.

Edited by Arthur M. Usmani, Naim Akmal.

p. cm.—(ACS symposium series, ISSN 0097–6156; 556)

Includes bibliographical references and indexes.

ISBN 0–8412–2908–2

1. Biosensors—Congresses. 2. Polymers in medicine—Congresses. 3. Polymers—Diagnostic uses—Congresses. 4. Piezoelectric polymer biosensors—Congresses.

I. Usmani, Arthur M., 1940– . II. Akmal, Naim, 1962– . III. American Chemical Society. Division of Industrial and Engineering Chemistry. IV. American Chemical Society. Meeting (205th: 1993: Denver, Colo.) V. Series.

[DNLM: 1. Polymers—diagnostic use—congresses. 2. Electrodes—congresses. 3. Biosensors—congresses. QT 34 1994]

R857.B54D53 1994
616.07'56—dc20
DNLM/DLC
for Library of Congress 94–895
CIP

The paper used in this publication meets the minimum requirements of American National Standard for Information Sciences—Permanence of Paper for Printed Library Materials, ANSI Z39.48–1984. ∞

PRINTED IN THE UNITED STATES OF AMERICA

Foreword

THE ACS SYMPOSIUM SERIES was first published in 1974 to provide a mechanism for publishing symposia quickly in book form. The purpose of this series is to publish comprehensive books developed from symposia, which are usually "snapshots in time" of the current research being done on a topic, plus some review material on the topic. For this reason, it is necessary that the papers be published as quickly as possible.

Before a symposium-based book is put under contract, the proposed table of contents is reviewed for appropriateness to the topic and for comprehensiveness of the collection. Some papers are excluded at this point, and others are added to round out the scope of the volume. In addition, a draft of each paper is peer-reviewed prior to final acceptance or rejection. This anonymous review process is supervised by the organizer(s) of the symposium, who become the editor(s) of the book. The authors then revise their papers according to the recommendations of both the reviewers and the editors, prepare camera-ready copy, and submit the final papers to the editors, who check that all necessary revisions have been made.

As a rule, only original research papers and original review papers are included in the volumes. Verbatim reproductions of previously published papers are not accepted.

M. Joan Comstock
Series Editor

Contents

Preface

THE TECHNOLOGY that is responsible for the way we live today is a result of polymer applications. Polymers are used in electronics, in medicine, in space—in fact, everywhere. Advances in polymer science and technology have resulted in many successful implants, therapeutic devices, and diagnostic assays. Today's medicine, including diagnostic medicine, is placing demands on technology for new polymeric materials and new application techniques. Polymers are thus finding ever- increasing use in diagnostics, immunodiagnostics, and biosensors.

Polymer-enhanced biosensors and dry chemistries are very important now. They will certainly take a quantum jump by the turn of the century. Polymers are used to enhance speed, sensitivity, and versatility of biosensors and dry chemistries to measure glucose, cholesterol, urea, dissolved oxygen, and other important analytes.

This volume has been organized into four sections, namely, biosensors, biosensor polymers and membranes, biocompatibility and biomimetics, and immobilization and stabilization methods. This book should be useful to biochemists, chemists, and chemical engineers doing or planning to do research and development work in diagnostics, specifically biosensors and dry chemistries. Polymer scientists should also find this volume worthwhile to read.

We wish to express our appreciation to Boehringer Mannheim, Teledyne, and Bridgestone/Firestone for support given to us.

ARTHUR M. USMANI
Firestone Building Products Company
525 Congressional Boulevard
Carmel, IN 46032–5607

NAIM AKMAL
Teledyne Brown Engineering
Analytical Instruments
16830 Chestnut Street
City of Industry, CA 91749–1580

December 21, 1993

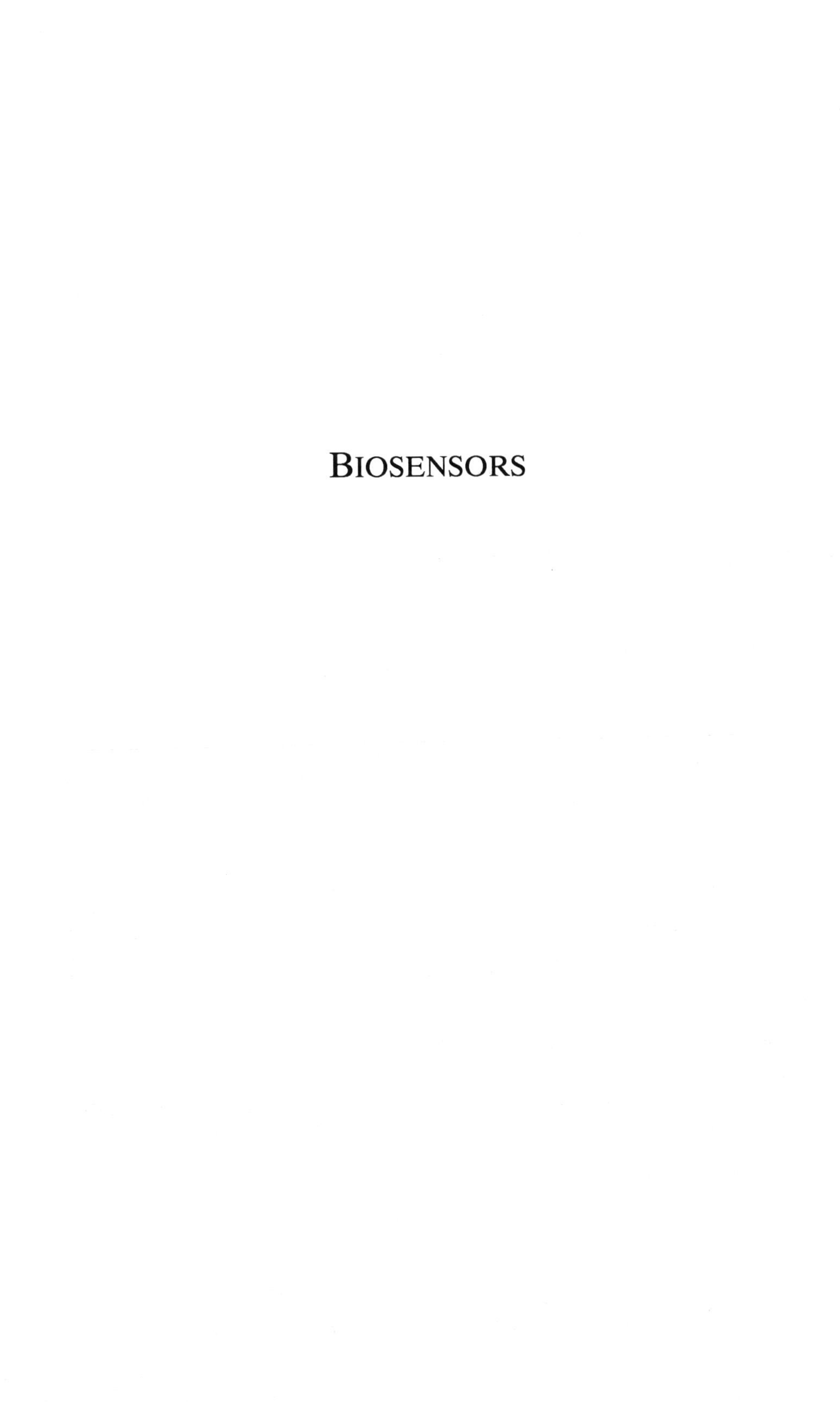

BIOSENSORS

Chapter 1

Diagnostic Polymers and Coatings

Chemistry, Technology, and Applications

Arthur M. Usmani

Firestone Building Products Company, 525 Congressional Boulevard, Carmel, IN 46032–5607

Advances in materials science and technology, notably polymers, have resulted in many successful implants, therapeutic devices and diagnostic assays in the medical field. Diagnostic polymers and coatings play a very important functional role in dry chemistries, immunodiagnostics, and biosensors. Enzymes, also polymeric in nature are extensively used to measure or amplify signals of many, but specific, metabolites. Furthermore, polymers, plastics and other polymeric materials are invariably utilized in such devices for functions other than chemical, biochemical, and electrochemical reactions. We will describe chemistry, technology, design, applications and future prospects of diagnostic polymers and coatings in this report.

Materials, notably polymers (fibers, films, rubber products, molded plastics, coatings, polymeric carbon and composites), have resulted in many successful implants, therapeutic devices and diagnostic assays in the medical field (*1*). In contrast to other less important biomaterials, e.g., metals and ceramics, polymers can be manufactured in a wide range of compositions. Their properties can be regulated by suitable composition and process modifications. Furthermore, they can be shaped into simple as well as complex configurations. Polymer–enhanced biosensors and dry chemistries are very important now – they will certainly take a quantum jump by the turn of the century. Polymers are being used to enhance speed, sensitivity and versatility of both electrical and optical biosensors as well as dry chemistries in diagnostics to measure glucose, urea, dissolved oxygen, and other vital analytes. The polymer–enhanced biosensors will find ever increasing applications in diagnostic medicine, automated drug delivery systems, environmental and industrial process monitoring, and in basic neurobiology research.

Today's medicine including diagnostic medicine is placing demands on technology for new materials and new application techniques. Polymers are thus finding every increasing usage in diagnostics, immunodiagnostics and bisensors (*2*). The volume of such biotechnological polymers used in diagnostic medicine is extremely low, yet they are highly value–adding in nature.

In this paper we will describe diagnostic polymers and coatings. Background on dry chemistry, and medical aspects will also be addressed for completeness.

0097–6156/94/0556–0002$08.00/0

Dry Chemistry History

Dry or solid–phase chemistry traces its origins back to the ancient Greeks. Approximately 2000 years ago copper sulfate, or verdigris, was a vital ingredient in the tanning and preserving of leather; so valuable, in fact, that dishonest tradesmen were adulterating copper sulfate with iron salts. Pliny described the procedure for soaking reeds of papyrus in a plant gall infusion or a solution of gallic acid that would turn black in the presence of iron. This is the first recorded use of a dry–chemistry system. In 1830, a filter paper impregnated with silver carbonate was used to detect uric acid.

Self–testing or home monitoring is likewise an old concept; we all measure our own weight and body temperature. Even before the discovery of insulin, Elliot Joslin, some 75 years ago, advocated that diabetics should monitor their glucose levels by frequent urine testing with Benedict's qualitative test (*3*). In 1922, insulin became available and the self–testing of urine became a necessity. Around the mid–40s, Compton and Treneer developed a dry tablet containing sodium hydroxide, citric acid, sodium carbonate, and cupric sulfate (*4*). When this tablet was added to a small sample of urine in a test tube, the solution boiled, resulting in reduction of blue cupric sulfate to yellow or orange color if glucose were present. Around 1956, glucose urine strips, impregnated and based on enzymatic reactions using glucose oxidase, peroxidase and indicator, were introduced by Miles and Eli Lilly (*5*). Dry chemistry blood glucose test strips coated with a semipermeable membrane to which blood could be applied and wiped off were introduced in 1964 by Miles. Low–cost, lightweight plastic–housed reflectance meters with data management capabilities and memories are now available as well.

Paper–matrix dry reagents have a coarse texture, high pore volume, and uneven large–pore size, resulting in nonuniform color development in the reacted strips. In the early 1970s, Boehringer Mannheim in Germany developed a coating–film– type dry reagent by applying an enzymatic coating onto a plastic support. This gave a smooth, fine texture and therefore uniform color development (*6*) and is still the workhorse of the diagnostics industry. Nonwipe dry chemistries in which the device handles the excess blood by absorption or capillary action have recently appeared in the marketplace.

Discrete multilayered coatings, developed by the photographic industry, were adapted in the late 1970s by Eastman Kodak to coat dry–reagent chemistry formats for clinical testing (*7,8*). Each zone of a multilayered coating provides a unique environment for sequential chemical and physical reactions. The basic multilayered coating device consists of a spreading layer, separation membranes, reagent, and reflective zone, coated onto the base support. The spreading layer wicks the sample and applies it uniformly to the next layer. The separation membranes holds back certain sample constituents, such as red blood cells and allows only the desired metabolites to pass through to the reagent zone that contains all the necessary reagent components. Other layers may be included for enzyme immobilization, selective absorption, filtering, reflecting media, or elimination of interfering substances.

Dry Chemistry Background

Millions of people with diabetes are required to check their glucose and obtain results in a matter of minutes. Science has not yet invented an insulin formulation that will respond to the body's senses. Since injected insulin does not automatically adjust, the dose required to mimic the body's response must be adjusted day to day depending upon diet and physical activity. Self–monitoring of blood glucose levels is essential to people with diabetes, and this has become possible for the past 20 years due to the advent of dry chemistries (*9*). By regular and accurate monitoring of her blood glucose level by dry chemistry, an expectant mother can have a normal pregnancy and give birth to a healthy child. Athletes with diabetes also face significant problems and must monitor their blood glucose level that could vary with the level of physical exertion. Dry chemistries are not only useful to diabetics, but to patients with various other medical problems.

Enzymes are marvelously designed, though very specific catalysts, and are essential in dry chemistries. In a typical glucose measuring dry reagent, glucose oxidase (GOD) and peroxidase (POD) enzymes, along with a suitable indicator, e.g., tetramethylbenzidine (TMB), are dissolved and/or dispersed in a latex or water soluble polymer. The enzyme–containing coating is applied to a lightly pigmented plastic film and dried to a dry thin film. The coated plastic, cut to about 0.5 cm x 0.5 cm size is the dry reagent. The user applies a drop of blood and allows it to react with the strip for about 60 seconds or less. The blood is wiped off and the developed color is then read by a meter or visually compared with the pre–designed printed color blocks to precisely determine the glucose level in the blood. Thus, dry chemistries are users friendly.

Dry reagent test kits are available in the form of thin strips. Strips are usually disposable either coated or impregnated and mounted onto a plastic support or handle. The most basic diagnostic strip thus consists of a paper or plastic base, polymeric binder, and reacting chemistry components consisting of enzymes, surfactants, buffers and indicators. Diagnostic coatings or impregnation must incorporate all reagents necessary for the reaction. The coatings can be single or multi–layers in design. A list of analytes, enzymes, drugs and electrolytes assayed by dry chemistry diagnostic test kits is given in Table I (*10,11*).

Dry chemistry systems are very widely used in physician's offices, hospital laboratories and many homes worldwide. They are used for routine urinalysis, blood chemistry determinations, immunological and microbiological testing. Today dry chemistry systems are also used for assays in applications ranging from chlorine in swimming pools to coolant levels in cars and trucks (*12*).

The advantage of dry chemistry technology is that it eliminates the need for reagent preparation and many other manual steps common to liquid reagent systems, resulting in greater consistency and reliability of test results. Each test unit contains all the reagents and reactants necessary to perform analyses.

The Reactions

Dry chemistry reagent diagnostic tests are used for the assay of metabolites by concentration or by activity in a biological matrix. Many clinical chemistry diagnostic tests are based on the principle of reacting a body fluid, e.g., blood or urine containing a specific analyte with specified enzymes. The reactive components are usually present in excess, except for the analyte being determined. This is done to ensure that the reactions will go to completion quickly. Other enzymes or reagents are also used to drive the reactions in the desired directions (*13*). Glucose and cholesterol are the most commonly measured analytes. Glucose concentrations, deviating from the normal, are considered clinically significant, indicating chronic, hyper– or hypoglycemia.

Table I. List of Diagnostic Test Kits

Substrates	Enzymes	Drugs	Electrolytes
Glucose		Phenobarbitone	
Urea		Phenytoin Theophyline	Sodium ion
Urate		Carbamazepine	Chloride
Cholesterol (Total)	Alkaline Phosphate (ALP)		CO_2
Triglyceride	Lactate dehydrogenase (LDH)		
Amylase (total)	Creatine kinase MB isoenzyme (CK–MB)		
Bilirubin (total)	Lipase		
Ammonium ion Creatinine Calcium Hemoglobin HDL cholesterol Magnesium(II) Phosphate (inorganic) Albumin Protein in cerebrospinal fluid			

Glucose is determined by conversion to gluconic acid and hydrogen peroxide using glucose oxidase. Hydrogen peroxide is then coupled to an indicator, e.g., 3,3′,5,5′–tetramethylbenzidine (TMB) by the enzyme peroxidase. The rapidly (< 60 s) developed color is then measured at or around 660 nm. The biochemical reactions in dry chemistries in two important analytes are shown in Figure 1 (*14*).

Dry Chemistry Constructions

The basic components of typical dry reagents that utilize reflectance measurement are a base support material, a reflective layer, and a reagent layer either single or multi–layer. The functions of the various building blocks are now described. The base layer serves as a building base for dry reagent and usually is a thin, rigid thermoplastic film. The function of the reflectant layer is to reflect light not absorbed by the chemistry to the detector. Typical reflective materials are white pigment filled plastic film, coating, foam, membrane, paper and metal foil. The reagent layer contains the integrated reagents for a specific chemistry. Typical materials are paper matrix, fiber matrix, coating–film as well as hybrids.

An example of a single–layer coating reagent that effectively excludes red blood cells (RBC) is shown in Figure 2. In this case, a coating usually an emulsion type that contains all the reagents for a specific chemistry, is coated onto a lightly TiO_2 filled thermoplastic film and dried. For glucose testing, the coating will contain glucose oxidase, peroxidase, and an indicator, e.g. TMB. It may also contain buffer for pH adjustment, minor amounts of ether–alcohol type organic coalescing agent, and traces of a hindered phenol type antioxidant to function as a color signal ranging compound. A drop of blood is applied to this coating and allowed to react for a short pre–determined time (usually 60 seconds), the excess blood is wiped off with cotton or other absorbent material, and then read either visually or by a meter. Paper impregnated type dry chemistry is shown in Figure 3.

The schematic of a typical multi–layer coating dry reagent for example blood urea nitrogen is shown in Figure 4. It consists of a spreading and reflective layer. Sample containing BUN is spread uniformly by this layer. The first reagent layer is a porous coating–film containing the enzyme urease and buffer (pH 8.0). Urease reduces BUN to NH_3. A semi–permeable membrane coating allows NH_3 to permeate while excluding OH^- from the second reagent layer. The second reagent layer is comprised of porous coating–film containing a pH indicator wherein the indicator color develops when NH_3 reaches the semi–permeable coating–film. Typically, such dry reagents are slides (2.8 x 2.4 cm) with an application area of about 0.8 cm^2 and the spreading layer thickness is about 100 μm.

Touch and drain dry chemistry construction, shown in Figure 5 has been designed in our laboratories (*15*). It has found use in the evaluation of diagnostic coating–films. This construction should also find acceptance by users.

Diabetes Considerations

While researchers are opening new pathways to understand diabetes, medicinal chemists are exploring them to discover new drugs to treat the disease and its complications. At the same time, biotechnologists are discovering better ways to manage the disease.

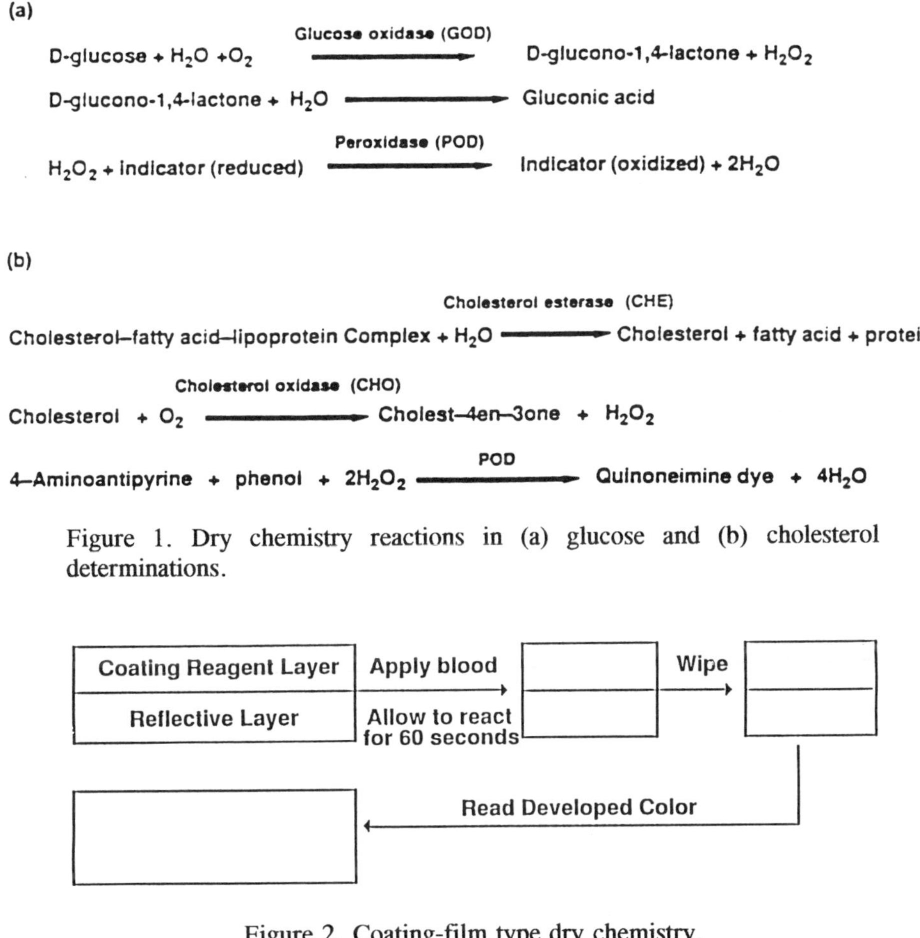

Figure 1. Dry chemistry reactions in (a) glucose and (b) cholesterol determinations.

Figure 2. Coating-film type dry chemistry.

Film Membrane
Reagent Impregnated Paper
Reflective Support Layer

Figure 3. Paper impregnated and overcoated type dry chemistry.

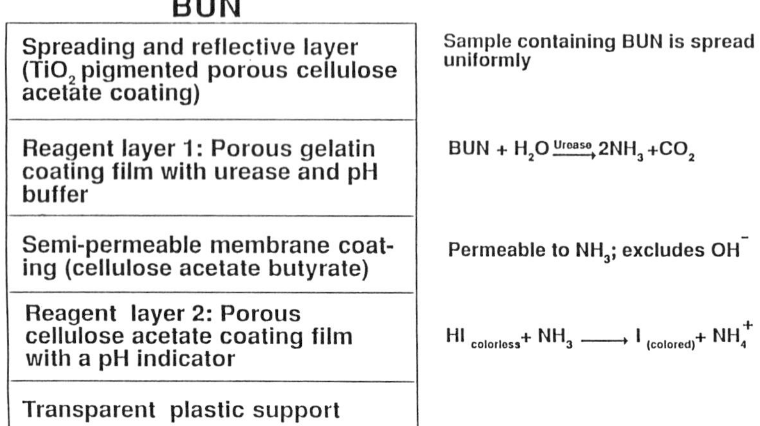

Figure 4. Example of a multi-layer coating dry chemistry for blood urea nitrogen.

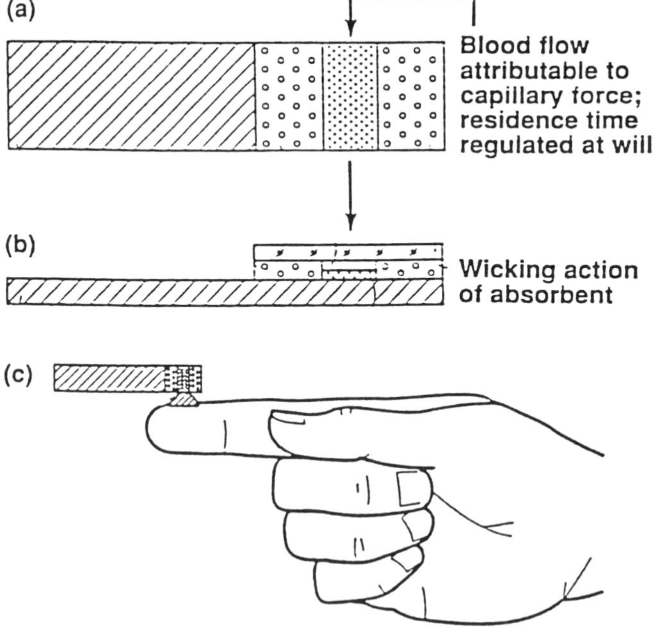

Figure 5. Touch and drain dry chemistry construction. (a) Dry coated surface, (b) cross-section of dry coated surface, adhesive, and cover piece, and (c) contact with blood drop results in blood filling the cavity. After desired reaction time, blood is drained off by touching end of cavity with absorbent material.

There are about 15 million estimated diabetics in the USA, half of whom have been diagnosed with the disease. More than 1.5 million diabetics are treated with injected insulin and the rest with weight loss, diet, and oral antidiabetic drugs, e.g., sulfonylureas (Tolbutamide, Tolazamide, Chloro–propamide, Glipizide, Glyburide) (*16*).

The current US market for drugs to control blood glucose is about $1 billion, equally divided between insulin and antidiabetic drugs. Insulin will grow by about 10% annually whereas antidiabetic drug market will shrink at an annual rate of 3%. Blood glucose monitoring market is over $600 million in the US and expected to grow about 10% annually.

Diagnostic Polymers and Coatings

In most dry chemistries, polymers account for more than 95 by weight of the strips. Polymers are therefore important and must be selected carefully. We have linked polymer chemistry with biochemistry that has lead to better understanding and improved dry chemistries. General considerations of polymers in dry chemistries are described in paragraphs following.

The polymer binder is necessary to incorporate the system's chemistry components in the form of either a coating or impregnation. The reagent matrix must be carefully controlled to mitigate or eliminate non–uniformity in reagent concentrations due to improper mixing, settling, or non–uniform coating thickness. Aqueous based emulsion polymers and water soluble polymers are extensively used. Table II provides a list of commonly used matrix binders (*17,18*).

Polymers must be carefully screened and selected to avoid interference with the chemistry. The properties of polymers, e.g., composition, solubility, viscosity, solid content, surfactants, residual initiators, film forming temperature, glass transition temperature, and particle size should be carefully considered (*19*). Generally, the polymer should be a good film–former with good adhesion to the support substrate and should have minimal or no tack for handling purposes. The coated matrix or impregnation must have the desired pore size and porosity to allow penetration of the analyte being measured, as well as the desired gloss, swelling characteristics, and surface energetic. Swelling of the polymer binder due to absorption of the liquid sample may or may not be an advantage depending on the overall system. Emulsion polymers have distinct advantages over water soluble polymers due to their high molecular weight, superior mechanical properties, and potential for adsorbing enzymes and indicators due to micellar forces.

Polymeric binders used in multi–layered coatings include various emulsion polymers, gelatin, polyacrylamide, polysaccharides, e.g., agarose, water soluble polymers, e.g., polyvinyl pyrrolidone, polyvinyl alcohol, co–polymers of vinyl pyrrolidone and acrylamide, and hydrophilic cellulose derivatives, e.g., hydroxyethyl cellulose and methyl cellulose.

During the past several years, we have researched new diagnostic polymers and novel dry chemistries therefrom. These include water–borne tough coating films, nonaqueous coatings and molded plastic systems. These chemistries are described below.

Water–borne Coatings

In a tough coating–film, the whole blood is allowed to flow over a tough coating–film, (Figure 5). About 100 emulsion and water soluble polymers were researched to determine their suitability in a tough coating–film. A styrene–acrylate emulsion (50% solid) was found to be most suitable. Diagnostic coatings were made using this polymer. A coating comprising of 51.0 g styrene–acrylate emulsion, 10.0 g 15% linear alkylbenzene sulfonate (LABS), 0.1 g GOD, 0.23 g POD, 0.74 g TMB indicator, 10.0 g hexanol and 13.2 g 1–methoxy–2–propanol gave good dose response (Figure 6). This coating–film also gave linear response, in transmission mode, up to 1600 mg/dL glucose. The coating was slightly tacky and was found to be highly residence time dependent.

Inclusion of ultra fine mica eliminated tack. Coated plastic could be rolled for automated strip construction. Coating–films residence time dependency would create a serious problem if the whole blood residence time cannot be regulated by strip construction. To make the coating–film residence time independent, continuous leaching of color from films during exposure time is essential. Enzymes and indicator can leach rapidly near the blood/coating–film interface and develop color which can be drained off. The color formed on the coating–film is then independent of the exposure time. We found that this gave a coating–film with minimal residence time differences.

A number of water soluble polymers were investigated. A low molecular weight polyvinyl pyrrolidone (e.g., PVP K–15 from GAF or PVP K–12 from BASF) gave coatings with no residence time dependency. This diagnostic coating comprises of 204.0 g of styrene–acrylate emulsion, 40.0 g of 15% solution of LABS, 0.41 g GOD (193 U/mgm), 1.02 g POD (162 U/mgm), 2.96 g TMB, 7.4 g PVP K–12, 32.0 g micromica C–4000, 20.0 g hexanol and 6.0 g of nonionic surfactant, e.g., Igepal CO–530. The nonionic surfactant serves as a surface modifier to eliminate RBC retention.

Excellent correlation was found when results at 660 nm and 740 nm were correlated with a reference hexokinase glucose method. The dose response was excellent up to 300 mg/dL glucose. The water–borne coatings, however, do not lend to ranging, coatings could be only slightly ranged by antioxidants.

Non–aqueous Coatings

For the past 20 years, dry reagent coatings have exclusively been water– borne because of the belief that enzymes function effectively only in water medium. We have researched nonaqueous enzymatic coatings for dry chemistries to which red blood cells will not adhere in addition to giving quick end–point reaction. These new coatings have superior thermostability as well. The new non–aqueous coatings can easily be ranged by a variety of antioxidants whereas water–borne enzymatic coatings are very difficult to range. The differences are explained using organic reaction mechanisms.

The non–aqueous diagnostic polymer must impart particular desirable properties to the resulting coating–film. Most importantly, hydrophilicity and hydrogel character to the film must be imparted, thus allowing intimate contact with aqueous whole blood sample. Furthermore, non–aqueous polymer may also provide some enzyme stabilizing effects.

An acrylic resin comprising of 65 wt% 2–hydroxyethyl methacrylate (HEMA), 33 wt% butyl methacrylate (BMA) and 2 wt % dimethylaminoethyl methacrylate (DMAEMA) was made by solution polymerization at 40% solid in xylene/1–methoxy–2–propanol (1/1). The polymerization was done at 90°C for 6 h using 1% azobisisobutyronitrile initiator. Another acrylic polymer,

Table II. Common Matrix Binders

Emulsion Polymers	Water Soluble Polymers
Acrylics	Polyvinyl alcohol
Polyvinyl acetate: homo and co–polymers	Polyvinyl pyrrolidone Highly hydroxylated acrylic
Styrene acrylics	Polyvinylethylene glycol acrylate
Polyvinyl propionate: homo and co–polymers	Polyacrylamide Hydroxyethyl cellulose
Ethylene vinyl acetate	Other hydrophilic cellulosics
Lightly crosslinkable acrylics	Various co–polymers
Polyurethanes	

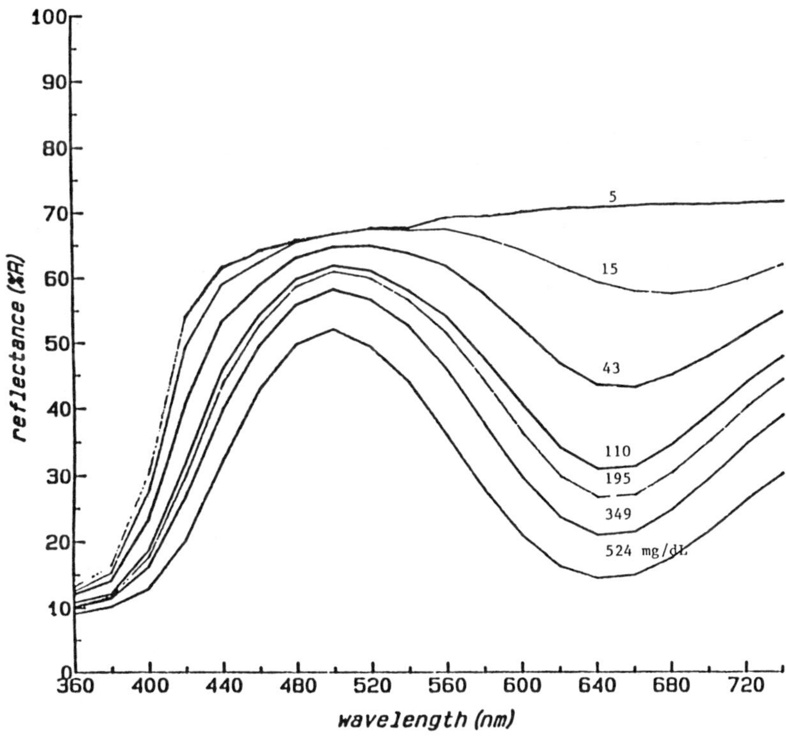

Figure 6. Dose response of water-borne tough coating-film for 120 s residence time.

HEMA/BMA/t–butylaminoethyl methacrylate (65/34/1) was also made similarly. In fact, a very large number of polymers were made – the above compositions are typical among the best.

In organic solvents, GOD and POD enzymes become extremely rigid and can be dispersed with ease. Dispersions were made by grinding GOD and POD in xylene/methoxy propanol with or without a surfactant using an Attritor mill (2–4 h) or a ball mill (24 h). Grinding was continued until dispersion of <1 μm were obtained. The composition of a typical dispersion is 1.876 g GOD, 4.298 g POD, 11.79 g sodium dodecyl sulfate, 41.06 g xylene and 41.06 g 1–methoxy–2–propanol.

A generalized method for preparation of the non–aqueous enzymatic coating involved adding polymer solution, TMB, mica, surface modifiers and solvents to the enzyme dispersion. Ranging compound can be post–added to the coating. Only slight mixing on a ball mill is required after ingredient addition to complete enzymatic coating preparation. The composition of a typical non– aqueous coating useful for low range blood glucose detection by wt% is 33.29 HEMA/BMA/DMAEMA (65/33/2) (40% solid), 2.38 TMB, 1.17 GOD, 2.68 POD, 3.28 sodium dodecyl benzene sulfonate, 26.53 xylene, 26.53 1–methoxy–2–propanol, and 1.96 cosmetic grade C–4000 ultrafine mica. Many surfactants and surface modifies were investigated that eliminated RBC retention.

In order to improve resolution in measurement of glucose levels in the high range (220–800 mg/dL), many hindered phenol antioxidants were found useful. These antioxidants that function as ranging compounds include 3–amino–9–(aminopropyl)–carbazol dihydrochloride (APAC), butylated hydroxy toluene (BHT), and BHT/propyl gallate (85/15). The ranging compounds were effective in the ranging compound: TMB molar ratios of 1:2.5 to 1:20.

The coatings were applied on a lightly TiO_2 pigmented polycarbonate plastic film at a wet film thickness of 100 μm and dried in an air–forced oven at 50^{0}C for 15 minutes. A lab. coater was used for applying coating onto polycarbonate at 2 m/min rate.

For dose response and other studies with whole blood, the coated–films were incorporated into a touch and drain test device, (Figure 5). In this device, the blood is caused to flow across the surface of the coating–film. The residence time can thus be regulated at will. The color signal of reacted coating–films were monitored by diffuse reflectance spectrophotometry. A McBeth 1500 visible spectrometer was used for measuring color signals.

The characteristics and performance of aqueous and non–aqueous diagnostic coatings differ. The composition and structural differences are now discussed. In aqueous coatings, the continuous phase comprise of dissolved GOD and POD enzymes and dispersed TMB indicator that are absorbed onto latex particles due to micellar forces. The continuous phase in non–aqueous coatings is organic solvents and dispersed GOD, POD and dissolved TMB which coates enzyme dispersion uniformly constitute the discontinuous phase. Antioxidants, useful as ranging compounds, are soluble in organic solvents whereas they are dispersed and adsorbed onto latex particles in aqueous coatings.

Figure 7 shows the dose response curves for blood glucose samples from 30 mg/dL to 231 mg/dL glucose for the above described coating–film. At 660 nm, there is a total change in reflectance (%R) of over 30, thus allowing accurate measurement of glucose at these levels. The dose response for the same coating–film in the high range blood glucose (171 to 786 mg/dL) gave 10% R change. This will preclude an accurate measurement of glucose at these levels.

To improve dynamic range and resolution, a large number of antioxidants can be used in nonaqueous coatings. A typical example is coating to which BHT and PG have been added. The dose response of the ranged coating is illustrated in Figure 8. A difference of more than 30% R between lowest and highest blood glucose has been accomplished by the ranging compounds.

The water–borne enzymatic coatings do not lend themselves to ranging easily. Most hindered phenols are ineffective. Only APAC is effective in some aqueous coatings (*20*). In contrast, almost all hindered phenols are effective in non–aqueous coatings. There are several reasons for this disparity between the aqueous and non–aqueous coatings. The physical reason is due to the solubility of the hindered phenols in organic solvents. In non–aqueous coatings, the ranging compounds uniformly coat the enzyme particles. In a similar vein, the indicator TMB is uniformly coated onto enzyme particles. H_2O_2 generated due to the reaction of blood glucose in the coating–film is thus readily available to the antioxidant. In aqueous coatings, the antioxidant will be found discontinuously as aggregate, large insoluble particles and adsorbed onto emulsion particles. Much of the generated H_2O_2 does not find the antioxidant.

In dry chemistries, end–point reactions are preferred over kinetic reactions. The new nonaqueous chemistry has not only quick end–point, but several other desirable characteristics. The chemistry is extremely fast reacting and the generated color signal is independent of the blood residence time.

The diagnostic polymers used in the non–aqueous coatings are "hydrogel" in nature and are fully wetted by blood almost instantaneously. Microscopic dye penetration experiments have shown that the liquid penetrates the polymer as well as the coatings in less than 2 s. The generated colors remain stable immediately after reaction and also for an extended time.

In general, uric acid is known to produce interference by lowering color signals (increasing percent R) in dry chemistries. Blood glucose containing a high level of uric acid (30 mg/dL) produced lowering of color signal in non–aqueous coating–films. The interference was, however, prevented by buffering the coatings to 5.5 pH.

Long–term stability of non–aqueous coating–films under elevated temperature (up to 65^{0}C) and moderate humidity (up to 50% RH) was very good. Stability of these coating–films were much better than comparable aqueous films.

The color resolution and sensitivity of reacted coating–films were found to be very good. The % R does response data for a high range, high solid coating were fitted to the best nonlinear function using equation $y = Ae^{-bx} + CX + D$ wherein y is %R, X is the concentration of glucose in blood in mg/dL, A is 66.93, b is –0.087, C is –0.09 and D is 4.28. Coating–films that deviate from such fittings lack accuracy and precision.

Molded Dry Chemistry

Enzymes are polymeric in nature and very specific catalysts. They are extensively used in diagnostics, immunodiagnostics, and biosensors to measure or amplify signals of many, but specific, metabolites. Typical properties of select diagnostic enzymes are presented in Table III.

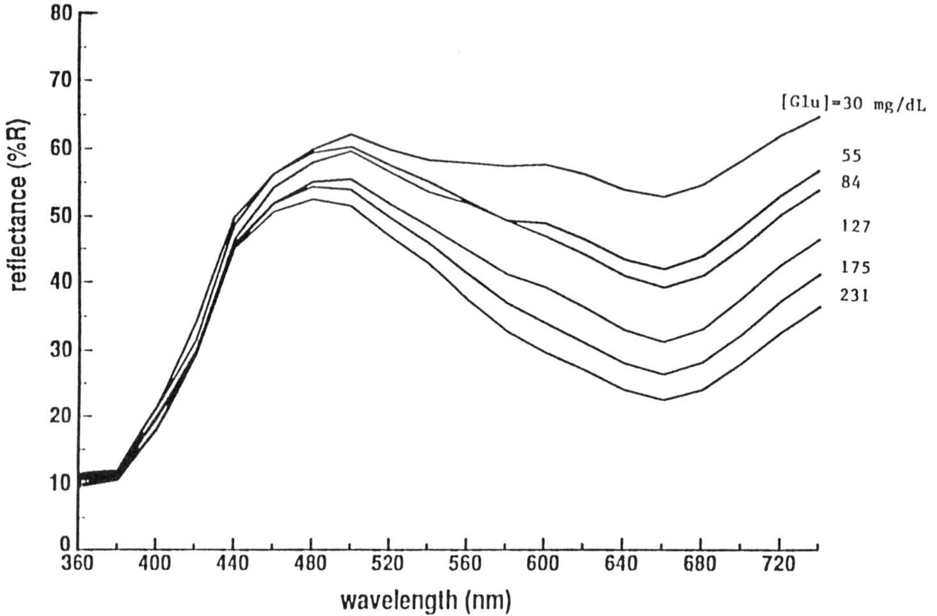

Figure 7. Dose response of non-aqueous coating in low glucose range.

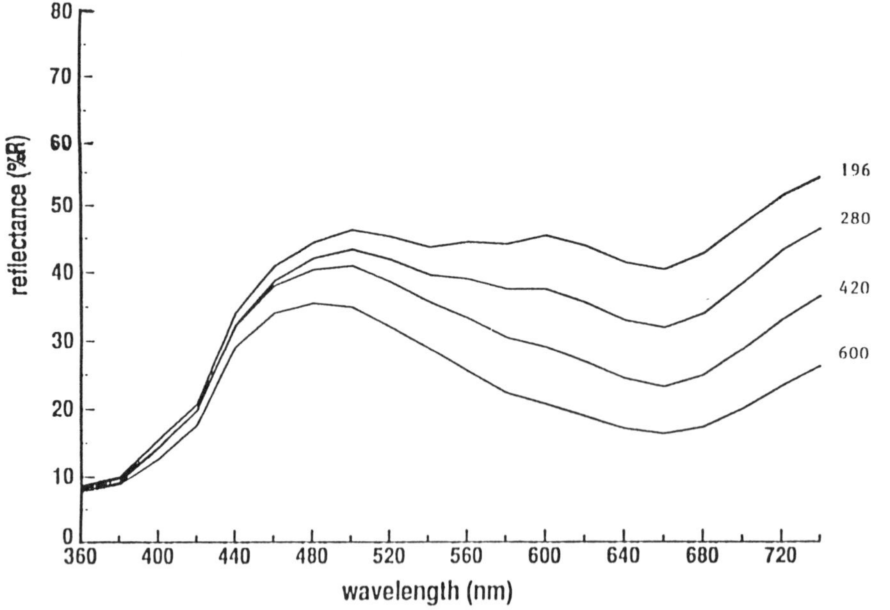

Figure 8. Coating ranged by BHT and propyl gallate.

In general, most enzymes are very fragile and sensitive to pH, solvent, and elevated temperatures. The catalytic activity of most enzymes is reduced dramatically as the temperature is increased. Typical enzymes used in diagnostics, e.g., GOD and POD are almost completely deactivated around 65^{0}C in solid form or aqueous solution. Despite wide and continued use of such enzymes in diagnostics for more than 30 years, limited or no thermal analysis work on these biopolymers has been reported until recently (*21*). Our DSC analysis results indicating glass transition temperature (Tg), melting temperature (Tm), and decomposition temperature (Td), are shown in Table IV. Below the glass transition temperature (Tg), the enzymes are in a glassy state and should be thermally stable. Around Tg, onset of the rubbery state begins, and the enzyme becomes prone to thermal instability. When the enzymes melt around Tm, all the tertiary structures are destroyed, thus making the enzyme completely inactive. The presence of chemicals can considerably influence enzyme stability.

In GOD enzyme, the redox center (FAD/FADH2) that can conduct electrons, is catalytically relevant. To keep or sustain the enzyme's activity, the redox centers must remain intact. The bulk of the enzyme, polymeric in composition, is an insulator – altering it will not affect the enzyme's catalytic activity.

An enzymatic compound containing GOD, POD, tetramethyl benzidine indicator, a linear alkylbenzene sulfonate, and PHEMA with weight composition similar to a typical water–borne or non–aqueous coating composition was ballmilled for 48 h. PHEMA was prepared by mass polymerization using 1% benzoyl peroxide at 125^{0}C/16 h. PHEMA was crushed, pulverized and finally ground in a ball mill to –80 mesh ($< 177\ \mu$m). Using the molding compound, strips were made by compression molding. The molded strips showed much improved thermal stability. This is schematically shown as follows:

$$\text{(GOD, POD) Coating, Dry film} \xrightarrow{65^0\text{C}} \text{Rapid Inactivation}$$

$$\text{(GOD, POD) Molding Compound} \xrightarrow[\text{Mold}]{\text{Heat/Pressure}} \text{Molded Strip}$$

Molding Schedule

Temp. ^{0}C	Time,Min.	Activity(Response to Glucose)
105	1	yes
105	5	yes
125	1	yes
150	1	yes
200	1	no

Why should the enzymes be stable in polymer melt whereas they quickly deactivate in water or dry form? We postulate a proposed mechanism, shown in Figure 9, for excellent thermal stability of molded strips. We speculate the enzyme is surrounded very tightly by coils of PHEMA at room and elevated temperatures. Temporary melting of the crystalline insulator regions of the

Table III. Typical Properties of Select Diagnostic Enzymes

	Cholesterol Oxidase (CO)	Cholesterol Esterase (CE)	Glucose Oxidase (GOD)	Peroxidase (POD)
Source	Streptomyces	Pseudomonas	Aspergillus	Horseradish
EC	1.1.3.6	1.1.13	1.1.3.4	1.11.1.7
Molecular Weight	34,000	300,000	153,000	40,000
Isoelctric Point	5.1 ± 0.1 & 5.4 ± 0.1	5.95 ± 0.05	4.2 ± 0.1	
Michaelis Constant	$4.3 \times 10^{-5}M$	$2.3 \times 10^{-5}M$	$3.3 \times 10^{-2}M$	
Inhibitor	Hg_{++}, Ag_{+}	Hg_{++}, Ag_{+}	Hg_{+},Ag_{+},Cu_{++}	CN_{-}, S_{--}
Optimum pH	6.5 – 7.0	7.0 – 9.0	5.0	6.0–6.5
Optimum Temp.	45 – 50°C	40°C	30 – 40°C	45°C
pH Stability	5.0 – 10.0 (25°C,20h)	5.0 – 9.0 (25°C,20h)	4.0 – 6.0 (40°C,1h)	5.0–10.0 (25°C,20h)

Table IV. Thermal Analysis of Enzymes

Enzyme	Source	Tg(°C)	Tm(°C)	Td(°C)
Cholesterol Oxidase	Nocardia	50	98	210
Cholesterol Oxidase	Streptomyces	51	102	250
Cholesterol Esterase	Pseudomonas	43	88	162
Glucose Oxidase	Aspergillus	50	105	220
Peroxidase	Horseradish	50	100	225

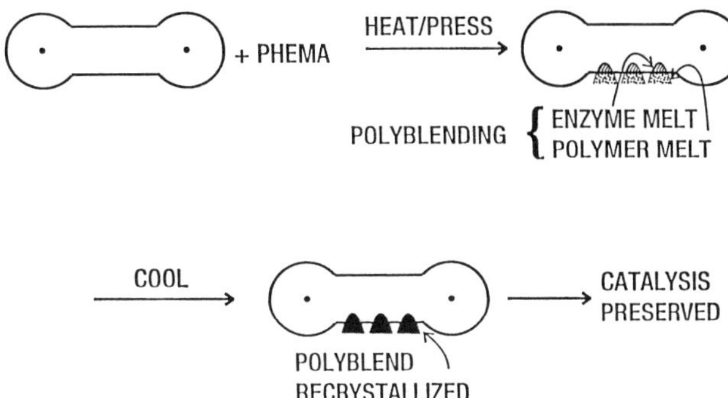

Figure 9. Preservation of catalyst in molded strips.

protein up to 150°C results in polyblending with PHEMA, the polyblending does not disturb the redox centers. When the molding is cooled, the enzyme/PHEMA blend crystallizes without destroying the redox centers. The catalytic activity of GOD is thus preserved during and after molding, and therefore, the molded strip responds to glucose. At 200°C the melt viscosity of the blend is lower with increased mobility. The redox center is disturbed at 200°C and the cooled strip, therefore, becomes inactive. Certainly, the stability of the enzymes will depend on the concentration of PHEMA and the melt viscosity of the enzymatic compound (*22*).

Molding of strips using RIM may lead to useful chemistries. Modification of enzymes for added thermal stability by our technique is a distinct possibility. Extension of our concept in coating–films for biosensors is also indicated.

Miscellaneous Uses of Polymers in Diagnostics

Dry chemistry diagnostics often take advantage of the immobilization capabilities of enzymes. Immobilization is a useful tool in the design of multi–layered coatings to reduce interaction between layers. Enzymes can be bound by the following methods: covalent coupling, adsorption onto the polymer surface, entrapment within the polymer matrix, or by crosslinking (*23–26*). Some immobilized enzymes show higher thermostability and lower sensitivity to pH and oxidation, which allows more flexibility in the design and processing of reagent strips. Enzyme immobilization can alter the reactivity of the enzyme depending on the site and type of binding.

Enzymes are usually bound to a variety of adsorbents such as diethylaminoethyl cellulose, agarose, polysaccharides, carrageenan, poly–acrylamide, polyacrylates, polystyrene, polyvinyl pyrrolidone, polyvinyl alcohol, polyvinylethylene glycol acrylate, collagen, and gelatin. Adsorption of antibodies and antigens onto polystyrene and other latex particles is widely used in solid phase immunoassay (*25*).

Enzyme stability is essential in dry chemistry diagnostics. Changes in enzyme activity can decrease the performance and shelf life of a product. Some of the basic approaches used to stabilize enzymes are by addition of soluble compounds and immobilization. The chemical binding of the enzymes onto polymers reduces the freedom to undergo conformational changes and thus reduces the chances of denaturation (*27*). Another stabilization technique involves dispersing certain hydrophobic polymers into a carrier material permeable to liquid and oxygen. Although the mechanism is not understood, the inclusion of certain dispersed polymers improve the stability of reagents such as peroxidase. Preferred classes of polymers are copolymers consisting of hydrophobic recurring units (styrene, alkyl acrylates, etc.) and ionic recurring units (acrylic acid, sulfonates, methacrylic acid, etc.).

The surface properties of a coated matrix can have a significant effect on the test's performance. Water soluble polymers and polymeric surfactants are used considerably to achieve the desired results. Water soluble polymers and surfactants can be used to control the rate of wetting and the release of reactants from different zones in multi–layered systems. Faster wetting of the strips could also mean a reduction in the overall test time. Polyvinyl pyrrolidone, polyvinyl alcohol, polyacrylamides, polyethylene glycol, ethyl cellulose, and hydroxypropyl methyl cellulose are some of the commonly used water soluble polymers.

Polymeric fibers such as polyesters, polypropylene, nylon, and the natural cellulosics are used extensively in the diagnostic field for sample application, reagent impregnation, matrix carrier, and removal of excess sample.

Diagnostic test strips vary considerably, but all have some of the basic building materials. Most dry reagent chemistries are built on a base support material usually made of a polyolefin coated paper or rigid plastic such as polyethylene terephthalate, polycarbonates, polymethyl methacrylate, polystyrene, or cellulose acetate. The base material is often used as the reflecting media and may contain titanium dioxide or barium sulfate. Polymer adhesives in the form of tapes and hot melts are used for assembling the test strips. Careful screening of the polymers used are necessary to avoid interference with the chemistry. Optically, clear foils and adhesives are also used when required. Polymeric based inks for screen printing and polymeric cosmetic foils are often used to apply trademarks and logos.

Redox Polymers for Biosensors

Biosensors and DNA probes are a few of the new trends in dry reagent diagnostics (*28*). Near infrared (NIR) method will be a reagentless system and non–invasive, although prospects for this method is uncertain at this time.

There are about three detection approaches in the glucose enzyme electrode. Oxygen consumption is measured in the earliest method. A reference, nonenzymatic electrode is required, however, to provide an amperometric difference signal. The second approach detects H_2O_2 but requires an applied potential of about 650 mV and an inside permseletive membrane. In the third generation biosensor, advantage is taken of the fact that the enzymatic reaction requires two steps. The enzyme (GOD) is reduced by glucose then the reduced enzyme is oxidized by an electron accepter, i.e., a mediator specifically redox polymer. Direct electron transfer between GOD and the electrode occurs extremely slowly; therefore, an electron accepter mediator is required to make the process proceed rapidly and efficiently (*29*).

Recently, two groups have explored and developed redox polymers that can rapidly and efficiently shuttle the electrons. The groups have "wired" the enzyme to the electrode with a long polymer having a dense array of electron relays. The polymer penetrates and binds the enzymes, and is also bound to the electrode.

Heller et al have done extensive work on Os–containing polymers. They have synthesized a large number of such Os–containing polymers and evaluated their electrochemical characteristics (*30*). Their most stable and reproducible redox polymer is a poly(4–vinyl pyridine) to which $Os(bpy)_2Cl_2$ has been attached to 1/6th of the pendant pyridine groups. The resultant redox polymer is water insoluble. To make it water soluble and biologic compatible, Heller et al have partially quaternized the remaining pyridine pendants with 2–bromoethyl amine. The redox polymer is water soluble and the newly introduced amine groups can react with a water soluble epoxy e.g., polyethylene glycol diglycidyl ether and GOD to produce a cross–linked biosensor coating–film. Such coating–films produced high current densities and a linear response to glucose up to 600 mg/dL.

Skotheim et al have used a flexible polymer chain, in contrast, to put relays. Their polymers provide communication between GOD's redox centers and electrode. No mediation was found when ferrocene was attached to a non–silicone backbone. Their ferrocene–modified siloxane polymers are stable and non–diffusing (*31*). Therefore, biosensors based on these redox polymers gave good response and superior stability.

Literature Cited

1) Azhar A.F. and Usmani A.M., "Degradation of Polymers in Medicine", in Handbook of Polymer Degradation, Maadhah A.G. (Ed), Marcel Dekker, New York, 1992.
2) Skarstedt M.T. and Usmani A.M., Polymer News, 1989, 14, 38.
3) Joslin E.P., Root H.P., White P. and Marble A., "The Treatment of Diabetes", 7th ed. Lea and Febiger, Philadelphia, 1940.
4) Compton W.A., Treneer J.M., U.S. Patent 2, 387, 244, 1945.
5) Free A.H., Adams E.C., Kercher M.L., Clin. Chem. 1957, 3, 163.
6) Rey H.G., Rieckmann P., Wielinger H., Rittersdorf W., U.S. Patent 3, 630,957, assigned to Boehringer Mannheim, 1971; also see Vogel, P. Braun, H.P., Berger, D., and Werner, W. U.S. Patent 4,312,834, assigned to Boehringer Mannheim, 1982.
7) Shirey T.L. Clin. Biochem. 1983, 16, 147.
8) Przybylowicz E.P., Millikan A.G., U.S. Patent 3, 992, 157, assigned to Eastman Kodak, 1976.
9) Schrenk W.J. and Usmani A.M., "History of Diagnostic Coating"., in History of Coatings, Mark H.F. and Seymour R.B. (Eds), Elsevier, New York, 1990.
10) Spiegel H.E., Kirk–Othmer Encyclopedia of Chemical Technology; 3rd Edition, Wiley, New York, 1985.
11) Walter B. Anal. Chem., 1983, 55, 449A.
12) Tabb D., Rapkin M and Usmani A.M. Boehringer Mannheim, Unpublished work, 1988.
13) Tietz N.W., "Fundamentals of Clinical Chemistry", W.B. Saunders, Philadelphia, 1976.
14) Diebold E., Rapkin M. and Usmani A.M., Chem. Tech. 1991, 21, 462.
15) Azhar A., Burke A., DuBios J. and Usmani A.M., Polymeric Mater. Sci. Eng., 1992, 66, 432.
16) Stinson S.C., Chem. and Eng. News, Sept. 30, 1991, 635.
17) Usmani A., Biotechnology Symposium, Plenum, Atlanta, 1992.
18) Bruschi B.J., US Patent 4, 006, 403, assigned to Eastman Kodak, 1978.
19) Scheler W., Makromol. Chem. Symp., 1987, 12, 1.
20) Azhar A., Burke A., DuBios J. and Usmani A., J. Appl. Polym. Sci., 1993, 47, 2083.
21) Kennamer J.E. and Usmani A.M., J. Appl. Polym. Sci., 1991, 42, 3073.
22) Kennamer J.E., Burke A.D. and Usmani A.M., "Molded Dry Chemistry", in Biotechnology and Bioactive Polymers, C.G. Gebelein and C.E. Carraher (Eds), Plenum, New York, 1992.
23) Schaeffer J.R., Burdick B.A., and Abrams C.T., Chem. Tech. (Sept. 1988) 546.

24) Carraher Jr C.E. and Sperling L.H., Polymer News, 1988, 13, 101.
25) Scouten W.H., Methods in Enzymology, 1987, 135, 30.
26) Boguslawski R.C., Smith R.S., and Mhatre N.S., "Applications of Bound Biopolymers in Enzymology and Immunology", Current Topics in Microbiology and Immunology, 1981.
27) Freeman A., Trends in Biotechnology, 1984, 2, 147.
28) Burke A., Azhar A., DuBios J. and Usmani A., Chem. Tech., 1991, 21, 547.
29) Reach G. and Wilson G.S., Analy. Chem., 1992, 64, 381A.
30) Gregg B.A. and Heller A., J. Phys. Chem., 1991, 95, 5970.
31) Boguslavsky L., Hale P., Skotheim T., Karan H., Lee H.and Okamoto Y., Polym. Mater. Sci. Eng., 1991, 64, 322.

RECEIVED January 21, 1994

Chapter 2

Fiber-Optic Sensors Based on Degradable Polymers

Venetka Agayn and David R. Walt

Max Tishler Laboratory for Organic Chemistry, Chemistry Department, Tufts University, Medford, MA 02155

A major deficiency in the use of sensors based on irreversible chemistries is that they cannot be regenerated and are therefore limited to a single use. Our laboratory has developed methods to use slow release polymers as a way to renew the supply of reagent(s) available at the fiber tip. Previous work has demonstrated this approach with two examples: a pH sensor based on a pH sensitive dye entrapped in ethylene vinyl acetate copolymer (EVA) and a sensor based on an immunoassay employing an antibody and an antigen entrapped in EVA matrices. We investigated the preparation of fiber-optic sensors using reagents confined in degradable microspheres that release over extended time periods. This paper describes the use of poly(lactide-glycolide) copolymers for the preparation of reagent containing microspheres. The release rates, sensor configurations and response times of various sensors prepared with these microspheres are described.

Recently, sensors have become accepted tools in the analytical arsenal. With the introduction of new materials and improved instrumentation and computer control, new avenues are available to solve different analytical problems. In order for an analytical technique to be considered a sensor it has to be continuous hence, allow repeated use. This requirement mandates that the chemistry employed be reversible. Unfortunately, despite the variety of sensing schemes available for various substances of interest many of them are essentially thermodynamically irreversible. In the area of medical and environmental immunoanalysis, vast progress has been accomplished in the last decade to develop assays for various analytes. However, due to the slow off-rates of antibody-antigen complexes, the reagents utilized in immunoassays can be used only once. Another irreversible reaction is the complex formation of different metal ions and some fluorescence or colorimetric indicators. A way to overcome these limitations without employing frequent sampling, is to use a pumping device to replenish reagents. This approach, which we can call "active", is widely applied in enzymatic and immunoassays, but requires substantial instrumentation and control for accurate delivery. Apart from this limitation, a constant delivery of reagent solution is required in substantial volumes over a long period of time posing challenges to reagent instability.

0097-6156/94/0556-0021$08.00/0

Another approach for the long-term delivery of reagents at the sensor tip is the use of " passive" systems. Reagents can be entrapped in polymer matrices and be released over time. In previous work we prepared a long-lasting pH sensor in which fluorescent pH sensitive dyes were entrapped in ethylene vinyl acetate copolymers[1]. The resulting polymer plugs were positioned around the fiber tip and the fluorescence signal was monitored as a function of pH. The response time was longer than 10 minutes and was largely dependent on the sensor configuration because of significant bulk fluid transfer delays in equilibrating with the surrounding environment. In another demonstration of the method, we prepared immunosensors using the same EVA polymer-fluorescent labeled antibody and antigen were embedded in two separate polymer plugs positioned in a chamber[2]. Upon release, the two reagents combined in the space around the fiber tip, resulting in a fluorescent signal that corresponded to the concentration of competing unlabeled analyte. This sensor was used for over a month, however it suffered from the same limitations as the pH sensor described above, mainly mass transfer and diffusional problems affecting response time. The immunosensor could distinguish only between the presence and absence of competing analyte, hence it was useful as a threshold rather than a quantitative sensor. We believe long response times associated with these sensors could be reduced by breaking the polymer into small particles, thus increasing the surface-to-volume ratio and thereby increasing the reagent release rate[3].

To assure steady and uniform release over extended periods of time we chose to employ microspheres to maximize the surface-to-volume ratio. We have investigated the use of degradable poly(lactide-glycolide) copolymers of different molecular weights for the entrapment of pH sensitive dyes. The sensors prepared were tested for both sensitivity to pH and response time.

Sustained Release Polymers Useful for Sensors Construction

Polymeric materials for continuous long-term release of entrapped substances (excipients) have been utilized extensively in the last two decades in drug delivery systems[4]. These polymers can be classified into two major groups as shown in Table I. The non-erodible carriers, such as polyacrylamide, polyvinyl alcohol and poly(2-hydroxy methacrylate) have been used widely in sensor preparation mainly as supports for physical or chemical immobilization of fluorescent molecules[5] or enzymes[6]. As discussed above, EVA has been shown to be appropriate as a reservoir polymer for sensor development.

A second family of slow-release polymers we have employed[7] in conjunction with fiber-optics are some of the degradable polymers listed in Table I. Polylactide, polyglycolide and their co-polymers are polyesters which hydrolyze slowly in water. They are both hydrophobic and water insoluble polymers. Characteristics such as crystallinity and degradation rate vary depending upon molecular weight, monomer ratio and formulation procedure. Both lactic and glycolic acids are bifunctional α–hydroxy acids that may be condensed to form polyesters. Polylactic acid is optically active and is used in polymer formation in either its D(+), L(-) or racemic forms. Upon degradation it releases lactic acid, which is part of the metabolism of many organisms; hence the D(+) form can be metabolically utilized, while the L(-) form is inactive and can be safely excreted. The glycolide is a polymer of glycolic acid, which is utilized by higher plants in photo respiration. The polymers of both lactic and glycolic acids are FDA approved for use in implants, sutures and microspheres as drug carriers. These properties make them safe to implant in patients and do not pose toxicity threats if employed in the environment.

Table I. Polymers used for controlled release

Nonerodible carriers	Degradable polymer carriers
Hydrogels	
polyacrylamide	polylactic acid (polylactide)
polyvinyl alcohol	polyglycolic acid (polyglycolide)
poly (2-hydroxyethyl methacrylate)	poly (lactic acid-co-glycolic acid)
poly N-(2-hydroxypropyl methacrylamide)	poly (hydroxybutyric acid-co-hydroxyvaleric acid)
poly vinylpyrrolidone	poly (ε-caprolactone):polyvalerolactone
poly methyl methacrylate	polyorthoesters
	poly alkylcyanoacrylates
Hydrophobic polymers	natural polymers:albumin, gelatin, starch
ethylene vinyl acetate copolymer	cross-linked polypeptides and proteins
silicone elastomers	polyanhydrides: co-polymers of sebacic acid, bis(p-
microporous polypropylene	carboxy-phenoxy)- propane and dodecanedioic acid

Other polyesters amenable to slow degradation are the co-polymers of hydroxybutyric and hydroxyvaleric acid. These polymers are obtained from bacterial fermentation and their degradation products can be metabolized. Other polymers of interest are the polyorthoesters, polyalkylcyanoacrylates, and natural polymers such as gelatin, starch, albumin and synthetic peptides. Other polymers with attractive characteristics are the polyanhydrides, such as co-polymers of sebacic acid and dodecanedioic acid but their degradation is strongly affected by variations in pH.

A different approach for sensor preparation can be envisioned by the use of sensitive polymers in conjunction with insensitive dyes. In this case the dye simply provides a convenient way to monitor the response of the polymer, which is the actual active sensing element. Candidates for such materials are the pH sensitive co-polymers of cellulose, e.g. hydroxypropyl methylcellulose. They are used for medical preparations as coating materials to protect against the acidic environment in the stomach and to deliver the drug to the basic environment in the intestines where they dissolve rapidly.

Structures for Microparticles, Enclosure of Substances and Release Pattern

A variety of different formulations for controlled release polymers exist including microparticles, microcapsules and microspheres. Microparticles range in size from 1-200 μm, while particles with a diameter smaller than 1 μm are called nanoparticles. Microcapsules are microparticles which have the substance of interest enclosed in a shell of degradable polymer. Microcapsules however are characterized by a relatively fast release of large amounts of the enclosed substance. Microspheres (Figure 1), on the other hand, are monolithic in structure, i.e. have the substance uniformly distributed within the polymer layer. This distribution results in a more uniform release over longer periods of time. We selected such microspheres prepared from poly(lactide-glycolide) copolymers to develop our sensors.

Release pattern from slow-release polymers are generally divided into two major types: homogeneous (bulk) and surface erosion. The glycolide-lactide copolymers display bulk degradation, i.e. they exhibit significant degradation in the polymer matrix interior. Ideally, a polymer should exhibit only surface erosion as for the polyanhydrides and poly(ortho)esters. Unfortunately, due to their high pH sensitivity these materials cannot be used for all application.

Experimental

Preparation of Microspheres

Microspheres can be prepared by two approaches. Controlled emulsion polymerization[8] starts with the monomers and achieves the desired particle size by emulsion formation followed by polymerization. Solvent evaporation[9] starts from the polymer and involves a series of steps as outlined in Figure 2. Obtaining good shapes and sizes of microspheres with a relatively high load of enclosed reagents presents a challenge in designing the optimal procedure for the specific physico-chemical characteristics of the polymer and the entrapped substance. Various modifications of the basic solvent evaporation procedure have been employed to obtain particles of polymer containing the entrapped reagent. Spray drying[10] can be utilized with or without vacuum to facilitate the solvent evaporation process. An alternative to the emulsification technique is used to prepare a series of lactic and glycolic acid polymers and was found appropriate for hydrophilic species[11]. The substance to be entrapped is dissolved in water and added to the dissolved polymer to form the so called "inner emulsion" water/oil ($\mathbf{w_1/o}$). This emulsion is then added with stirring to a high viscosity water solution to form a second emulsion: $\mathbf{w_1/o/w_2}$. This complex emulsion

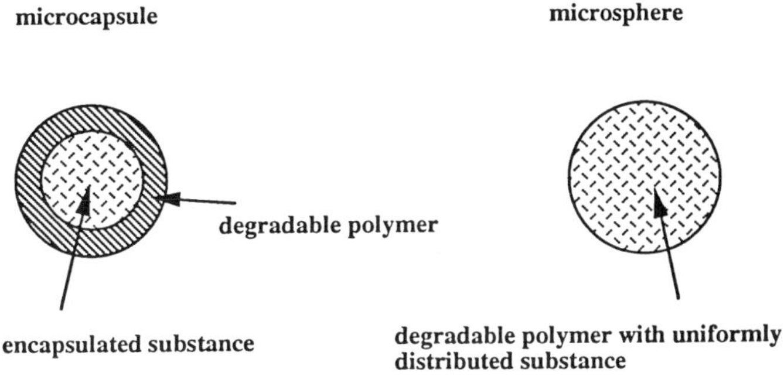

Figure 1. Devices for slow release of substances based on degradable polymers

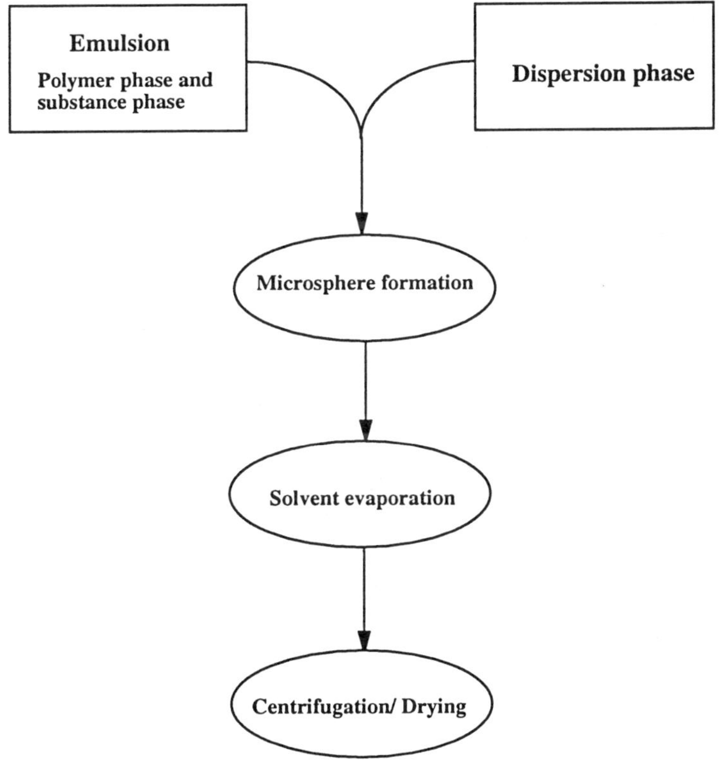

Figure 2. Preparation of microspheres.

is finally added to a water solution to allow for solvent evaporation and microsphere hardening.

In our work we utilize different procedures based on emulsion formation followed by the solvent evaporation method[12]. The poly (lactide-glycolide) polymer (100 mg, Medisorb, DE) is dissolved in a solvent (acetone or methylene chloride, 1 ml) to form the "oil phase" and to this solution the core material is added. Different approaches can be used depending on the hydrophobicity of the substance. When a lipophilic material like fluorescein diacetate is used, it can be added (10-30% load on the polymer weight) directly to the dissolved polymer. The resulting clear viscous solution then is dispersed in a water solution with an emulsifying agent (polyvinyl alcohol, 0.1%) and the resulting emulsion is stirred for several hours to allow solvent evaporation. The microspheres are then separated and washed by decantation, filtration or centrifugation and subsequently dried.

We prepared microspheres containing hydrophilic substances such as fluorescein and 8-hydroxypyrene 1,3,6 trisulfonate (HPTS) utilizing the complex emulsification approach. Fluorescein or HPTS (15% w/w of polymer) was dissolved in 35 µl of water and was added to the polymer solution (1ml). A $\mathbf{w_1/o}$ emulsion is formed by sonication and 2-4 ml of 1% polyvinyl alcohol are then added. The solution is vortexed to form a $\mathbf{w_1/o/w_2}$ emulsion. The emulsion is poured into a 250 ml solution of 0.1% polyvinyl alcohol and stirred for several hours to allow solvent evaporation. The microspheres were separated as described above.

Sensor Configuration

The initial sensor design is shown in Figure 3. Microparticles with entrapped reagent were dispersed in a hydrophilic polymer layer, such as polyacrylamide or polyHEMA. The porous polymer contains the microspheres which upon degradation release reagents, diffusing into the polymeric matrix. Alternatively, HPTS was dispersed in a solution of polylactide-glycolide polymer and this mixture was applied to the fiber tip to provide a coating uniformly distributed on the sides and the distal end of the fiber.

Results and Discussion

The preparation of microspheres involves surface chemistry phenomena which are not well quantified and are hard to generalize. Therefore the preparation conditions and procedures are unique to each case and depend on both the polymer used and the characteristics of the reagent/substance to be entrapped. The sizes of the microspheres can be controlled by changing the stirring rate. In general, higher rates result in smaller diameter spheres. The microspheres obtained in this way varied between 2 to 200 µm, depending on the preparation conditions. Figure 4 shows poly (lactide-glycolide) copolymer microspheres containing HPTS dye. The use of polymers with various molecular weights affects the release rate from the microspheres as seen from Figure 5. The polymer of 100,000 molecular weight released over three months, while the lower molecular weight polymer released only over three weeks.

Sensor Design

When using sensors with the configuration in Figure 6, the response of the sensor did not correlate with the changes in pH in a reproducible fashion. Therefore we performed experiments to determine how the presence of entrapped dye at the tip of the fiber would affect the sensor response. Using a composite of PLGA and HPTS, we tested two different sensor configurations. The initial design had the tip covered with polymer, while later designs had polymer surrounding, but not covering, the tip (Figure 6). The response of each sensor was tested in different 0.1 M phosphate buffers. HPTS is a pH sensitive dye with both acid and base sensitive peaks. As seen

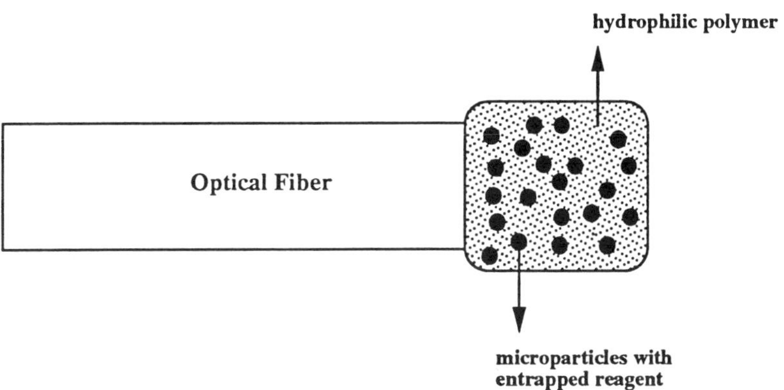

Figure 3. Initial sensor configuration of a fiber-optic sensor based on degradable microparticles.

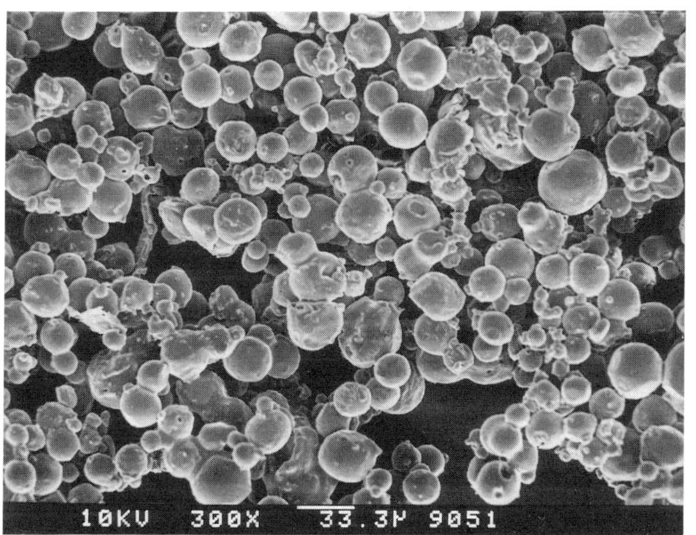

Figure 4. Microspheres of PLGA containing HPTS. The SEM shows a uniform distribution of sizes in the range 20-60 µm.

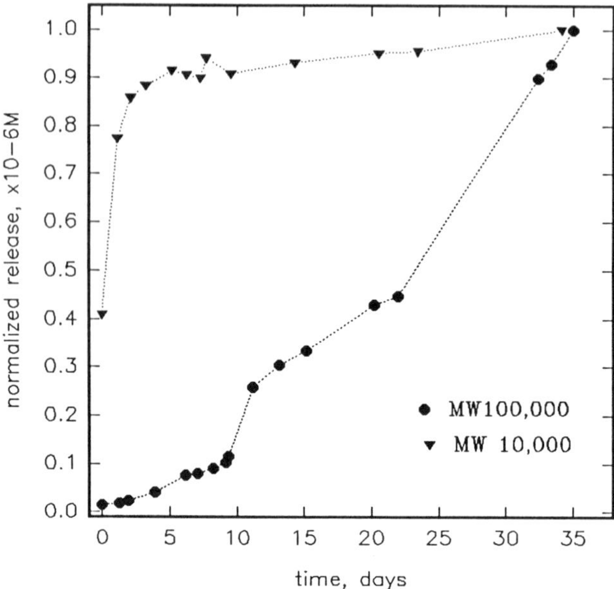

Figure 5. Release of fluorescent reagents from microspheres of PLGA of different molecular weights.

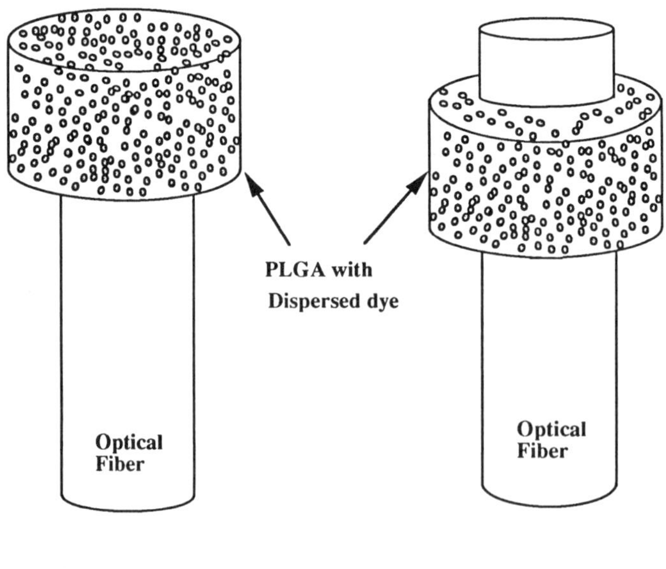

Figure 6. Configuration of pH sensors based on HPTS/ poly (lactide-glycolide) composite.

from the excitation spectra of the two sensor configurations (Figure 7 A and B), when the tip is covered, the acid peak at 405 nm possesses a high intensity at both acidic and basic pH, while the base peak at 460 nm remains virtually unchanged under both conditions. We attribute this result to the capability of the dye dispersed in the polymer to fluoresce, resulting in a relatively constant background signal corresponding to the acid peak of HPTS in the acidic micro- environment in the polymer. If micropores are formed in the polymer matrix and erosion proceeds, the pH in this environment will be acidic because of the character of the degradation products. This also will contribute to

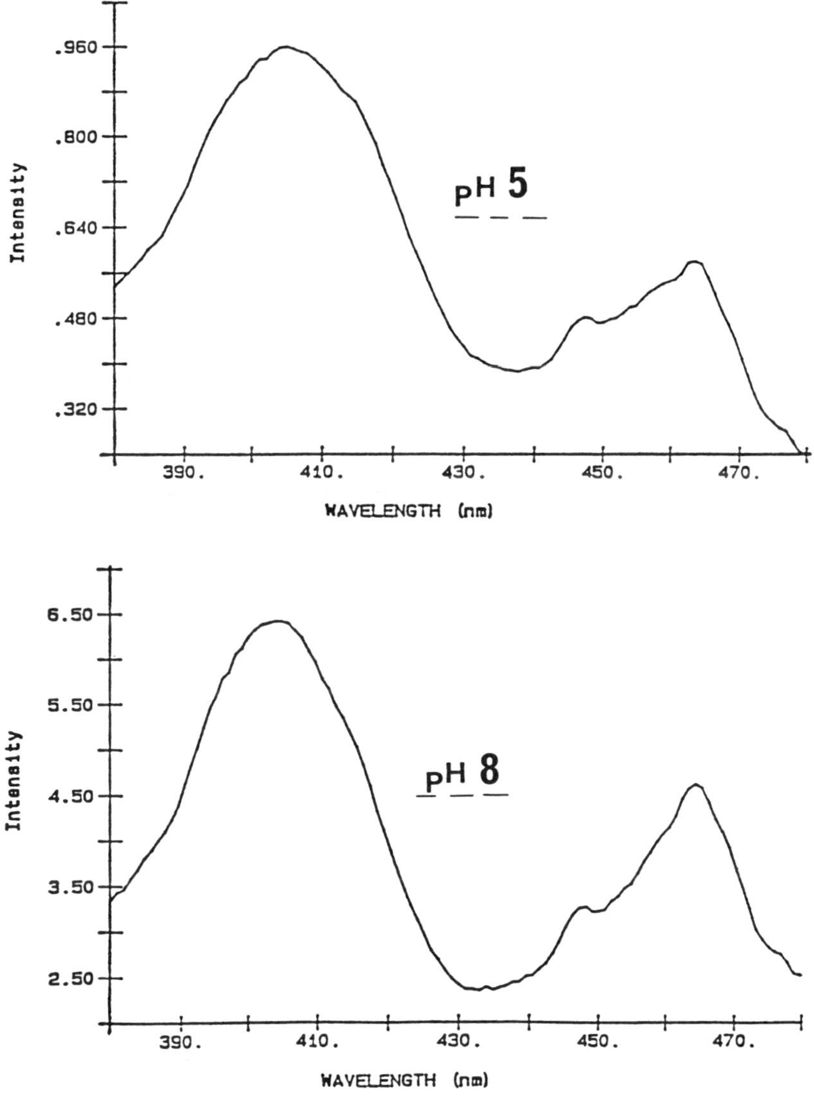

Figure 7A. Excitation spectra of HPTS sensor with a polymer/dye layer covering the sensor tip.

a high background of signal corresponding to HPTS in acidic pH. The sensor with no polymer on the tip responds as expected for HPTS: at pH 5 the peak at 405 nm is of high intensity while the peak at 460 nm is low; in pH 8.0 the situation is reversed (Figure 7 B). Taking the ratios of 460/405 nm we obtain the typical pH dependent titration curve of HPTS (Figure 8).

A pH sensor was tested based on fluorescein entrapped in PLGA of 10,000 molecular weight. Its response to pH is shown in Figure 9. It appears that the greatest sensitivity is in the pH range of 7.5-8.0. This can be attributed to the effect of the degradation products, i.e. lactic or glycolic acid which seem to create an acidic micro-environment and shift the pH optimum to more basic values.

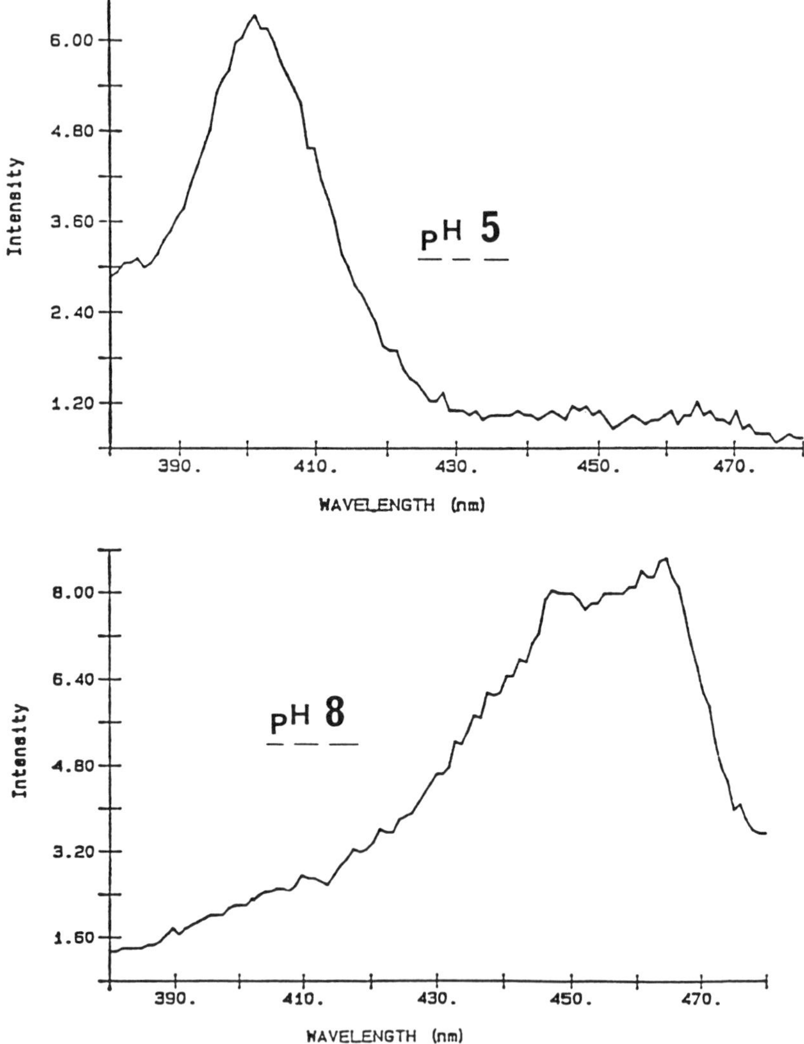

Figure 7B. Excitation spectra of HPTS sensor with fiber tip clear from polymer/dye layer.

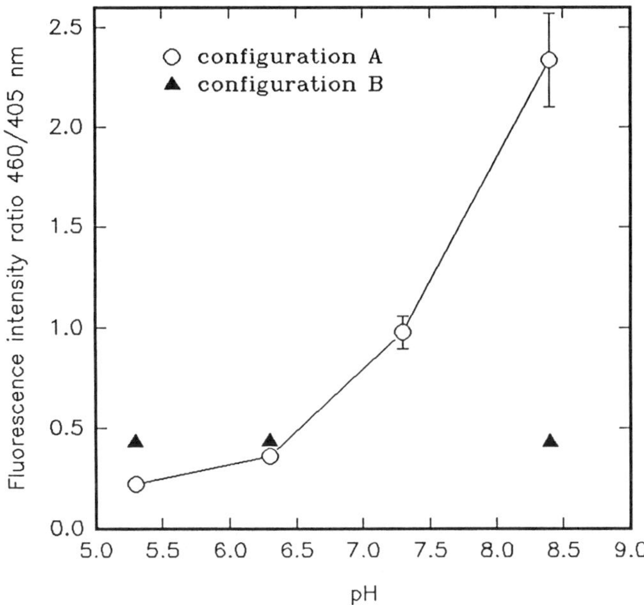

Figure 8. Comparison of responses of two sensors with different sensor configurations.

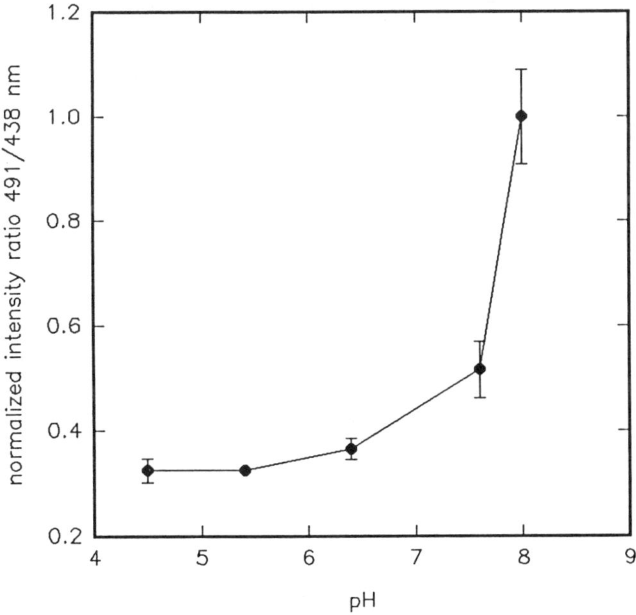

Figure 9. Response of pH sensor based on microspheres. The polymer was PLGA and the dye was fluorescein.

A

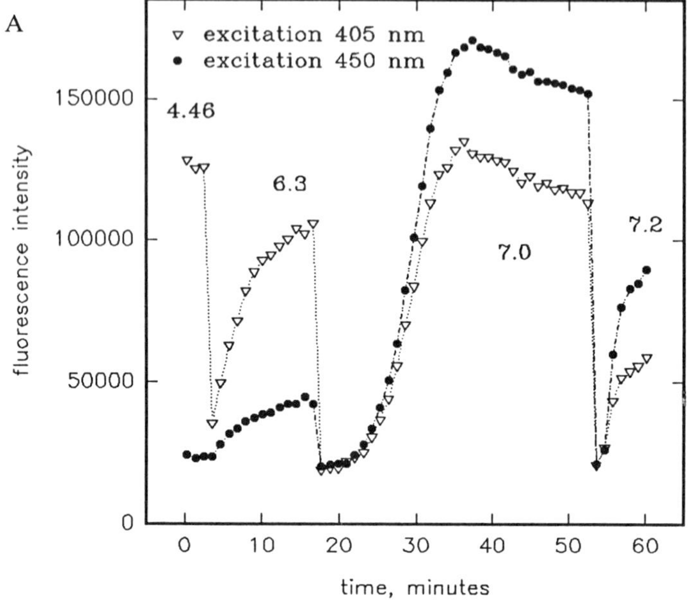

B

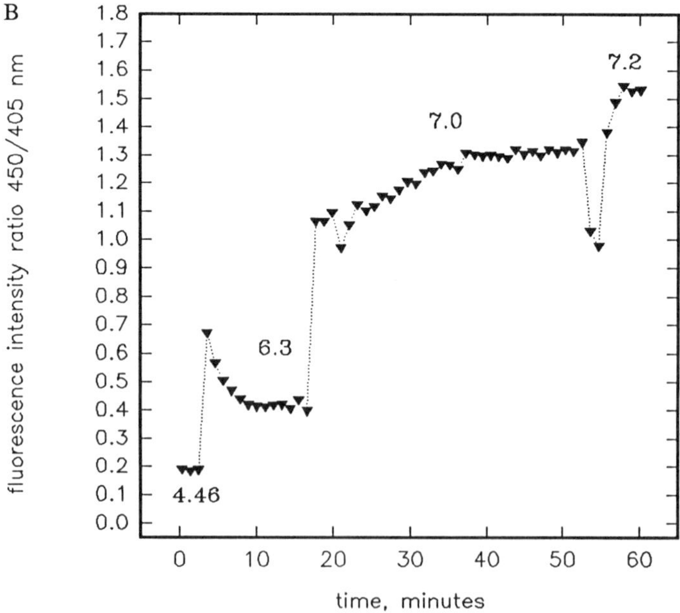

Figure 10. Response of pH sensor based on PLGA microspheres with entrapped HPTS. A. Fluorescence intensity at 520 nm when exciting at 405 and 450 nm. B. Changes in the intensity ratio when exciting at 450 and 405 nm with pH changes.

Experiments with fiber-optic sensors using HPTS microspheres embedded in the matrix of polyHEMA and deposited only around the sides of the fiber tip were performed. The response was tested in 100-200 µl 0.1 M phosphate buffer solutions (Figure 10, A and B). As can be seen, the response is essentially instantaneous. A problem however is maintaining an appropriate level of dye concentration around the sensor tip to elicit a response. Two competing design criteria are balancing the reagent release for rapid response with the necessity to build up sufficient dye concentration at the sensor tip to register a signal. These criteria require a high load of reagent inside the polymer to assure a useful response time. In large volumes the diffusion rate of the released dye away from the sensor tip is much faster than the degradation rate of the polymer replenishing the reagent. Further improvements of the sensor design to accomplish better balance between these two phenomena are under way.

Conclusions.

We have demonstrated that fiber-optic sensors can be prepared using degradable polymer microspheres which provide a continuous supply of reagent at the tip of the fiber. Using polymers of different molecular weights the release rate can be controlled to achieve the required response.

Acknowledgments.

This work was supported in part by a grant from the Environmental Protection Agency administered by Tufts Center for Environmental Management.

Literature Cited

1. Luo, S.; Walt, D.R. *Anal. Chem.* 1989, **61**, 174.
2. Barnard, S.M.; Walt, D.R. *Science* 1991, **251**, 927.
3. Luo, S.; Ph.D. Thesis, Tufts University, Medford, MA.
4. Langer, R.*Science* 1990, **249**, 1527.
5. Kulp, T.J., Camins, I., Angel, S.M., Munkholm, C., Walt, D.R. *Anal. Chem.* 1987, **59**, 2849.
6. Luo, S.; Walt, D.R. *Anal. Chem.* 1989, **61**, 1069.
7. Agayn, V.; Walt, D.R. *Proc. SPIE,* 1993.
8. Kreuter, 1985. In *Methods in Enzymology* **112**, 129.
9. J.Cowsar, D.R., Tice, T.R., Gilley, R.M., English, J.P. 1985. In *Methods in Enzymology* **112**, 101.
10. Forni, F., Coppi, G., Vandelli, M.A., Cameroni, R., *Chem. Pharm. Bull.* 1991, **39**, 2091.
11. Ogawa, Y., Yamamoto, M., Okada, H., Yasiki, T., Shimamoto, T., *Chem. Pharm. Bull.* 1988, **36**, 1095.
12. Benita, S., Benoit, J.P., Pusieux, F., Thies, C.J. *J. Pharm. Sci.* 1984, **73**,1721.
13. Cohen, S., Yoshioka, T., Lucarelli, M., Hwang, L.H., Langer, R. *Pharm. Res.* 1991, **8**, 713.

RECEIVED January 21, 1994

Chapter 3

Enhancement of the Stability of Wired Quinoprotein Glucose Dehydrogenase Electrode

Ling Ye[1], Ioanis Katakis[1], Wolfgang Schuhmann[2], Hanns-Ludwig Schmidt[2], Johannis A. Duine[3], and Adam Heller[1]

[1]Department of Chemical Engineering, The University of Texas at Austin, Austin, TX 78712
[2]Lehrstuhl für Allgemeine Chemie und Biochemie, Technische Universität München, Vöttingerstrasse 40, D–85350 Freising-Weihenstephan, Germany
[3]Department of Microbiology and Enzymology, Delft University of Technology, Delft, Netherlands

Quinotrotein glucose dehydrogenase (EC1.1.99.17) was electrically wired to a glassy carbon electrode with an osmiun complex containing redox polymer. The current output of the electrode reached 1.8 mA cm^{-2} at 70 mM glucose and it was insensitive to oxygen and nearly insensitive to pH change in 6.3 to 8.3 pH range. The half life of the electrode was about 8 hours. The main cause of the current decay was the leaching of the enzyme from the redox polymer/enzyme film. The stability of the electrode was improved by increasing the extent of the crosslinking of the film and/or overcoating the electrode. The overcoated electrodes had typical half life of 33 hours.

In our previous paper (*1*), we reported the high current density wired quinoprotein glucose dehydrogenase GDH electrode. In addtion to the extraordinarily high current density, the total insensitivity to oxygen and the wide pH operation range make the wired GDH electrode very attractive. However the reported half life time of the GDH electrode was only 8 hours. In the present paper, we report the results of the study for enhacing the operational stability of the wired GDH electrode.

Materials and Methods

Glucose oxidase (EC 1.1.3.4, from Aspergillus niger) and bovine serum albumin (BSA, fraction V) purchased from Sigma were used as received. 1 M glucose (from Sigma) stock solution was made with deionized water, allowed to mutarotate for more than 24 hours at room temperature and was stored at 4°C. Poly(ethylene glycol 400 diglycidyl ether) (PEGDE) was purchased from Polysciences and was used as received. The other chemicals were reagent or better grade. A pH 7.3 phosphate buffer solution (PBS), containing 0.15 M NaCl and 20 mM Na_2HPO_4, was used for all electrochemical measurements. Deionized water was used throughout the experiments. Potassium hexachloroosmate was purchased from Strem Chemicals and used as received. Polyallylamine (PAL) was purchased from Polysciences and used as received.

0097–6156/94/0556–0034$08.00/0

The redox polymers, POs-EA, was polyvinylpyridine partially N-complexed with [osmium bis(bipyridine) chloride]$^{+/2+}$ and quaternized with bromoethylamine. The synthesis followed the published procedures (*1*). The structure and the composition of POs-EA is shown in Fig. 1. The elemental analysis and UV-VIS spectroscopy showed that the repeating unit of the POs-EA had 1.1 ethylamine pendant functions and 1.8 unsubstituted pyridine rings per osmium complexed pyridine.

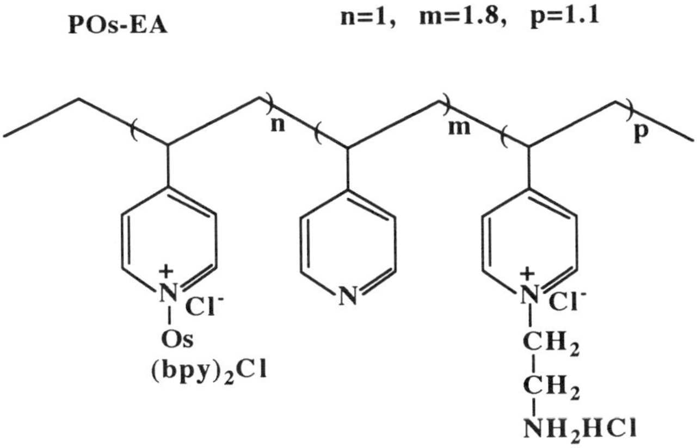

Fig. 1. Structure of the redox polymer POs-EA.

GDH apo-enzyme and PQQ were obtained as described by Van der Meer (*2*). 500 μl of the apo-enzyme (1mg ml^{-1} in a phosphate buffer, pH 7.0) was reconstituted by addtion of 10 μl of the PQQ (5 mg ml^{-1} in 10 mM HEPES with 3 mM $CaCl_2$, pH 8.0), and the mixture was incubated at room temperature for 2 hours. The enzymatic activity of the reconstituted holo-enzyme was assayed spectrophotometrically by monitoring the decoloration of Wurster's Blue (*3*). The reconstituted GDH (in its reconstitution solution) can be stored at 4°C for more than two months without measurable loss of activity. The isoelectric focusing electrophoresis was done using the procedure of Katakis, Davidson and Heller (Katakis, I.; Davidson, L.; Heller, A. in preparation for publication).

Rotating disk electrodes were made by press fitting 3 mm diameter glassy carbon rods into Teflon housing. Stationary disk electrodes were prepared by sealing the glassy carbon rods into glass tubings with a fast curing epoxy. The surfaces of the electrodes were polished with rough, then increasingly fine sand paper, then with 5 and 1 μm alumina to a mirror finish. The electrodes were sonicated for ca. 5 minutes and rinsed with deionized (DI) water after each polishing step. The enzyme electrodes were prepared by applying sequentially 1 μl of the polymer POs-EA (5 mg ml^{-1} in DI H_2O), 0.5 μl of GDH (1 mg ml^{-1} in the reconstituted solution) and 1 μl of freshly made solution of PEGDE 400 (2 mg ml^{-1} in DI H_2O) onto the electrode surface, unless otherwise indicated. The components were mixed, allowed to dry, then cured for 24 hours at room temperature. Before the electrochemical measurements, the electrodes were soaked in phosphate buffer solution (PBS) for 30 minutes and rinsed with DI water.

The electrochemical measurements were carried out in a conventional 3-electrode electrochemical cell. A Model 400 bipotentiostat from EG&G Princeton Applied Research was used for the amperometric measurements, and a KIPP&ZONEN BD101 strip chart recorder was used to record the data. The rotator used was from Pine Instruments. All steady-state currents were measured at 0.4 V vs SCE.

Results and Discussion

Figure 2 shows the current dependence of a GDH electrode on glucose concentration. The steady state current at saturating glucose concentration reached 1.8 mA cm^{-2}, three times the current density for the similar glucose oxidase electrode. To the best of our knowledge, this was the highest current density ever

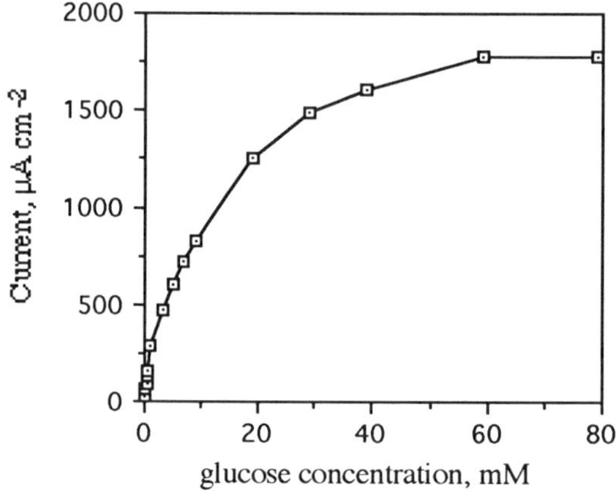

Fig. 2 Current dependence of the GDH electrode on glucose concentration. GDH loading=10% by weight; PEG concentration=12% by weight; E=0.4 V vs SCE; Rotation rate=1000 RPM; Argon atmosphere.

reported for mediated glucose electrodes. This was not very surprising if we notice that the natural electron acceptor for GDH is chytochrom b_{562}, a rather big molecule, while that for GOX is oxygen. From the x-ray crystal structure of GOX (*4*), we know that the FAD redox centers of GOX are buried about 15 Å deep in side the insulating glycoprotein shell. Since oxygen is such a small molecule, it can diffuse easily into the active site of GOX and take electrons from $FADH_2$ efficiently. But for the redox polymer used, a strong complex formation with GOX is required in order to shorten the distance between the osmium centers and $FADH_2$ for the electron exchange to occur (Katakis, I.; Davidson, L.; Heller, A. in preparation for publication). However, in the case of GDH, since its natural electron acceptor, chytochrome b_{562}, is a large molecule, the PQQ redox center of GDH is most likely loacated close to the surface of the enzyme molecule, i.e., the distance for the electron transfer between the osmium centers of the redox polymer and $PQQH_2$ is short. If the Marcus theory holds in this case, i.e., the electron transfer rate decreases exponentially with increasing distance, the electron transfer from $PQQH_2$ is faster than from $FADH_2$ to the osmium center of the redox polymer. This faster electron transfer is translated to the higher current acheived with wired GDH electrode. This agrees with the reported results for apparent bimolecular electron transfer rate constant from the reduced active centers to free diffusing mediators: this rate constant of electron transfer from reduced GDH to ferrocene monocarboxylic acid, 96.0x10^5 L mol^{-1} s^{-1}, is about 50 times that from reduced GOX, 2.01x10^5 L mol^{-1} s^{-1} (*5*). In fact, direct electron transfer from PQQ enzymes to the surface of a solid electrode has been reported (*6*). We also observed some direct electron transfer from GDH to high surface area graphite electrodes. The sensitivity of the GDH electrode was 120 μA cm^{-2} mM^{-1} at 6 mM glucose concentration. The apparent Km of the electrode obtained from the Eadie-Hofstee plot is 7 mM.

In addition to the extraodinarily high current density, another advantage of GDH electrode over GOX electrode is that GDH electrode is totally oxigen insensitive, as reported in our preveous paper (*1*). However the half life time of the GDH electrode was only 8 hours (dotted line in Fig. 4). We performed some preliminary studies to elucidate the mechanism responsible for the fast decay of the current.

Figure 3 shows the results of isoelectric focusing (IEF) electrophoresis. The electric field is indicated by the positive and negative signs. Samples were premixed and deposited directly on the gel. 10 μg of each compound was used for

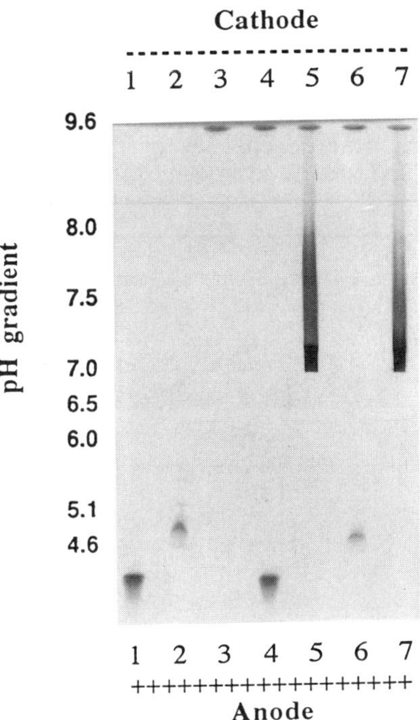

Fig. 3 Isoelectic focusing electrophoresis. Samples deposited on lane 1 to 7: 1, GOX; 2, BSA; 3, GDH; 4, GOX and GDH; 5, GOX, GDH and POs-EA(I); 6, BSA and GDH; 7, BSA, GDH and POs-EA.

each lane. From lane 1 to 7, the samples deposited on the gel are 1:GOX; 2: BSA; 3: GDH; 4: GOX and GDH; 5: GOX, GDH and POs-EA(I); 6: BSA and GDH; 7: BSA, GDH and POs-EA. As expected, the negative charged GOX and BSA migrated toward the anode (the positive pole), and the positive charged GDH and POs-EA migrated toward the cathode (the negative pole). When GOX and GDH are deposited together (lane 4), they migrated to the opposite direction and formed clear bands seperately, indicating that there was no detectable complex formation under the experimental conditions. However, when GOX, GDH and POs-EA were deposited together (lane 5), the GOX band disappeared, indicating that GOX and POs-EA formed a strong complex which was not separable by the electrophoresis. The clear band of GDH at the top of lane 5 indicates that GDH did not form a detecable complex with either GOX or POs-EA. The same results were obtained with BSA (lane 6 and 7), i.e., BSA formed complex with POs-EA but GDH did not complex with either.

As we see from Figure 3, GDH has an isoelectric point of 9.6. Because it is highly positively charged and all of the amino groups are protonated at neutrual pH, it can not react with the diepoxide which was used as the cross linker to form the redox polymer/enzyme hydrogel. It also does not complex with the redox polymer, as shown by the IEF results in Figure 3. Essentially, for a GDH electrode made with POs-EA and the diepoxide cross linker, GDH was simply entrapped in the three dimensional network of the cross linked POs-EA. If this entrapment is not sufficient, as it is the case for a losely cross linked, highly porous hydrogel, GDH is very likely to leach out from the redox polymer/enzyme hydrogel, especially when there is a repulsion force between the redox polymer and GDH since both are highly positively charged at pH 7.3. If our speculation is correct, we should be able to stabilize the current output of the GDH electrode by preventing the leaching of GDH.

Figure 4 shows the time dependence of the currents of different GDH electrodes. The dotted line was the decay curve of normal GDH electrode made

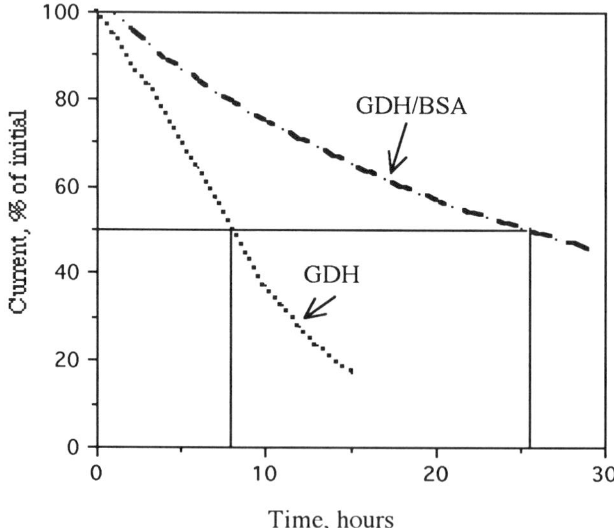

Fig. 4 Comparison of the operational stability of GDH electrodes at 21.3°C. E=0.4 V vs SCE; Rotation rate=1000 RPM; Argon atmosphere.

only with POs-EA, GDH and the diepoxide as a cross linker. The dashed line is the decay curve of a GDH electrode with BSA in the film. When 40% by weight BSA was added to the film, the half life of the electrode was prolonged from 8 hours to 26 hours. But the current of the electrode was decreased by about 58%. The possible reasons for the stabilization of the GDH electrode by coimmobilized BSA are the following: First, since BSA is negatively charged at pH 7.3, it may serve as a shield in the film to decrease the repulsion between GDH and the positively charged redox polymer. Secondly, we have seen from the IEF results that BSA formed a complex with POs-EA. This might result in a less porous or more crosslinked network so that the leaching of GDH is decreased. Finally, the increase in the stability by coimmobilization of BSA could be due to the decrease in current density since the decay of the current was associated with the electrochemical turnover of the sensor (*1*).

The solid line in Figure 4 is the decay curve for a GDH electrode which was overcoated by polyallylamine (PAL) as described else where (Katakis, I.; Ye, L.; Duine, J.; Schmidt, H.-L.; Heller, A. in preparation for publication). The stability was futher improved to a half life time of 33 hours because that the top layer of crosslinked PAL prevented the leaching of the enzyme. It is worthwhile to point out that the PAL overcoating increased the half life time of the GDH electrode by a factor of 4 at the expense of only 22% decrease in current output. This suggests that the main cause for the fast decay of the uncoated electrode was the leaching of the enzyme.

The results with different GDH electrodes are summarized in Table I. Here we can see that overcoating the GDH electrode with PAL did not only enhance the operational stability of the electrode, but also increased the apparent Km of the electrode to 14 mM from 7 mM of the corresponding uncoated electrode, resulting in a wider dynamic measurement range.

Table I. Properties for GDH Electrodes

Electrode composition (by weight)	Apparent Km mM	Initial current at 100 mM glucose $\mu A\ cm^{-2}$	Half life time hours
80% POs-EA 8% GDH 12% PEGDE	7 ± 1	1800 ± 200	8 ± 1
40% POs-EA 40% BSA 4% GDH 16% PEGDE	2 ± 0.2	413 ± 50	26 ± 2
80% POs-EA 8% GDH 12% PEGDE over coated with PAL	14 ± 2	1400 ± 100	33 ± 2

Conclusion

The high current density achieved with wired quinoprotein glucose dehydrogenase electrode was due to the fast electron transfer between the osmium centers of the redox polymer and the reduced redox center of GDH, $PQQH_2$ which is most likely located close to the surface of the enzyme. The main cause of the fast decay of the GDH electrode is the leaching of GDH. The leaching can be prevented by incorporating a negatively charged protein in the redox polymer/enzyme hydrogel, or overcoating the electrode. Overcoating the GDH electrode with polyallylamine not only increased the operational half life time of the GDH electrode from 8 to 33 hours, but also increased the apparent Km of the GDH electrode from 7 to 14 mM without significant loss of the current.

Aknowledgment

The work at the University of Texas at Austin was supported by the Office of Naval Research, the Welch Foundation and the National Science Foundation.

Literature Cited

(1) Ye, L.; Hämmerle, M.; Olsthoorn, A. J. J.; Schuhmann, W.; Schmidt, H.-L.; Duine, J. A.; Heller, A. *J. Anal. Chem.* **1993**, *65(3)*, 238-241.
(2) Van der Meer, R.A.; Groen, B.W.; Van Kleef, M.A.G.; Frank, J.; Jongejan, J.A. and Duine, J.A. *Meth. Enzymol.* **1990**, *188*, 260-283.
(3) Dokter, P.; Frank, J. and Duine, J.A. *Biochem. J.* **1986,** *239,* 163-167.
(4) Hecht, H. J.; Kalisz, H. M.; Hendle, J.; Schmid, R. D.; Schomburg, D. *J. Mol. Biol.* **1993**, *229*, 153-172.
(5) Davis, G. *Biosensors* **1985**, *1*, 161-178.
(6) Ikeda, T.; Matsushita, F.; Senda, M. *Biosens. Bioelectr.* **1991**, *6*, 299-304.

RECEIVED February 7, 1994

Chapter 4

Amperometric Glucose-Sensing Electrodes with the Use of Modified Enzymes

F. Mizutani, S. Yabuki, and T. Katsura

National Institute of Bioscience and Human Technology, 1–1 Higashi, Tsukuba, Ibaraki 305, Japan

Two kinds of modified glucose oxidases (GOD's), a polyethylene glycol-modified GOD and a lipid-modified GOD were prepared and used in the construction of enzyme electrodes. The polymer- modified GOD and a mediator (ferrocene) were incorporated into carbon paste (CP) to prepare a ferrocene-mediated glucose-sensing electrode. The polymer-modified enzyme exhibited higher activity than native enzyme in the CP matrix, owing to the enhanced affinity toward the hydrophobic matrix resulting from the enzyme modification. The higher enzyme activity resulted in an enhanced electrode response to glucose. Another glucose-sensing electrode was prepared by using the water-insoluble, lipid-modified GOD: the modified enzyme was immobilized on the surface of a glassy carbon (GC) electrode with a thin Nafion coating. The lipid-modified GOD-based electrode showed high performance characteristics such as rapid response, high sensitivity and superior stability.

The modification of enzyme by attaching ions or molecules is a suitable way for providing it with useful functions. For example, the use of modifiers such as polyethylene glycol (PEG) (*1,2*) and synthetic lipids (*3*) enhances the affinity for hydrophobic environments. PEG-modified enzymes are soluble and active in various organic solvents as well as aqueous solutions (*1,2*). Lipid-modified enzymes are insoluble in aqueous solutions, but show catalytic activities in both aqueous and organic media (*3*). These unique properties lead us to apply PEG- and lipid-modified enzymes in the construction of enzyme electrodes.

In this article, we describe the preparation and use of amperometric electrodes for sensing the analytically significant substrate glucose based on a PEG-modified glucose oxidase (GOD) (*4,5*) and on a lipid-modified GOD (*6*). The PEG-modified GOD has been incorporated into a carbon paste (CP) electrode: the modified enzyme exhibits higher activity in a hydrophobic CP matrix than hydrophilic, native GOD (*7-10*). The lipid-modified GOD has been immobilized on a glassy carbon (GC) electrode with a thin Nafion overcoat. The water-insoluble modified enzyme is far more stable between the electrode surface and the polymer layer than native GOD (*11*).

0097–6156/94/0556–0041$08.00/0

Experimental

Reagents. The enzymes used were GOD (E.C. 1.1.3.4, from *Aspergillus* sp., Grade II, Toyobo) and peroxidase (E.C. 1.11.1.7, from horseradish, Grade I-C, Toyobo). Methoxypolyethylene glycol activated with cyanuric chloride (activated PEG; Mwt., 5,000) was obtained from Sigma Chemical, a lipid, *N*-(α-trimethylammonioacetyl)didodecyl-L-glutamate chloride, from Sogo Pharmaceutical, and Nafion [5% (w/v) solution, 1,100 equiv. wt.], from Aldrich Chemical. F-kit (Boehringer Mannheim) was used for the spectrophotometric measurement of glucose. The kit uses the enzyme pair hexokinase/glucose-6-phosphate dehydrogenase. Other reagents were of analytical reagent grade. Deionized doubly distilled water was used throughout.

Modified GOD's. GOD modified with PEG was prepared as follows. The GOD, as received (native GOD, 20 mg), and activated PEG (200 mg) were dissolved into 3 ml of 0.1 M borate-KOH buffer (pH 10). The solution was then incubated at 37°C for 2 h, after which the reaction between GOD and activated PEG was stopped by adding acetic acid to the solution to pH 6. Unattached PEG was removed in an ultrafiltration cell (Millipore) using a dialysis membrane (PTTK membrane, Millipore; cut off molecular weight, 30,000) and 0.05 M $(NH_4)HCO_3$ as a dialyzing solution. Finally, a light yellow-colored powder of PEG-modified GOD (ca. 30 mg) was obtained by lyophilization.

GOD modified with the lipid was prepared according to the procedure of Okahata et al. (*3*). A buffer solution (5 ml, 0.02 M potassium acetate buffer plus 0.1 M KCl, pH 6) containing 25 mg GOD was mixed with an aqueous dispersion (50 ml) of 100 mg of the lipid. The precipitate formed after incubation of the mixture at 4°C for 24 h was lyophilized. A light yellow powder (ca. 70 mg) was obtained.

The GOD content in these modified enzymes were determined by measuring the adsorption by flavin adenine dinucleotide in solutions of the lyophilized products (the solvents used were water and benzene for the PEG- and lipid-modified GOD, respectively) at 450 nm. The enzyme activities of the native and modified GOD's were measured by using a peroxidase/phenol/4-aminoantipyrine chromogenic system. The solution (or dispersed medium in the case of the lipid-modified GOD) was 0.1 M potassium phosphate buffer (pH 7, 25°C).

Enzyme Electrode Systems. Two kinds of glucose-sensing CP electrodes, CPE I and II, were prepared by using the PEG-modified and native GOD, respectively. The modified or native GOD, 1,1'-dimethylferrocene, and carbon paste (CP-O; Bioanalytical Systems), 1:1:8 by weight, were thoroughly mixed together. A portion of the mixture was placed in a hole (3 mm diameter, 4 mm diameter) at the end of the electrode body (model 11-2010, Bioanalytical Systems).

Similarly two kinds of glucose-sensing GC electrodes, GCE I and II were prepared by using the lipid-modified and native GOD, respectively. The GC disk electrodes used (diameter, 3 mm; model 11-2012, Bioanalytical Systems) were first activated according to the procedure of Wang and Tuzhi (*12*). After activation, a drop of benzene solution containing the lipid-modified GOD or of an aqueous solution (pH 7) containing native GOD was placed on the GC electrode surface, and each solvent was allowed to evaporate at room temperature. The surface density of each GOD on GC electrode was 0.7 mg cm^{-2}. Finally, a Nafion membrane coating was made by dip-coating each electrode in 0.5% (w/v) Nafion solution, which was prepared by diluting the 5% solution as received with a mixture of 2-propanol [50% (v/v)] and water (*13*), and the electrode was allowed to dry with the surface facing down at room temperature for 1 h. The thickness of each GOD/Nafion layer was several microns.

A potentiostat (HA-502, Hokuto Denko) was used in a three-electrode configuration for amperometric measurements. The enzyme electrode thus prepared, an Ag/AgCl reference electrode, and a platinum auxiliary electrode were immersed in 10 ml of a test solution of 0.1 M potassium phosphate buffer solution (pH 7) in a cylindrical cell. The measurements were carried under argon atmosphere with CPE I and II and in air with GCE I and II. The solution was stirred with a magnetic bar, and the temperature of the solution was kept at 25°C.

Further, a GC electrode coated with a lipid-modified GOD/Nafion layer was incorporated into an electrochemical flow detection system (Bioanalytical Systems) and the response to glucose was recorded. A flow-cell electrode (Model 11-1000, Bioanalytical Systems; electrode area, 0.14 cm^2) was used as the base electrode, and a lipid-modified GOD layer and a Nafion overcoat were similarly prepared by the procedure described above. The sample volume used was 10 μl and the flow rate was 1 ml min^{-1}.

Glucose-Sensing Electrode based on PEG-Modified Enzyme

Properties of PEG-modified GOD. The GOD content in the lyophilized powder of PEG-modified enzyme was determined to be 50%. The PEG-modified GOD was soluble in organic solvents, such as benzene and hexane, as well as in aqueous media. The GOD activities of the modified- and native enzymes were 15 and 105 mU mg^{-1}, respectively. This indicates that the enzyme activity is considerably reduced by the modification process. On the other hand, when the PEG-modified GOD was doped into CP, it exhibited a much higher activity than the native enzyme. The GOD activities on the surfaces of CPE I and II were ca. 0.1 and 0.02 U cm^{-2}, respectively. The higher activity of modified GOD in the CP matrix is attributable to its enhanced affinity for the hydrophobic matrix, which can be proven by the solubility of the modified enzyme in organic solvents such as hexane. The PEG-modified GOD is highly dispersed in the CP matrix, and the modifier protects GOD from denaturation by the oil contained in CP.

Glucose Response of Enzyme Electrode. The increase in the GOD activity in CP by modifying the enzyme with PEG resulted in enhanced response toward glucose on CPE I, compared to CPE II. Figure 1 shows the current responses to 5 mM glucose on CPE I and II. The electrode potential, 0.4 V *vs.* Ag/AgCl, was sufficient for the oxidation of 1,1'-dimethylferrocene to ferricinium ion, and hence for obtaining the ferrocene-mediated current response for glucose (*14*):

$$\text{Glucose} + \text{GODox} \rightarrow \text{gluconolactone} + \text{GODred}$$
$$\text{GODred} + 2\text{FcR}^+ \rightarrow \text{GODox} + 2\text{FcR} + 2\text{H}^+$$
$$2\text{FcR} \rightarrow 2\text{FcR}^+ + 2e^- \text{ (at the electrode)}$$

In this scheme, GODox and GODred represent the oxidized and reduced forms of GOD, respectively, and FcR^+/FcR, the ferricinium ion/ferrocene couple. As shown in Figure 1, the glucose response on CPE I was about five times as large as that on CPE II.

CPE I gave a linear current response up to 10 mM, and a significant increase in the current was still observed with glucose concentrations between 10-50 mM. The detection limit was 0.1 mM (signal-to-noise ratio = 5). The relative standard deviation for ten successive measurements of 5 mM glucose on CPE I was ca. 2%.

The effect of storage (in the test solution at 4°C) of CPE I and II were then examined. On each electrode, the response to glucose gradually decreased and become ca. 60% of the initial value after 2 weeks. Long-term stability was not improved by modification of the enzyme. The decrease in the electrode response was caused by leaching of the (modified or native) enzyme out of the CP matrix, since the solution used for storing CPE I or II showed significant GOD activity and the current response was not reduced when the electrode surface was covered with a dialysis membrane (Viscase Seals).

Glucose-Sensing Electrode based on Lipid-Modified Enzyme

Properties of Lipid-Modified GOD. The GOD content in the lyophilized product of lipid-modified GOD was calculated to be 33%. The modified enzyme was soluble in nonpolar organic solvents, such as benzene and chloroform. The GOD activity of the modified enzyme was 38 U mg^{-1}-solid, which corresponds to 115 U mg^{-1}-GOD. This shows that the enzyme activity does not decrease during the modification process and that the enzyme substrates, glucose and oxygen permeate easily into the lipid layer on the GOD molecule to reach active site.

After immersing GCE I and II in the test solution for 1 h, the GOD activities on the electrode surfaces were measured. GCE I exhibited a GOD activity of ca. 0.2 U cm^{-2}, whereas GCE II exhibited an activity lower than 1 mU cm^{-2}. In the case of CPE II, the native GOD placed on the electrode surface easily leaked through the Nafion overcoat, and only a small amount of the enzyme remained immobilized. In contrast, the water-insoluble, lipid- modified GOD was stable between the electrode surface and the polymer overcoat which permitted the much higher enzyme activity.

Glucose Response of Enzyme Electrode. Figure 2 shows current-time curves for GCE I and II. The potential of each electrode was set at 0.9 V *vs.* Ag/AgCl. The current on each electrode increased immediately after the addition of glucose and reached a steady state within a few seconds. As the Nafion layer is thin, the added glucose is expected to diffuse quickly through the layer, and is available to be oxidized by the GOD reaction:

$$\text{Glucose} + O_2 \xrightarrow{\text{GOD}} \text{gluconolactone} + H_2O_2$$

The hydrogen peroxide produced near the electrode surface is immediately oxidized to give an anodic current response.

As shown in Figure 2, the glucose response on GCE I was far larger than that on GCE II. The much higher activity on the electrode surface with GCE I is responsible for the larger glucose response.

The relative standard deviation for ten successive measurements of 0.2 mM glucose on GCE I was 1.3 %. GCE I gave a linear current response up to 3 mM, and a significant increase in the current was observed with glucose concentration in the range of 3-10 mM. The detection limit was as low as 0.2 µM (signal-to-noise ratio, 5). The effect of storage (in the test solution at 4°C) of CPE I was then examined for 3 weeks: the response to glucose did not decrease during this period.

Flow Injection Measurement of Glucose. The GCE coated with a lipid-modified GOD/Nafion layer thus showed high performance characteristics such as rapid response, high sensitivity, and high stability. These characteristics are particularly promising for the use of the electrode in flow injection measuring systems. Therefore we examined the analytical potential of the modified GOD/Nafion-coated GC electrode by measuring glucose in beverages. Figure 3 shows typical responses by the electrode. The peak current was proportional to the glucose concentration up to 10 mM, under the present experimental conditions, and the detection limit was 10 µM. Table I gives the results for the determination of glucose in beverages. Each sample was diluted with 0.1 M potassium phosphate buffer (pH 7) by a factor of 50 before use. The results were compared with those given by the F-kit method. The agreement was excellent; the regression equation between the results obtained by the present electrode method (x) and those by the F-kit method (y) was $y = 0.993x + 0.572$ for the seven samples given in Table I. The long-term stability of the electrode was examined by measuring 10 mM glucose 30 times a day each day for 12 weeks. The average value of the electrode response for the 30 measurements did not decrease for 10 weeks. Such an accurate and stable measuring system is suitable for the use in a flow arrangement.

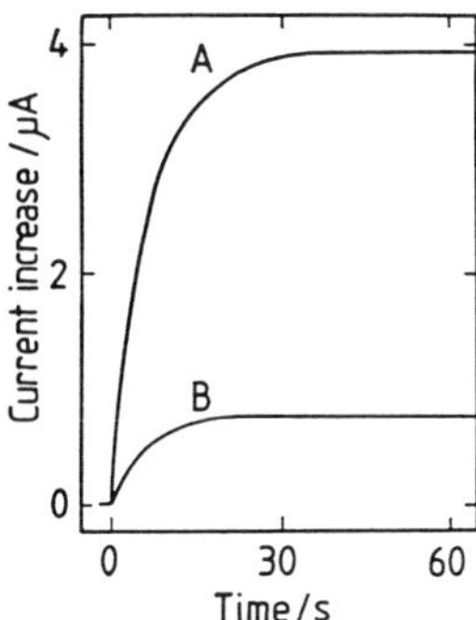

Figure 1. Response/time curves for (A) CPE I and (B) II to 5 mM glucose.

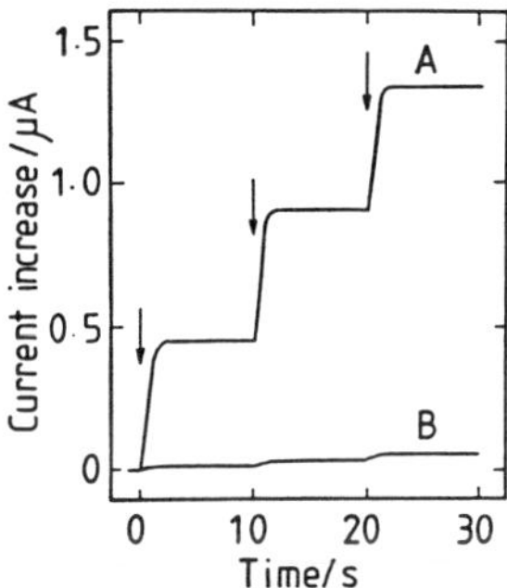

Figure 2. Response/time curves for (A) GCE I and (B) II obtained on increasing the glucose concentration in 0.5 mM steps.

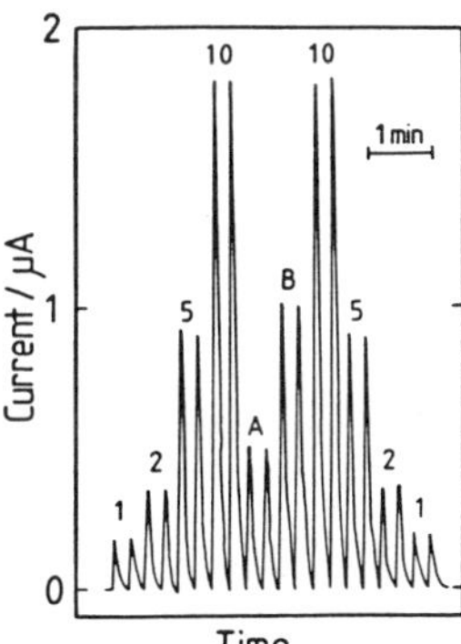

Figure 3. Recorded calibration and sample peaks for determination of glucose in beverages. 1, 2, 5 and 10, glucose concentrations of standard solutions given in mM; A and B, sample solutions (samples 1 [orange juice] and 2 [cola] in Table I, respectively). Each of the beverage was diluted by a factor of 50 with 0.1 M phosphate buffer [pH 7]).

Table I. Comparison of results obtained for glucose in beverages by different methods

Sample no.	Glucose concentration (mM)	
	proposed method (x)	F-kit method (y)
1	138	141
2	279	276
3	138	137
4	250	253
5	291	288
6	142	141
7	146	142

Conclusion

Modification of GOD with PEG was effective for enhancing the affinity of the enzyme for a hydrophobic CP matrix. A water-insoluble, lipid-modified GOD could be immobilized on the GC electrode surface far more tightly than the native enzyme. These modifiers produced high performance characteristics for GOD-based electrodes.

We have prepared and tested other modified enzymes. These include a choline-sensing CP electrode based on a PEG-modified choline oxidase, which exhibits a wide dynamic range (*15*), and a GC electrode, using a lipid-modified lactate oxidase (*16*), which shows a fast response to L-lactate.

Literature Cited

(*1*) Inada, Y.; Nishikawa, H.; Takahashi, K.; Yoshimoto, T.; Saha, A. R.; Saito, Y. *Biochem. Biophys. Res. Commun.* **1984,** *122,* 845-850.

(*2*) Takahashi, K.; Ajima, A.; Yoshimoto, T.; Okada. M.; Matsushima, M.; Tamura, Y.; Inada, Y. *J. Org. Chem.* **1985,** *50,* 3414-3415.

(*3*) Okahata, Y.; Tsuruta, T.; Ijiro, K.; Ariga, K. *Thin Solid Films,* **1989,** *180,* 65-72.

(*4*) Mizutani, F.; Yabuki, S.; Katsura, T. *Bull. Chem. Soc. Jpn.* **1991,** *64,* 2849-2851.

(*5*) Yabuki, S.; Mizutani, F.; Katsura, T. *Biosens. Bioelectron.,* **1992**, *7,* 695-700.

(*6*) Mizutani, F.; Yabuki, S.; Katsura, T. *Anal. Chim. Acta,* **1993,** *274,* 201- 207.

(*7*) Matsuszewski, W.; Trojanowicz, M.; *Analyst,* **1988,** *113,* 735-738.

(*8*) Gorton, L.; Karan, H. I.; Hale, P. D.; Inagaki, T.; Okamoto, Y.; Skotheim, T. A. *Anal. Chim. Acta,* **1990,** *228,* 23-30.

(*9*) Wang, J.; Wu, L.-H.; Lu, Z.; Li, R.; Sanchez, J. *Anal. Chim. Acta,* **1990,** *228,* 251-257.

(*10*) Amine, A.; Kauffmann, J.-M.; Patriarche, G. J. *Talanta,* **1991,** *38,* 107- 110.

(*11*) Wang, J.; Leech, D.; Ozsoz, M.; Martines, S.; Smyth, M. R. *Anal. Chim. Acta,* **1991,** *245,* 139-143.

(*12*) Wang, J.; Tuzhi, P. *Anal. Chem.* **1986,** *58,* 1787-1790.

(*13*) Harrison, D. J.; Turner, R. F. B.; Baltes, H. P. *Anal. Chem.* **1988,** *60,* 2002-2007.

(*14*) Cass, A. E. G.; Davis, G.; Francis, G. D.; Hill, H. A. O.; Aston, W. J.; Higgins, I. J.; Protokin, E. V.; Scott, L. D. L.; Turner, A. P. F. *Anal. Chem.* **1984**, *56,* 667-671.

(*15*) Yabuki, S.; Mizutani, F.; Katsura, T., unpublished work.

(*16*) Mizutani, F.; Yabuki, S.; Katsura, T., *Denki Kagaku*, in press.

RECEIVED September 14, 1993

Chapter 5

Electron-Transport Rates in an Enzyme Electrode for Glucose

Nigel A. Surridge[1], Eric R. Diebold[1], Julie Chang[2], and Gerold W. Neudeck[2]

[1]Boehringer Mannheim Corporation, P.O. Box 50457, Indianapolis, IN 46250–0457
[2]School of Electrical Engineering, Purdue University, West Lafayette, IN 47907

We describe here the kinetics and mechanism of electron transport in an enzyme electrode where glucose oxidase is connected to the electrode surface with an electron transfer relay consisting of an osmium polypyridyl complex pendent on a polypyridine backbone. The effective electron diffusion coefficient in this polymer hydrogel was measured in films containing various loadings of enzyme using a steady-state-method involving interdigitated array electrodes and was found to range between $1.6x10^{-9}$ and $<1.5x10^{-10}$ cm^2 s^{-1}. The behavior of this parameter as a function of enzyme loading was related to the mechanism proposed to be operating in these sensors.

With the introduction of commercial electrochemical micro-biosensors such as the Exactech system from Medisense and the silicon based, "6+" system from i-Stat, it is apparent that this technology will have a major impact on rapid blood chemistry determinations over the next 5 years. A classic example is that of glucose determination in small, whole blood samples, particularly those of self-monitoring diabetic patients. Here there are particular advantages to a completely electrochemical test, where the instrumentation can be designed in a more compact manner than an opto-electronic system, and can practically make determinations in 5 μL blood volumes obtained by finger-stick. As evidenced by the i-Stat system, there is clearly a market for small, disposable electrochemical tests in the emergency room, surgical and critical care environments as well as the home.

In most electrochemical systems used for glucose determination, a common element is the use of an electron transfer "shuttle" for transport of electrons between glucose oxidase and the electrode surface. This species can be a monomeric, freely diffusing mediator that is soluble in the analyte sample, or polymeric in nature, and immobilized on the electrode surface along with the enzyme in the form of a thin film or hydrogel. An example of the latter type has been pioneered by Gregg and Heller whereby a polymeric electron-transfer relay based on an osmium-substituted polyvinylpyridine polymer is co-immobilized with glucose oxidase on an electrode

0097–6156/94/0556–0047$08.72/0

surface, and crosslinked to form an insoluble hydrogel matrix which, nevertheless, can be rapidly hydrated and into which glucose diffusion is facile.(*1*,*2*,*3*)

In these particular systems, the polymer films are crosslinked and so stabilized on the electrode by the addition of poly(ethylene glycol) diglycidyl ether (PEGDGE) to the formulation prior to dispensing. A representation of the resulting matrix is shown in Figure 1, illustrating that the enzyme forms a tightly bound complex with the osmium polymer, and is embedded within it on the electrode.

In an electrochemical biosensor we are seeking to compete effectively with the enzyme's natural cofactor - O_2 in the case of glucose oxidase (GOx) - for the transfer of redox equivalents from the active site to the electrode surface. In the case of GOx particularly, a good choice for artificial electron transfer relay will be determined by a molecules ability to reach the reduced $FADH_2$ active site, undergo fast electron transfer and then transport this reductive equivalent to the electrode as rapidly as possible. The reaction sequence operative in these types of films may be summarized in Scheme I,

$$GOx(FAD) + G \underset{k_{-1}}{\overset{k_1}{\rightleftharpoons}} GOx(FAD).G \xrightarrow{k_2} GOx(FADH_2) + GL \qquad (i)$$

$$GOx(FADH_2) + 2Os(III) \xrightarrow{k_3} GOx(FAD) + 2Os(II) + 2H^+ \qquad (ii)$$

$$Os(II)_1 + Os(III)_2 \xrightarrow{k_e} Os(III)_1 + Os(II)_2 \qquad (iii)$$

$$Os(II)_2 \xrightarrow{fast} Os(III)_2 + e^- \qquad (iv)$$

SCHEME I

where Os(II) indicates the reduced form of the relaying complex and Os(III), the oxidized form, and GL is gluconolactone. Reactions (i) and (ii) describe the enzyme-substrate reaction according to the uni uni uni uni ping pong mechanism(*4*) where the natural cofactor is replaced by two Os(III) centers undergoing one-electron transfers with the $FADH_2$ sites.

In this work we seek to understand more fully that part of the sequence which transports the reducing equivalents from the enzyme to the electrode by physical diffusion and/or self-exchange reactions, characterized by k_e, Scheme I, using electrochemical techniques established with related redox polymer films.(*5*,*6*)

Hill and Cass have previously asserted that the important rate of electron transfer between GOx and, in their case ferrocene mediator, can be determined in homogeneous solution phase by the use of electrocatalytic currents obtained from the steady-state GOx-glucose reaction, and working curves derived by Nicholson and Shain relating diffusion limited and catalytic currents.(*7*,*8*) A priori, this approach

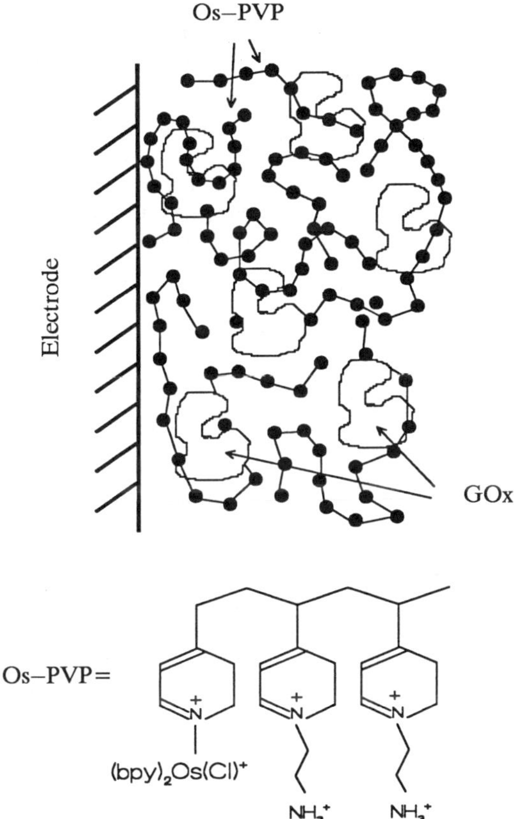

Figure 1. Schematic depiction of the Polymer/Enzyme hydrogel films.

is rather difficult to implement in these films given the requirement that the concentration of enzyme be substantially larger than that of the mediator.(*9*) In the operational configuration of these films the reverse is always true. In any case, Savéant and Bourdillon have lucidly demonstrated the limitation of assuming "pseudo-first-order" kinetics when using high concentrations of glucose and low concentrations of the enzyme.(*10*) In fact they have outlined a treatment of voltammetric data pertaining to the general case of coupled second-order reactions, such as those in Scheme I, which should also be applicable to the films described here.

Gregg and Heller have interpreted the activated processes involved in these enzyme-redox polymer films using a combination of Arrhenius plots and maximum current density behavior.(*2*) In that work, electron exchange between Os sites in the immediate vicinity of the enzyme active site, equation (iii) Scheme I, was clearly implicated in rate control over much of the formulation range studied. In other words, the active site was considered as a point source of electrons causing a locally low concentration of Os(III) around the enzyme, and slowing the escape of the catalytically produced reductive equivalents from the immediate polymer regions surrounding the $FADH_2$.

In this paper, we explore the behavior of the self exchange reaction over a formulation range where similar phenomena are occurring, using steady-state measurements of the apparent electron diffusion constant, D_e. Previously, values of D_e had only been obtained in films without inclusion of enzyme, thus yielding information only on the upper limit of electron transport.

The advantages of such measurements using Interdigitated Array Electrodes has been well documented, and includes decoupling of *macroscopic* ion motion from the rate of electron hopping when the ratio of oxidized to reduced sites in the film is constant.

The use of current measurements to determine both conductivity and ultimately intrinsic properties such as electron diffusion constants in polymer materials requires knowledge of the system geometry. This is especially true of Interdigitated Arrays, where the current carrying path is through material which lies both between and above the patterned metal fingers on the silicon. Theory has been developed to describe several geometric cases, the simplest of which deals with the case where the polymer film thickness, t, is less than or equal to the finger height, h, resulting in the simple form of equation 1,(*11*)

$$D_e = \frac{i_L dp}{Q} \cdot \frac{N}{(N-1)} \tag{1}$$

where p is the center-to-center distance between adjacent fingers, d the inter-finger gap or spacing, N the total number of fingers, Q the total charge under a slow scan (1 mV s^{-1}) cyclic voltammogram and i_L the steady-state limiting current flow between electrodes when one is set at a potential positive of the redox couple and the other at a potential more negative. (Q/F represents the number of moles of electroactive osmium centers on the IDA). This eqn has been shown to work well with thin, electropolymerized films just filling the gaps between fingers.(*5,12*) In addition,

Nishihara and Murray(*6*) have developed equation 2 which more appropriately describes a situation where a relatively thick film, ($t>h$), coats the IDA fingers, and contains a slowly diffusing redox couple.

$$D_e = \frac{i_L d}{nFCLt(N-1)} \tag{2}$$

L is the length of the fingers, and C the redox site concentration. This eqn holds equally well for films where redox sites are diffusionally immobile on the macroscopic scale, and conduction is due mainly to short range electron hopping by

$$D_{app} = D_{phys} + D_e \tag{3}$$

self-exchange reactions. In fact the Dahms-Ruff relationship, as written in equation 3, appears to hold along the continuum of site diffusivity in various types of redox conductive polymers,(*12,13,14*) where the apparent diffusion constant is a combination of physical site diffusion and diffusion by self exchange. In all films studied here, we assume that the epoxy crosslinked polymers form a matrix in which the long-range polymer diffusion contributions are negligible, and that $D_{phys}=0$. Under these conditions, electron transport is constrained to occur by self exchange reactions between Os sites. However, Blauch and Savéant have recently shown the importance of considering *short range* bounded motion of the redox sites on a flexible backbone as contributing to the self-exchange rate constant.(*15*) In the highly hydrated, hydrogel-like matrix of these films, we suggest that significant, local motion of the redox sites will occur in the process of forming activated complexes, and that these motions may be rapid on the timescale of the electron transfer reaction. Under conditions of local or "bounded" diffusion, a modification of the Dahms-Ruff and Laviron-Andrieux-Savéant(*16,17*) theory has been suggested relating the apparent diffusion coefficient to the activated self-exchange rate constant as shown in equation 4.

$$D_e = \frac{1}{6} k_e (\delta^2 + 3\lambda^2) C \tag{4}$$

where δ is the site-to-site distance at electron transfer, and λ characterizes the mean displacement of the site from its equilibrium position.(*15*) Under these conditions $k_e \approx k_{act}$, the *activated, bimolecular rate constant* rather than a combination of diffusion- and activation-controlled rate constants.

In this paper we attempt to analyze measured values of D_e according to some of these models as a function of enzyme content, and to rationalize the behavior of these same films when producing catalytic currents, according to the observed electron transport behavior.

Experimental

Electrode Fabrication. Due to its material versatility, flatness, dielectric properties (ϵ=3.9 at 300K), and processing simplicity, silicon was used as the base substrate for these highly resolved electrodes. 4000 Å of silicon dioxide was thermally grown on 3" wafers and borderlines etched with buffered hydrofluoride to mark an area for a maximum of 54 replicates of the IDA electrode dice. Rather than subtractive processing involving a metal etch, a liftoff procedure was employed to pattern the metal regions. Two levels of photoresist are spun onto the wafer, exposed and pattern defined on the SiO_2. 300 Å of titanium as an anchor layer was then evaporated onto the pattern, followed by 2000 Å of gold. (At this step, the masking in the evaporator allowed for metallization of only 35/54 possible dice). The photoresist and overcoated metal was then removed by acetone wash, and the dice cut out using a diamond saw. The highest yield achieved was 30 out of 35 metallized IDA's. Each die measures 12 mm by 5 mm with two contact pads of 2x2 mm at one end with contact rails of 100 μm width running to the interdigitated area at the other end of the die. Here, a total of 200 fingers are interlaced with a finger width and gap of 5 μm each.

Reagents. The Redox Polymer (Os-PVP) was synthesized in-house according to literature method.*(1)* The wt% of $(bpy)_2Os(vpy)Cl$ sites pendent on the PVP backbone was determined to be 14% from visible spectroscopy*(18)* and ICP analysis. This corresponds to about one Os complex for every three pyridine units on the PVP backbone. Glucose oxidase (Boehringer Mannheim Corp, Grade I, assayed at 200 units/mg) was used as received. Poly(ethylene glycol 400 diglycidyl ether) (PEGDGE) cross-linker (Polysciences #08210), and (4-(2-Hydroxyethyl)-1-piperazine-ethanesulfonic acid) (Hepes) (Boehringer Mannheim Corp) were also used as received.

The following stock solutions were prepared: 5 mg/ml Os-PVP in Nanopure water, 1.25 mg/ml PEGDGE in Nanopure water, and several GOx concentrations between 0-10 mg/ml in 5 mM Hepes buffer, pH 7.9. Formulations with varying enzyme concentrations were made by mixing the above stock solutions in a 2:1:2 ratio, respectively.

Film Fabrication. IDA electrodes were cleaned in a March Instruments Plasmod equipped with a GCM-100 gas manifold. They were subjected to an Ar plasma at 0.8 Torr and RF power of 50 watts for 3 minutes to remove organic contaminants, and also to increase surface hydrophilicity. 2 μL of the mixed reagents were then carefully dispensed onto the electrode fingers using a 2 μL Hamilton syringe. Extreme care was taken to avoid scratching the electrode fingers which would result in electrical shorting of the IDA electrodes. The reagents usually spread considerably beyond the dimensions of the fingers. In cases where the reagent did not cover all of the fingers, the drop was carefully manipulated with the syringe tip to cover all the fingers. 2 μL of reagent were also dispensed onto the 0.04 cm^2 working electrode of non Ar plasma treated planar electrodes. In the case of the planar electrodes, the drops were not spread beyond the dimensions of the working electrode. The electrodes were air dried at room temperature then placed in a forced air oven at 45°C for 15 minutes. Electrodes were further cured for a minimum of 48 hrs in a desiccated cabinet at room temperature.

Sections of the dry reagent films were removed around the perimeter of the IDA electrodes with the use of a micro-manipulator tip and a Cambridge stereo zoom microscope (see below). A Dektak IIA step profiler was used to determine the approximate film thickness of the IDA electrodes. Electrodes were tested for shorts between the electrode fingers using a Fluke 87 multimeter prior to electrochemical testing.

Electrochemical Measurements. Electrochemical experiments with the IDA electrodes were performed with an Ensman 400 bipotentiostat which is capable of independently controlling the applied potential at two electrodes versus a common reference, and measuring the currents i_A and i_B, flowing through each channel. This system was interfaced with a Keithley 575 data acquisition control system. The working electrodes comprised the two sets of Au fingers on an IDA, while a Pt wire counter electrode was wrapped around an ABTECH RE 803 Ag/AgCl mini-reference electrode. Measurements were performed in a solution of 10 mM sodium phosphate buffer containing 150 mM sodium chloride (pH 7.2). Cyclic voltammograms (E=100-600 mV) were run by shorting together both sets of fingers at scan rates of 10 mV/s (to equilibrate films) and then at 1 mV/s for data collection. Integrating the charge, Q, under the 1 mV/s cyclic voltammogram was used to determine the Os surface coverage on the IDA electrodes. i_L measurements were performed by monitoring i_A and i_B while independently controlling the potentials of both fingers. E_A was stepped from 100 mV to 600 mV while E_B was kept constant throughout the experiment at 100 mV. i_L was taken to be equal to i_B where it began to plateau vs time.

In the case of planar electrodes, a Bioanalytical Systems 100B Electrochemical Analyzer was used to collect cyclic voltammograms at various scan rates. A multi-channel potentiostat designed and custom built in house was used for amperometric measurements of sensor response to glucose. The planar 3-electrode design of these electrodes comprises a 0.04 cm^2 square gold working, gold counter, and Ag quasi-reference electrode on silicon, and these were used as received from Teknekron Sensor Development Corp (CA).

Results

Film Configuration and Homogeneity. A micrograph of a typical dried film of the Os-PVP/GOx/PEGDGE mixture on a 5μm IDA electrode (A) and (C), along with a bare IDA prior to casting the film (B) is shown In Figure 2. Contact rails (100 μm wide) can just be seen descending from the IDA area. By cleaning the Si/IDA electrodes in the Ar plasma prior to film deposition, the electrode surface becomes sufficiently hydrophilic to cause the drop to spread further on the surface, resulting in a thinner, more even, dry polymer film. This step proved important in generating films with a thickness that was both reproducible and of a value appropriate for application of the theory described (*vide supra*). Also seen in Figure 2A and 2C are

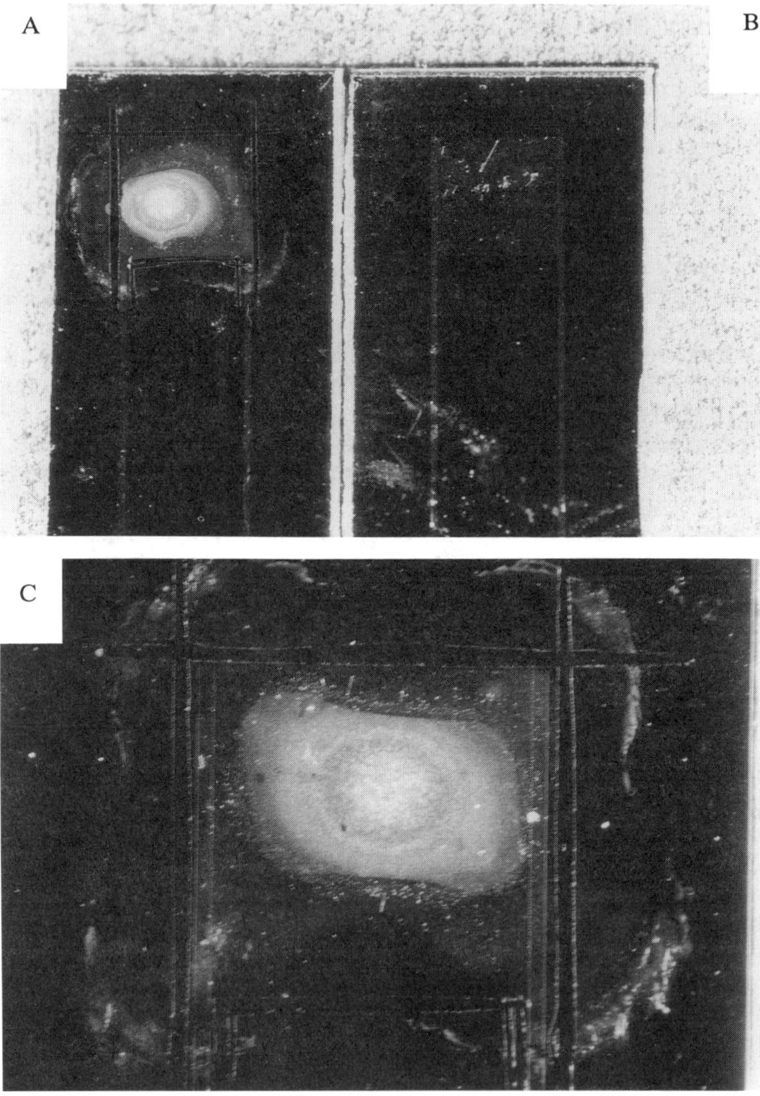

Figure 2. Photo-micrographs of Si/IDA electrodes with polymer films cast onto the electrode area (A and C), and blank electrode without film (B). (At this magnification the individual IDA fingers are not resolved).

sections of polymer film that have been removed around the perimeter of the electrodes using a micro-manipulator tip, thus forming a trench down to the underlying silicon. This was done to electrically isolate that region of the polymer that does not lie immediately over the gold fingers. (This polymer does not contribute to the measured film conductivity since it does not lie between any adjacent pair of fingers, and should not be included in the electrochemical determination of Os surface coverage, Γ, *vide infra*).

A step profiler was used to determine an approximate film thickness for each electrode studied by profiling a single section of the film along a line about halfway between the contact rails, and across the entire diameter of the polymer film. A typical profile is shown in Figure 3, where the silicon wafer at the bottom of the trench in the film can be seen at either end of the scan (these points indicated in Figure 3 were used as a reference for the film thickness, and planarity of the underlying wafer). In general, the films show considerable surface roughness at feature sizes on the order of $\leq 50\mu m$ horizontally, with peak-to-valley dimensions on the order of 5-25% of film thickness. This roughness is confined to the outer regions of the film, well removed from the fingers, and is expected to affect conductivity measurements to a lesser extent. However, several films also display a slight bowl shape in their profiles upon drying, although this is not pronounced for the film in Figure 3. The finger height of these IDA's is 2000Å, or about a third of the total film thickness shown in Figure 3. We can expect that the film conductivity in the center of the film may be less than that near the edges of the bowl where the film is thicker. Nevertheless, we feel justified in using these surface profiles to assign an *average thickness, t*, to the films used for conductivity studies, and to calculate the concentration of Os sites in a *dry* film (vide infra). These average thicknesses are shown in Table I.

Determination of Surface coverage of Os sites. Prior to measuring the limiting current between the finger sets of the IDA, all films were subjected to two slow-sweep cyclic voltammograms in phosphate buffered saline. This had two purposes: 1) To bring the films to equilibrium with the bathing electrolyte by oxidizing all possible Os sites. It is known that some regions of these types of polymers do not undergo reversible oxidation, and that once oxidized can not be re-reduced. This is confirmed by the fact that the second and subsequent scans are consistently smaller than the first. Many reasons have been postulated for this phenomena, including an irreversible phase transition within selected regions of the film. 2) To quantitate the total number of electroactive Os sites present on the electrode. This is done by integrating the charge, Q, under the *second* voltammogram, when the system is no longer changing. Figure 4 shows three different voltammograms of films with increasing amounts of GOx. These values are shown in Table I. Figure 4A and 4B show very similar shaped redox waves for films containing 40% and 47% GOx respectively, and yield similar values for total charge. In fact, all the film formulations with the exception of that containing the highest proportion of enzyme (64%), show similar electrochemical characteristics for the osmium oxidation-reduction. The highest enzyme level (Figure 4C) results in noticeably slower electron diffusion on the already slow timescale of the voltammetric measurement,

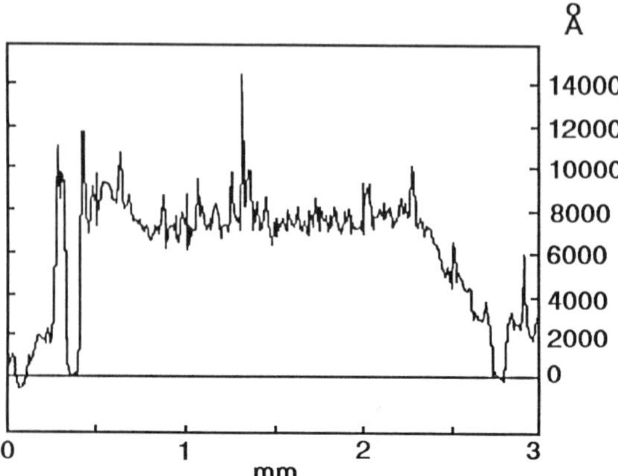

Figure 3. Typical Surface Profile (12 μm tip) taken of a film on an IDA. With reference to Figure 2, the line of scan is parallel to, and halfway between the contact rails.

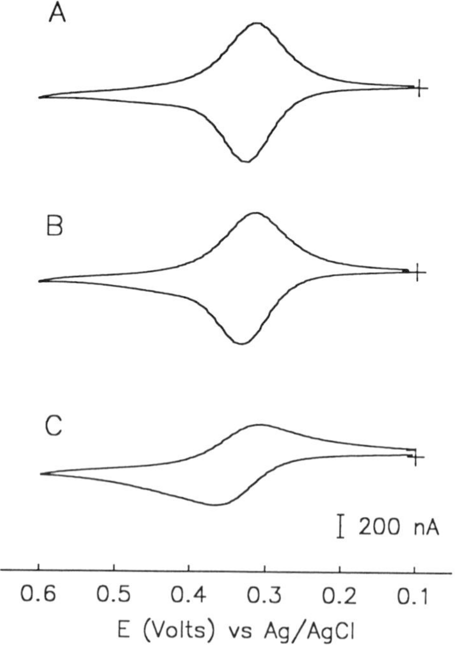

Figure 4. Cyclic voltammograms at 1 mV s^{-1} of Polymer/Enzyme films containing: (A) 40, (B) 47 and (C) 64 wt% of GOx in 0.15 M saline (buffered with phosphate, pH 7.2). Films were supported on the IDA electrodes as in Figure 1, and both finger sets were connected to the working electrode input of the potentiostat.

TABLE I. Film Parameters at various levels of GOx Enzyme loading[a]

wt% GOx	i_L (amps)	Q (C)[b]	Γ (mols cm^{-2})	t (Å)	C_t (mols cm^{-3})[c]	$D_{e,1}$ (cm^2s^{-1})	$D_{e,2}$ (cm^2s^{-1})
0 [d,e]	1.2×10^{-6}	5.0×10^{-5}	1.1×10^{-8}	-	-	1.2×10^{-8}	-
20 [d,e]	4.1×10^{-7}	6.4×10^{-5}	1.4×10^{-8}	-	-	3.3×10^{-9}	-
33 [e]	2.8×10^{-7}	5.5×10^{-5}	1.2×10^{-8}	-	-	2.5×10^{-9}	-
40 [f]	1.6×10^{-7}	5.2×10^{-5}	1.1×10^{-8}	5800	2.1×10^{-4}	1.6×10^{-9}	1.6×10^{-9}
40	1.4×10^{-7}	5.9×10^{-5}	1.3×10^{-8}	5300	2.6×10^{-4}	1.2×10^{-9}	1.2×10^{-9}
44 [f]	1.5×10^{-7}	7.1×10^{-5}	1.5×10^{-8}	7300	2.3×10^{-4}	1.1×10^{-9}	1.1×10^{-9}
44	1.0×10^{-7}	7.0×10^{-5}	1.5×10^{-8}	6600	2.5×10^{-4}	7.4×10^{-10}	7.4×10^{-10}
47 [f]	1.1×10^{-7}	7.5×10^{-5}	1.6×10^{-8}	6000	2.9×10^{-4}	7.3×10^{-10}	7.2×10^{-10}
47	8.0×10^{-8}	6.3×10^{-5}	1.3×10^{-8}	6000	2.5×10^{-4}	6.3×10^{-10}	6.3×10^{-10}
47	6.3×10^{-8}	6.3×10^{-5}	1.3×10^{-8}	5000	3.0×10^{-4}	5.0×10^{-10}	5.0×10^{-10}
52 [f]	6.2×10^{-8}	7.2×10^{-5}	1.6×10^{-8}	5900	2.9×10^{-4}	4.3×10^{-10}	4.3×10^{-10}
52	6.0×10^{-8}	5.5×10^{-5}	1.2×10^{-8}	5800	2.2×10^{-4}	5.5×10^{-10}	5.4×10^{-10}
52	5.5×10^{-8}	7.2×10^{-5}	1.5×10^{-8}	6100	2.8×10^{-4}	3.8×10^{-10}	3.8×10^{-10}
57	3.9×10^{-8}	6.6×10^{-5}	1.4×10^{-8}	6600	2.3×10^{-4}	3.0×10^{-10}	3.0×10^{-10}
64 [f]	1.5×10^{-8}	$>4.9\times10^{-5}$	$>1.1\times10^{-8}$	8200	$>1.4\text{x}10^{-4}$	$<1.5\times10^{-10}$	$<1.5\times10^{-10}$
64	1.4×10^{-8}	$>5.5\times10^{-5}$	$>1.2\times10^{-8}$	6400	$>2.0\text{x}10^{-4}$	$<1.3\times10^{-10}$	$<1.3\times10^{-10}$

[a] All films contain 4 μg of OS-PVP and 0.5 μg PEGDGE. [b] Area under a cyclic voltammogram with sweep rate = 1 mV/s. [c] Calculated from an area of 0.048 cm^2 for the IDA electrode and the value for t. [d] Average value for 3 IDA electrodes. [e] Experiment run at separate time from the remaining entries in the table. [f] Where there are multiple entries per GOx level, values of i_L were measured on consecutive days following film fabrication reading down the table.

as evidenced by increased diffusional tailing of the wave, and lower apparent surface coverage. *(It is noteworthy that the electrochemical behavior is so similar for all but the latter film, indicating that we are not grossly affecting the macroscopic film structure over this limited range of enzyme content).*

Using the value for surface charge and the values for average film thickness, t, it is possible to calculate an average concentration of osmium sites in each film using equation 5,

$$C_t = Q/nFLptN \quad (5)$$

where C_t is the concentration of sites in a dry film. These values along with surface coverages for each film, Γ, are shown in Table I.

i_L Measurements. Limiting current measurements were made immediately after obtaining slow sweep voltammogramms by poising both sets of IDA fingers (Channels A & B, Figure 5) at 100 mV vs Ag/AgCl with a bi-potentiostat until the current through both fingers had equilibrated close to zero, (usually less than 2 minutes). Baseline current was collected at this time, followed by application of a 500 mV step to the finger set on Channel A (t=0, Figure 5), thus poising each finger set on opposite sides of E° for the $Os^{II/III}$ couple (0.32 V vs Ag/AgCl). The current response was immediate on Channel A ($i_{L,A}$) as Os sites were oxidized adjacent to those fingers, this charge being represented by the peaks seen on Channel A in Figure 5. Current response on Channel B, (held constant at 100 mV), was slower to respond, but gradually increased to an equal and opposite value to that seen on Channel A as concentration gradients are generated within the redox polymer. The rate at which the "crossed concentration gradients" can be generated is, of course, dependent on the rate of charge redistribution which can occur in the polymer, and is indicated by the time taken to reach equilibrated currents on both channels. It is evident from Figure 5 that films higher in GOx content take longer to establish quasi-steady-state concentration gradients which is consistent with the voltammetric behavior described above.

Generally current-time traces for the films reached a quasi-plateau as seen in Figure 5 for three of the formulations, and it was possible to assign a limiting current value, i_L, at a point between 10 and 20 minutes after application of the step voltage to Channel A. These values are reported in Table I and used in equation 1 and 2 to generate estimates for the effective electron diffusion constant, $D_{e,1}$ and $D_{e,2}$ respectively, and these values are also reported in Table I. ($D_{e,1}$ - calculated from equation 1 - is the diffusion coeffecient calculated without theoretical reference to film thickness, while $D_{e,2}$ - calculated from equation 2 - does allow for the measured film thickness). It is interesting to note that the calculated values are independent of the theoretical approach taken, indicating that the vast majority of current is carried along flux lines that lie *parallel* to the surface of the IDA electrode rather than along semi-cylindrical paths above the fingers as described by Aoki.(*19*) (ie. D_e is independent of t in the range of t studied here, ≈5000-8000Å). Calculated values of $D_{e,2}$ are plotted on a log scale in Figure 6 from 0 to 70 wt% GOx in the formulation,

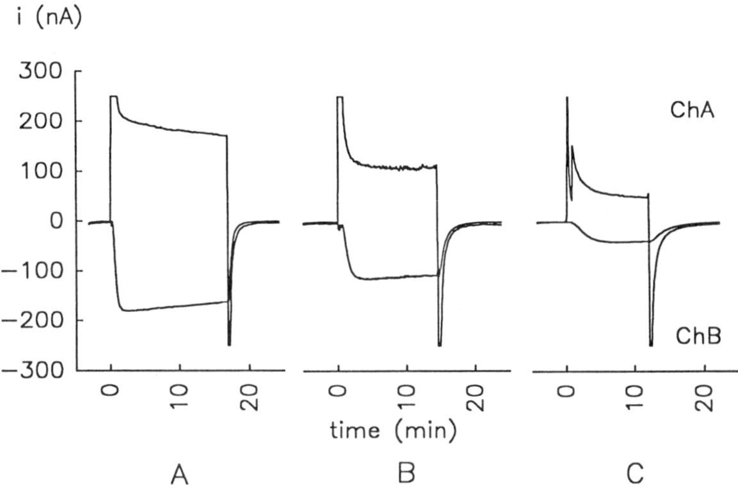

Figure 5. Limiting current (i_L) plots for IDA electrodes coated with films containing (A) 40%, (B) 47% and (C) 64% GOx by weight immersed in the same solution as Figure 4. Channel A stepped from 0.1 V to 0.6 V, Channel B held at 0.1 V (vs Ag/AgCl). Data obtained immediately after obtaining cyclic voltammograms in Figure 4.

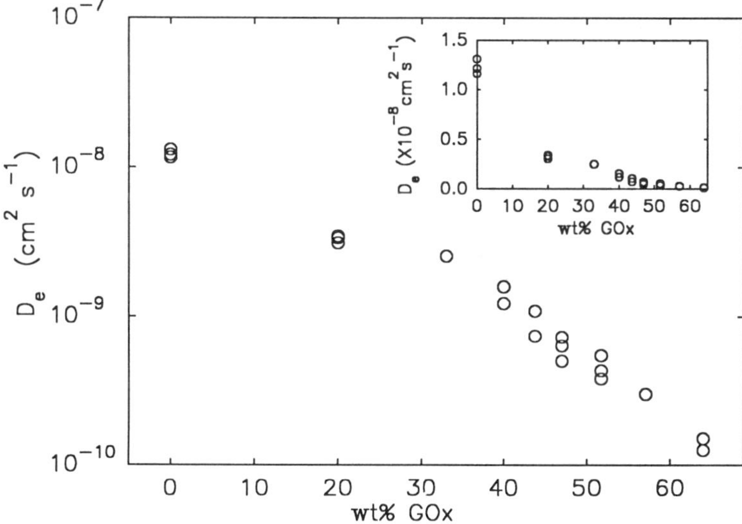

Figure 6. Values for $D_{e,2}$ as listed in Table I on logarithmic scale (linear scale on the inset).

and on a linear scale in the inset. The pure Os-PVP polymer, with only cross-linker added, yields values of ≈$1.2x10^{-8}$ cm^2s^{-1} which compares with a value of $4.5x10^{-9}$ cm^2s^{-1} for a batch of polymer synthesized and studied earlier.(*20*) These values also compare to $1.8x10^{-8}$ cm^2s^{-1} for an *electropolymerized* film of Os(poly-vbpy)$_3$, (vbpy=4-methyl-4'-vinyl-2,2'-bipyridine), as measured by IDA in acetonitrile solution.(*5*)

Cyclic Voltammetry at Planar Electrodes. The effect of counter-ion diffusion on measurements of electron mobility within redox polymers where ions are required for local electroneutrality has been the topic of much discussion in the literature.(*21,22,23*) The effect of reduced ion-mobility on fast or transient measurements of D_e has been well described and is often referred to as the electrostatic coupling, or migration effect.(*21*) The net result of this effect in polymer films such as those described herein, is to cause an overestimate when a transient technique is used to measure D_e, or for that matter, when any technique is used where ions are undergoing restricted motion while the measurement is in progress. It is this consideration that led to the use of IDA electrodes to measure limiting currents *after* steady state has been achieved in the film, and ions are no longer moving on the macroscopic scale.

To determine if this migration effect may be operative in this class of redox polymers, we conducted a study of the peak current dependency in cyclic voltammograms on sweep rate. Films were cast onto 0.04 cm^2, planar gold electrodes at the same time, and from the same formulations as those used in the IDA experiments. Given the relative diameter of the dried films, and the amount of material applied we know that these films are at least as thick as those cast on IDA's. After curing, they were immersed in phosphate buffered saline and the potential cycled several times between 0 and 0.5 V vs Ag/AgCl to equilibrate the films. Voltammograms were then obtained on at least two films per formulation at scan rates between 0.05 and 0.50 V s^{-1}. The anodic peak currents ($i_{p,a}$) are plotted in Figure 7 vs sweep rate and also the square root of sweep rate. Given the non-linear behavior seen in Figure 7A we can conclude that classical thin layer voltammetry is not an appropriate model in this experiment, which implies that at least at the scan rates employed here, the film acts more like a semi-infinite reservoir of redox sites. ie, diffusion profiles during the scans do not extend to the film/solution interface. Although the fact that the intercepts of the $i_{p,a}$ vs $v^{1/2}$ plots equal zero does support this assumption, close inspection of the plots also reveals a degree of non-linearity, and this may be an indication of inhomogeneity in the films. In any case, the least squares slopes of these plots are shown in Table II along with calculated values for $D_e^{1/2}C$.(*24*) Using average values for C calculated from the appropriate formulations in Table I we have an estimate for the diffusion coefficient derived from a non-steady-state method, and these values are also shown in Table II as $D_{e,3}$.

Calculation of Rate Constants. Using the values of D_e shown in Tables I and II and equation 4, it is possible to calculate the approximate bimolecular rate constants for self-exchange in the films. From the concentration of Os sites, and assuming a simple cubic lattice we can calculate the number of Os centers/cm^3, $N_T=N_AC_t$, and

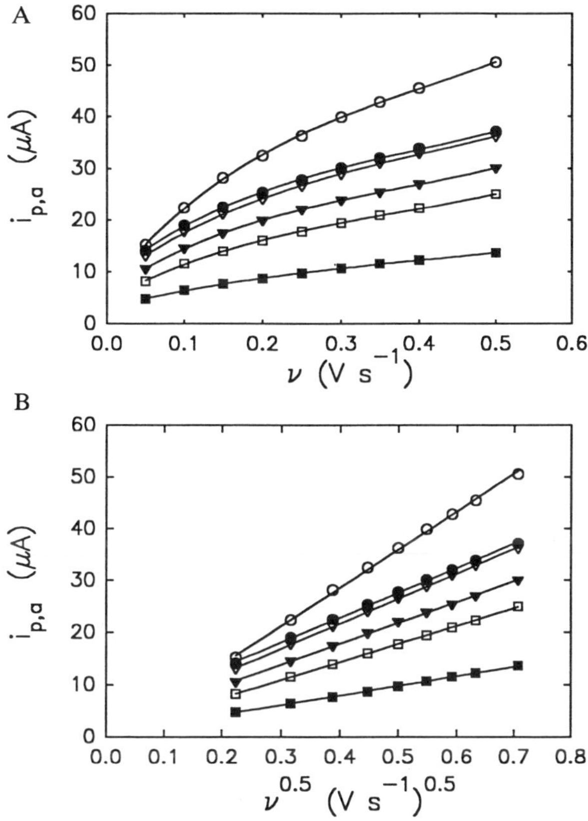

Figure 7. Anodic peak currents as a function of (A) scan rate ν, and (B) square root of scan rate ($\nu^{1/2}$), for films coated on 0.04 cm^2 planar gold electrodes containing, (○) 40%, (●) 44%, (▿) 47%, (▾) 52%, (□) 57% and (■) 64% GOx by weight. Solution as in Figure 4.

Table II. Cyclic Voltammetry and Rate Constants for Os-PVP/GOx Films

wt% GOx	Slope (Fig 7) ($\mu A\ V^{-1/2}\ s^{1/2}$)	$D_{e,3}^{1/2}C$ ($mols\ cm^{-1}s^{-1}$)	$D_{e,3}$ (cm^2s^{-1})	$k_{e,2}$ (Ms^{-1})	$k_{e,3}$ (Ms^{-1})
40	7.4×10^{-5}	7.2×10^{-9}	9.4×10^{-10}	2.6×10^{5}	1.7×10^{5}
44	4.8×10^{-5}	4.7×10^{-9}	3.8×10^{-10}	1.7×10^{5}	6.9×10^{4}
47	4.8×10^{-5}	4.7×10^{-9}	2.8×10^{-10}	9.7×10^{4}	4.4×10^{4}
52	4.0×10^{-5}	3.9×10^{-9}	2.3×10^{-10}	7.6×10^{4}	3.9×10^{4}
57	3.5×10^{-5}	3.4×10^{-9}	2.2×10^{-10}	5.7×10^{4}	4.2×10^{4}
64	1.8×10^{-5}	1.8×10^{-9}	1.1×10^{-10}	$<3.6\times10^{4}$	$<2.8\times10^{4}$

obtain an estimate for the spacing between adjacent Os sites in the film, $(N_T)^{-1/3} \approx 4$nm. Simplistically, we can estimate that the mean displacement of one Os site prior to encountering another during electron transfer is about half the equilibrium site separation which leads to the approximation, $\lambda \approx 2$nm (see equation 4) for all the films in Table I where C_t is evaluated. This is necessarily an oversimplification, but nevertheless, we have used this value along with both $D_{e,2}$ and $D_{e,3}$ and values for the concentration of Os sites derived from the dry film thickness measurements to calculate $k_{e,2}$ and $k_{e,3}$. These are both shown in Table II. It is likely that the dry film thickness values give an overestimate of the Os site concentration when the film is fully hydrated as it is when diffusion constants are measured, and this is discussed further below.

Sensor Response at Maximal Rates. In a parallel study to those measurements described above, we cast films of the same formulations again onto planar 2x2 mm gold electrodes, and cured in the same way. These electrodes were immersed in air equilibrated, phosphate buffered saline solutions containing either 2000 or 20000 mg/dL glucose, and potentials of 350 mV vs Ag applied for 40 seconds in a three electrode mode. The current response was recorded during this time, and at all formulations displayed a peak at early times corresponding to the oxidation of Os^{II} to Os^{III}. The current at longer times tended asymptotically to a steady state value corresponding to driving the glucose oxidase/glucose reaction (Scheme I, *vide infra*) at maximal rate, ie saturating glucose concentrations. The current value at 30 seconds is plotted in Figure 8 for two different experiments where [glucose]=2000 mg/dL, Figure 8A, and 20000 mg/dL, Figure 8B. In both cases, the response to glucose increases with increasing GOx, to a maximum value where GOx comprises about 45-50 wt% of the film. This confirms earlier studies on previous batches of the same polymer where the response rose as function of GOx in a pseudo first-order fashion up to point where the same sudden and rapid downturn in response occurred, see Figure 6 in reference 2. It is interesting that the higher glucose level actually results in depressed currents. We have explored the concentration range from 0 to 2000 mg/dL in detail, and response increases monotonically up to this point. The fall in current at higher substrate concentrations may be an indication of product inhibition of the system, although further work will be required to confirm this. In any case we will take the results at 2000 mg/dL as representative of the limiting behavior.

Discussion

The value for the electron diffusion constant calculated from these steady-state measurements for an Os-PVP film without enzyme, 1.2×10^{-8} $cm^2 s^{-1}$, Table I, compares to approximately 4×10^{-9} $cm^2 s^{-1}$ as previously determined for the same type of film by chronoamperomoetry,(*2*) and by IDA based measurements.(*20*) However, in the previous studies, the degree of Os complex substitution of the PVP backbone was approximately 1 in 5 of available pyridine rings. In this case, substitution is approximately 1 in 3, and it is possible that a slightly higher Os concentration in our films results in a faster rate of electron diffusion. Additionally we are measuring D_e

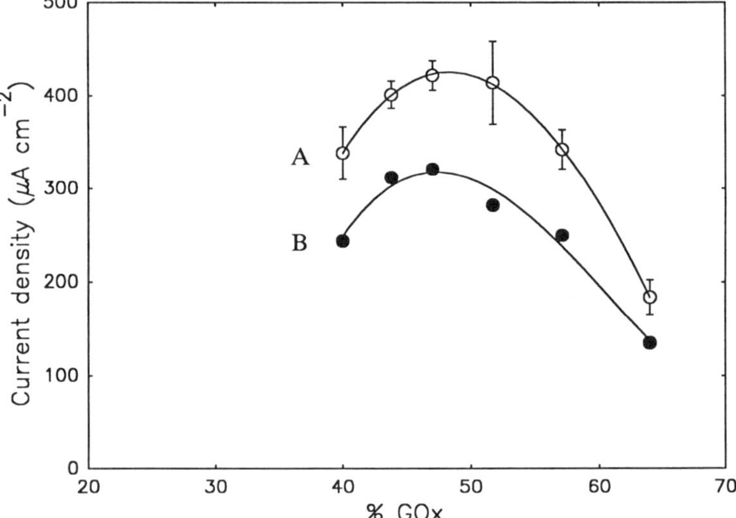

Figure 8. Current response of films with E_{app}=0.35 V vs Ag/Ag/Cl, coated on 0.04 cm^{-2} planar gold electrodes immersed in 0.15 M saline (pH 7.2) and (A) 0.111 M (○) and (B) 1.11 M (●) glucose. All solutions are air equilibrated.

over a much smaller distance, 5μm, than previously attempted. The first point is perhaps supported by the fact that D_e, as measured in films of Os(poly-vbpy)$_3$, where site concentration is higher, (see Results Section), is closer to the value reported here.(*5*)

As the concentration of enzyme in the film is increased from 0 to 65 wt%, the diffusion constant falls by 2 orders of magnitude with a function that is close to exponential. Simplistic application of the Dahms-Ruff theory (which implicitly requires that a "mean-field" approximation holds) would predict a linear dependence on concentration. The latter approximation would require that physical diffusion of the Os sites be rapid compared to electron hopping, which is clearly not the case here. Theory based on rigorously fixed site redox molecules and *extended electron transfer* (ie. static percolation), would indeed predict an exponential decrease in electron hopping with concentration.(*25*) However, simulations by Blauch and Saveant for the case of tethered redox sites also leads to a behavior part-way between that predicted by "static percolation" and "mean-field" approximations, resulting in a functional form of D_{ap} close to that seen in the inset of Figure 6. (See Figure 3B of reference 15 where t_e/t_p=0.1). It must be pointed out that the weak dependence of the film thickness, t, on enzyme content (Table I), leads to little electrochemically measurable decrease in site concentration as enzyme is increased. This of course makes a quantitative application of any of the models mentioned above rather difficult.

It is perhaps more constructive to consider the nature of the polymer-enzyme complex formed in these films. Recent work on this system has shown that a very strong, irreversible complex is formed between the two, dominated to a great extent by electrostatic interactions between the cationic polymer and (at pH 7) anionic enzyme.(*26*) It seems more likely that as the level of enzyme is increased in the films, greater electrostatic crosslinking of the polymer chains results in higher rigidity and slower electron diffusion. In other words, the average distance moved by the tethered Os sites, λ, is decreasing as GOx loading increases. This would effectively move the system progressively closer to "static percolation" with a concomitant fall in electron mobility.(*15*) A similar drop in electron transport rate has been observed in electropolymerized redox films containing either 2 or 3 covalent, crosslinking bonds per Os monomer.(*5*)

Over most of the range of GOx loading studied, it is unlikely that we approach the limiting case of static percolation, where at some critical concentration of enzyme, most routes for site-site electron hopping in the redox polymers become blocked by insulating protein. This is supported by the fact that the cyclic voltammetry and apparent surface coverages indicate that at least from 0 to 57% GOx, the macroscopic properties of the redox polymer are relatively undisturbed. That D_e does not fall to zero even at 64 wt% GOx further indicates that we are short of the static percolation threshold concentration(*15*) (if such a concentration exists in this system).

An additional consideration here is the intrinsic heterogeneity of the system. GOx is not iso-structural compared to the polymer, and the *microscopic* concentration of Os sites in the film volume that actually contains these sites may well be unchanged over a wide range of GOx content. Only when these regions or clusters of Os polymer become disconnected due to the presence of enzyme should a dramatic change in electrochemical behavior be observed.

In fact, the only discontinuity of electrochemical behavior occurs at 64% GOx, where the cyclic voltammogram becomes diffusional, Figure 4C, and where precipitation is actually observed in the solution prior to casting films. At this point there is clearly a change in the polymer-enzyme complexation reaction. Again, recent work(*27*) on a very similar system indicates that at approximately 70% GOx, no more enzyme is incorporated into the macromolecular complex, after which, additional enzyme may act simply to "disconnect" the redox conducting regions. Given the wave shape seen at this enzyme content, it is not possible to estimate with any accuracy the surface coverage of Os sites at a sweep rate of 1 mV/s, and so the values of Q and D_e shown in Table I must be regarded as lower and upper limits respectively.

The cyclic voltammetric data yield values of $D_{e,3}^{1/2}C$ that also decrease with increasing GOx, and the value at the lowest level of enzyme determined here(40%), $7.2x10^{-9}$ mols $cm^{-1}s^{-1}$, Table II, compares with a value of $2.5x10^{-8}$ determined previously for the same type of film with *no* enzyme present.(*2*) This is entirely consistent with the 1 order of magnitude drop seen in $D_{e,1}$ between these two levels of enzyme in Table I. However, when we use the values for C_t in Table I to calculate $D_{e,3}$ explicitly, Table II, we obtain values that are consistently lower than $D_{e,2}$. Again the possible overestimate of C_t in a *solvent swollen* film could well lead to an underestimate of $D_{e,3}$ here. Nevertheless, the fact that D_e as estimated by a dynamic technique ($D_{e,3}$) is *smaller* than that estimated by the steady-state technique ($D_{e,2}$) does argue against limited ion mobility dominating the dynamic measurement. The theory concerning coupled electron-ion conduction processes predicts a higher value of D_e using the dynamic technique.(*21*) Whilst it is possible that non-linear concentration gradients in the IDA experiments could lead to the appearance of migration effects, and an *overestimate* of $D_{e,1}$, recent and elegant work has demonstrated how remarkably linear the gradients actually are in highly charged redox gels based on polyvinylpyridine.(*28,29*) In addition, the same relationship between steady-state and transiently derived values for D_e has already been reported in the case of films with no enzyme present.(*2*)

Considering the sources of potential error in both types of experiment, it is likely that the real value of D_e can only be estimated to an order of magnitude anyway, and the relatively close agreement in Tables I and II is quite gratifying.

So taking averages of $D_{e,2}$ at each level of GOx, we calculated equivalent bimolecular rate constants for electron self-exchange in these films (Table II) using equation 4 (see results section). Over the range of GOx in the films from 40% to 64%, this value falls about one order of magnitude from $2.6x10^5$ to $<3.6x10^4$ $M^{-1}s^{-1}$ (as calculated from $D_{e,2}$). Previous studies of thin films of the electropolymerized polymer, poly-vbpy)$_3$Os, (where vpy=4-vinylpyridine), using similar steady-state techniques on an IDA electrode, yielded values for k_e of $\approx 6x10^6$ $M^{-1}s^{-1}$.(*5*) The major difference between the studies was the absence of non-conductive enzyme in the case of the electropolymerized film. In addition, the value derived for k_e in the case of the electropolymerized film was not corrected for a possible value of λ. However, this may not be inappropriate for the electropolymerized class of polymers where the cross-links between Os sites are bound directly to the Os metal centers, and may indeed not allow much motion of the sites. In this case $\lambda \rightarrow \delta$, and the "bounded diffusion" model approaches the static percolation model.(*15*) Given that the value for k_e of the Os-PVP polymer without enzyme is likely to be much larger than with 40% GOx, (see below), our estimated values appear to concur with those earlier measurements of electropolymerized osmium polypyridyl complexes.

Not surprisingly, the value of k_e obtained here is still less than that measured for $Os(bpy)_3$ freely diffusing in solution, $\approx 2.2 \times 10^7$ $M^{-1}s^{-1}$.(*30*) Inspection of $D_{e,1}$ in Table I indicates that we could expect perhaps a factor of five or ten increase in k_e were we to reduce the enzyme level to zero. This would put the rate constant within a factor of ten of the solution value, quite remarkably fast for an immobilized redox system. It is this behavior that makes this system developed by Heller et al such a good candidate for enzyme electrodes without having to resort to diffusionally mobile couples. Certainly this same polymer matrix with another enzyme does support very high current densities,(*31*) ($\approx$1.8mA cm^{-2}), that compare rather favorably with similar systems based on other polymeric redox couples.(*32*)

We turn now to the maximum current density generated by these films under quasi-saturating glucose concentrations and at an applied voltage sufficient to drive the oxidation of Os^{II} sites produced as a consequence of glucose oxidation, Scheme I, Figure 8, (and Figure 6 of reference 2). We observe the same behavior as reported previously for this system, where current increases with increasing GOx content up to a level of ≈35-45 wt%, after which it falls off rapidly. Increasing the GOx content beyond 65% in our case led to large scale precipitation of the enzyme-polymer complex, and a further decrease in current response.

Invoking any type of ping pong or bimolecular mechanism for the reaction as shown in Scheme I it is easy to explain the rising portion of the current response as GOx is increased. Under these circumstances, rate control for the entire process could lie with the association of the enzyme and substrate or cosubstrate or one of the dissociation steps or indeed in the electron transfer steps involving the Os complex and enzyme. It seems unlikely, in this region of film composition, that rate control is associated *entirely* with electron transfer events within the polymer alone, as k_e is in fact falling during this time, Table II. However, the earlier result that the activation energy for sensor response at low GOx content is the same as that measured for charge transport within a film containing no enzyme does beg the question as to whether similar events are involved in the formation of the activated complex in both processes.(*2*) This would be possible if the same polymer chain flexing motions are required to achieve the activated complex for electron transfer between Os sites and also for the approach of a polymer-Os site to the active site of a diffusionally immobile enzyme, while not requiring that the molecular rate constants be equal.

With the radical reversal of the maximum current trend at 45% GOx, it is enticing to ask whether the same processes control the sensor response after the maximum, as do before it. It is possible to argue that under limiting glucose conditions (Figure 8) and at low enzyme loadings (<40%), the system is kinetically limited by the enzyme-substrate-cosubstrate reaction (reactions i and ii, Scheme I). There are several pieces of evidence that tend to support this assumption. Previously published Eadie-Hofstree plots at 4.2 wt% GOx are certainly linear indicating a lack of diffusion control.(*2*) Catalytic cyclic voltammograms of a film containing 36 wt% GOx in 50 mM glucose solution yields perfectly flat limiting currents, with no sign of diffusional sloping, see Figure 1, reference 2. D_e and k_e are falling rapidly while the current response is rising at GOx < 40%. Cyclic voltammetric waveshapes do not change appreciably from 40-57 wt% GOx, indicating that heterogeneity of the film is not substantial in this range.

Beyond the turnover point at ≈45% GOx, it can also be argued that the conductivity of the redox polymer has been so reduced by addition of enzyme that rate control now passes to the self-exchange processes that transport the electrons to the electrode surface. A similar phenomena has been observed in a system where the catalytic site is a photocatalyst ($Ru(2,2'\text{-bipyridine})_3$) immobilized in a methyl-viologen redox polymer which accepts electrons from the Ru excited state. In this system when the mol fraction of the viologen is lowered enough, relative to the Ru sites, the photocatalytic current falls off, presumably as transport of the electrons through the poly-viologen chain becomes harder.(*33*)

Against this argument is the fact that these experiments were carried out in air where there is a competition between O_2 and Os^{III} for the reduced $FADH_2$ sites. As more enzyme is added it is possible that it becomes less well connected to the polymer and that oxygen competition becomes more effective at higher enzyme loadings, leading to a decrease in current density. In addition, current work underway in Heller's laboratory involving simulation of current response as a function of enzyme loading does indicate the possibility of a maximum in the sensor response when k_3 in Scheme I falls to low values, $<100\ M^{-1}s^{-1}$ even when electron transport through the polymer is assumed to be fast.(*27*)

Also it is possible that electron diffusion through the polymer away from the point source of the enzyme is *always* rate limiting as suggested earlier,(*2*) and supported by the equivalence of the activation energies for the sensor response and for charge transport through the film (*vide supra).* It is possible that the location of the GOx active site deeply buried within a narrow protein channel(*34*) allows only one or two polymer strands to approach sufficiently close to allow efficient electron transfer. In this case it is possible that the Os self-exchange reactions close to $FAD/FADH_2$ are restricted to intra- rather than inter-chain, which could also reduce the local rate constant. It is obvious that none of the interpretations can be ruled out on the basis of this data alone, and further work is needed to elucidate the mechanisms operating in different film regimes.

However, we can estimate a maximum allowable current density for the films from the measured diffusion constant at the film composition corresponding to the maximum sensor response (Figure 8) and the appropriate film thickness. Integration of Fick's law for the film thickness over which catalytic current is generated gives a maximum current density for such a film of nFD_eC_t/d. In a separate experiment, we fabricated films on planar gold electrodes equivalent to those used in Figure 8, where we measured the film thickness by step profilometer yielding $d \approx 11400$ Å. From these same films we obtained maximum current densities of 350 $\mu A\ cm^{-2}$ for oxidation of glucose solutions. However, the calculated maximum possible current for these films is 210 $\mu A\ cm^{-2}$, based on transport limitation through the bulk of the film. There is clearly a flaw in our calculations, and again we suspect that the intrinsic inhomogeneity of the films may be partly responsible. The Os site concentration, C_t, is calculated from the charge under the voltammogram and the physically measured film thickness and so assumes homogeneous mixing of the enzyme and polymer in the film volume. If the phases are essentially separate, the *local microscopic* concentration of Os sites may be significantly higher than the homogeneous estimate. This would lead to an underestimate of transport limited maximum current density.

Nonetheless, it is apparent that films with GOx > ≈40% are certainly operating quite *close* to the maximum permissible current flux based on the values of D_e measured here. The fall in D_e would then continue to cause rate limitation due to slow electron transport in the polymer.

This is an important consideration in the design of efficient amperometric biosensors where sensitivity ultimately defines the detection limits and influences the precision of the analysis. When attempting to optimize such a system for a commercial product we need to gain some understanding of the system elements so that efforts for improvement are directed at relevant problems. One issue in the rational design of polymeric enzyme electrodes is the question as to whether it is important to aim at more rigid polymeric matrices, with closely spaced redox sites that undergo fast exchange, or whether it is more important to accept slower self-exchange with a flexible polymer backbone where approach of the redox site to the enzyme active site is faster. It seems likely that the optimum system lies somewhere between the extremes, and these measurements of D_e should allow us to screen a number of polymeric candidates and predict the likelihood of electron transport limited behavior in any given regime of polymer/enzyme composition. Work is currently underway in both our laboratory and that of Heller to modify these types of polymer systems to achieve faster electron transfer without compromising their ability to act as good electron relays.

Acknowledgments

We would like to extend our gratitude to Ioanis Katakis and Prof. A. Heller for very helpful discussions during the preparation of this manuscript. We would also like to thank Dr David Deng and Dr Art Usmani for the synthesis and characterization of the Os polymers.

Literature Cited

1. Gregg, B. A.; Heller, A. *J. Phys, Chem.*, **1991**, *95*, 5970.
2. Gregg, B. A.; Heller, A. *J. Phys. Chem.*, **1991**, *95*, 5976.
3. Heller, A. *J.Phys. Chem.*, **1992.**, *96*, 3579.
4. Katakis, I.; Heller, A. *Anal. Chem.*, **1992**, *64*, 1008.
5. Surridge, N. A.; Zvanut, M. E.; Keene, F. R.; Sosnoff, C. S.; Silver, M.; Murray, R. W. *J. Phys. Chem.*, **1992**, *96*, 962.
6. Nishihara, H.; Dalton, F.; Murray, R. W. *Anal. Chem.*, **1991**, *63*, 2955.

7. Hill, H. A. O.; Sanghera, G. S. *in Biosensors a Practical Approach*, Cass, A. E. G., Ed., Oxford University Press, **1990**, NY.
8. Cass, A. E. G.; Davis, G.; Francis, D. G.; Hill, H. A. O.; Aston, W. J.; Higgins, I. J.; Plotkin, E. V.; Scott, L. D. L.; Turner, A. P. F. *Anal. Chem.*, **1984**, *56,* 667.
9. Nicholson, R. S.; Shain, I. *Anal. Chem.*, **1964**, *36*, 706.
10. Bourdillon, C.; Demaille, C.; Moiroux, J.; Savéant, J-.M. *J. Am. Chem. Soc.*, **1993**, *115*, 2.
11. Chidsey, C. E. D.; Feldman, B. J.; Lundgren, C.; Murray, R. W. *Anal. Chem.*, **1986**, *58*, 2844.
12. Feldman, B. J.; Murray, R. W. *Anal. Chem.*, **1986**, *58*, 2844.
13. Oyama, N.; Anson, F. *J. Electrochem. Soc.*, **1980**, *127*, 640.
14. Buttry, D. A.; Anson, F. C. *J. Am . Chem. Soc.*, **1983**, *105*, 685.
15. Blauch, D. N.; Savéant, J-.M. *J. Am. Chem. Soc.*, **1992**, *114*, 3323.
16. Laviron, E. J. *J. Electroanal. Chem.*, **1980**, *112*, 1.
17. Andrieux, C. P.; Savéant, J-.M. *J. Electroanal. Chem.*, **1980**, *111*, 377.
18. Forster, R. J.; Vos, J. G. *Macromolecules*, **1990**, *23*, 4372.
19. Aoki, K. *J. Electroanal. Chem.*, **1990**, *284*, 35.
20. Aoki, A. ; Heller, A. *J. Phys. Chem.*, manuscript submitted.
21. Andrieux, C. P.; Savéant, J-.M. *J. Phys. Chem.*, **1988**, *92*, 6761.
22. Surridge, N. A.; Jernigan, J. C.; Dalton, E. F.; Buck, R. P.; Watanabe, M.; Zhang, H.; Pinkerton, M.; Wooster, T. T.; Longmire, M. L.; Facci, J. S.; Murray, R. W. *Faraday Discuss. Chem. Soc.*, **1989**, *88*, 1.
23. Buck, R. P. *J. Phys. Chem.*, **1988**, *92*, 4196.
24. Calculated using equation 6.2.19 in: Bard, A. J.; Faulkner, L. R. *Electrochemical Methods Fundamentals and Applications*, **1980**, John Wiley and Sons, Inc., New York.
25. Fritsch-Faules, I.; Faulkner, L. R. *J. Electroanal. Chem.* ***1989***, *263*, 237.
26. Katakis, I.; Davidson, L.; Heller, A. manuscript in preparation.
27. Katakis, I. private communication.
28. Fritsch-Faules, I.; Faulkner, L. R. *Anal. Chem.*, **1992**, *64*, 1118.
29. Fritsch-Faules, I.; Faulkner, L. R. *Anal. Chem.*, **1992**, *64*, 1127.
30. Chan, M.-S.; Wahl, A. C. *J. Phys. Chem.*, **1982**, *86*, 126.
31. Ye, L; Hämmerle, M.; Olsthoorn, A. J. J.; Schuhmann, W.; Schmidt, H-.L.; Duine, J. A.; Heller, A. *Anal. Chem.*, **1993**, *65*, 238.
32. Hale, P. D.; Boguslavsky, L. I.; Inagaki, T. I.; Karan, H. I.; Lee, H. S.; Skotheim, T. A. *Anal. Chem.*, **1991**, *63*, 677.
33. Downard, A. J.; Surridge, N. A.; Gould, S.; Meyer, T. J. *J. Phys. Chem.*, **1990**, *94*, 6754.
34. Hecht, H. J.; Kalisz, H. M.; Hendle, J.; Schmid, R. D.; Schomburg, D. *J. Mol. Biol.*, **1993**, *229*, 153.

RECEIVED February 10, 1994

Chapter 6

Sandwich-Type Amperometric Enzyme Electrodes for Determination of Glucose

S. Mutlu[1], M. Mutlu[2], P. Vadgama[3], and E. Pişkin[1,4]

[1]Department of Chemical Engineering, [2]Department of Food Engineering, University of Hacettepe, 06530 Ankara, Turkey
[3]Department of Medicine, Clinical Biochemistry Section, University of Manchester, Salford M6 8HD, United Kingdom

Sandwich type amperometric enzyme electrodes for glucose based on oxidases were studied. Polyethersulphone (PES) and polycarbonate (PC) membranes were used as inner and outer membranes, respectively. Glucose oxidase was entrapped between these two membranes. The selectivity of the inner membrane against a group of electroactive compounds in blood were tested. The PES membrane fulfilled all the requirements of a permselective membrane. The outer surface of PC membranes were further modified by deposition of hexamethyldisiloxane (HMDS) in a glow-discharge reactor or by coating with polyvinylalcohol (PVAL) to increase the linear range of the electrode while keeping or improving the blood-compatible properties of membrane. Linearity of these enzyme electrodes were investigated in different media (i.e., buffer solutions, blood plasma and blood). Blood-compatibilities of the PC membrane and its modified forms were obtained in in-vitro experiments. HMDS treatment significantly increased the linearity of enzyme electrode but did not affect their blood-compatibilities or significantly change the selectively. Also, the PVAL coating, considerably, extended the linear range of the electrode, slightly improved blood-compatibility but did not change the selectivity.

In the last decade, a rapid technological evolution took place in the field of chemical sensors in general, and enzyme based sensors in particular (*1*). This progress has been at least partly due to the commercial interest in sensor technology for diagnostic medicine, both for real-time analysis of metabolites in metabolically unstable patients and rapid bed-side monitoring (*2-3*).

Sandwich type amperometric enzyme electrodes are amongst the most widely studied electrodes for diagnostic medicine. A typical sandwich type enzyme electrode consists of three major layers, i.e., outer membrane, enzyme layer and inner membrane. The outer membrane, where the analyte diffuses through should have diffusion-limiting property which provides a means of extending linear range through their reduction in local substrate concentration while maintaining sufficient oxygen

[4]Current address: P.K. 716, Kizilay, 06420 Ankara, Turkey

0097–6156/94/0556–0071$08.00/0

for the enzymatic reaction. In the middle, the enzyme layer accepts the analyte and necessary oxygen to achieve the enzyme-substrate reaction and to produce electrochemically detectable species. The immobilization of the enzymes is generally maintained by gel-entrapment. The inner membrane with permselective property prevents the interference by other electrochemical species on electrode (*4-5*). A number of techniques to improve linearity and selectivity, and also biocompatibility and enzyme utility of sandwich type electrodes have been described elsewhere (*6*). However, There are only a few approaches in use in screening new electrodes for linearity, selectivity and biocompatibility that were proven to be of value and accepted by majority of researchers.

Recently, we have studied a variety of sandwich type electrodes for development of enzyme electrodes based on oxidases. In this paper, we focused on extending linearity and blood-compatibility of the outer membrane, i.e., polycarbonate (PC) membrane, without affecting the selectivity. We modified the outer surface of this membrane by plasma polymerization of hexamethyldisiloxane (HMDS) in a glow-discharge reactor or by coating with polyvinylalcohol (PVAL). The PVAL coated surfaces were further treated with an anticoagulant, i.e., heparin. We studied the linearity, selectivity and blood-compatibility of these membranes and related electrodes. Here, we present the results of these studies.

Construction of the Electrodes

Materials. Polycarbonate (PC) membranes (i.e., outer membrane) with nominal pore size ranging from 0.01 to 0.05 μm (rated by manufacturer) were supplied by Poretics (USA), hexamethyldisiloxane (HMDS) was obtained from Wacker (Germany). The enzyme, i.e., glucose oxidase (E.C.1.1.3.4. from Aspergillus Niger: 292 IU mg^{-1} protein) was purchased from Sigma (UK). Polyethersulphone (PES) and polyvinylalcohol (PVAL, with an average molecular weight of 50000, 100% hydrolysed) were supplied by ICI (UK) and by Aldrich (Germany), respectively. Gluteraldehyde (25% aqueous solution), bovine serum albumin (fraction V), glucose, solvents, buffer components and other standard reagents were obtained from Sigma (UK). Cellulose acetate (CA; 39.8% acetyl content) was obtained from Aldrich (Germany). Blood plasma was donated by the "Kızılay Central Blood Bank" (Ankara, Turkey). Human blood were taken from volunteers.

Electrode. The main body of the sandwich type of electrode that we used in this study was a Rank oxygen electrode system (Rank Brothers, Bottishom, Cambridge), which is schematically described in Figure 1. The electrode consisted of a central 2 mm diameter platinum working electrode with an outer 12 mm diameter silver ring as the reference. The electrode was polarized at 650 mV (versus Ag/AgCl) for hydrogen peroxide detection. The meter was linked to a strip-chart recorder.

Outer Membrane. The outer membrane was a track-etched polycarbonate membrane, which was purchased from Poretics. PC membranes with three different nominal pore sizes of 0.01, 0.03 and 0.05 μm were tested.

The outer surfaces of these membranes were further modified by following methods: (1) glow-discharge treatment and (2) chemical coating.

Glow-Discharge Treatment. HMDS was deposited by plasma polymerization on the PC membranes in a glow-discharge reactor system shown schematically in Figure 2. The reactor was a glass tube (inner diameter: 6.4 cm and length: 35 cm) with two copper electrodes (surface area of each electrode: 140 cm^2) mounted on the outside.

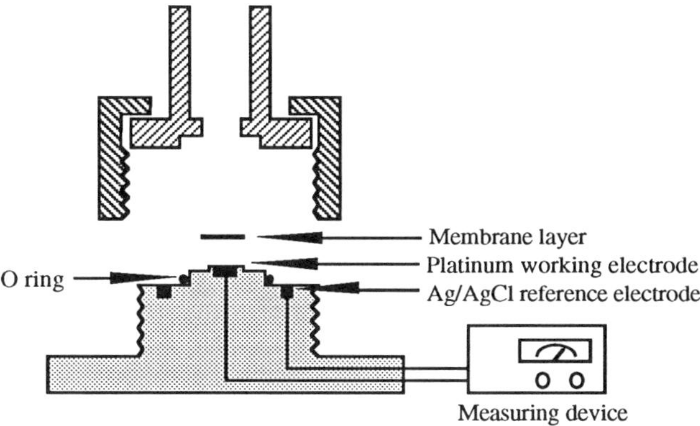

Figure 1. Schematic description of amperometric electrode system.

Radiofrequency generator (Tasarım Ltd. Co.-Turkey, frequency:13.6 MHz and power range: 0-100 W) was used to form the plasma in the reactor. Power losses were kept to a minimum by a matching network.

In a typical glow-discharge treatment, two membranes placed back to back (outer surfaces facing the plasma medium) were placed in the middle of the reactor. The reactor was evacuated to 10^{-3}-10^{-4} mbar. The monomer (i.e., HMDS) was allowed to flow through the reactor at a flow rate of 30 ml/min. The PC membranes were exposed to plasma medium for 5 min and 20 min at a discharge power of 10 W.

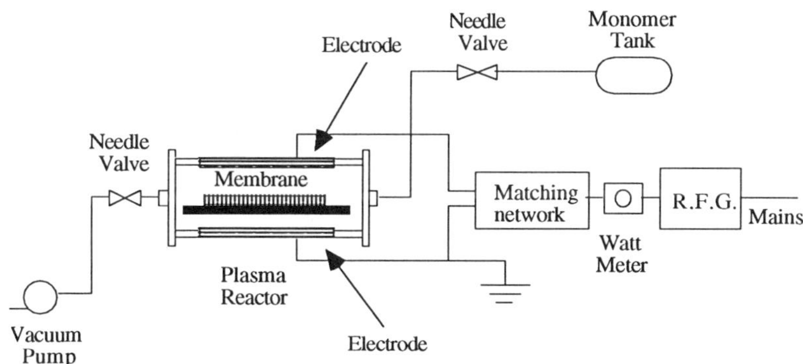

Figure 2. Glow-discharge apparatus.

Chemical Coating. The PC membranes were also coated with PVAL by a two-step procedure. Details of this procedure have been described previously (*7*). Briefly, a monolayer of PVAL molecules was first deposited on the outer surface of the PC membrane by a simple adsorption process. The PVAL concentration in the adsorption medium was 4×10^{-3} g/ml. The PC membrane (16 cm^2) was incubated with 130 ml of this solution for 2 h at room temperature (20°C). Next, PVAL molecules adsorbed on the PC membrane surface were crosslinked in order to have a non-water-soluble PVAL coating. Crosslinking was achieved chemically either by using terephtalaldehyde (at 80°C) or by exposing to γ-irradiation (0.2864 Mrad).

In order to include an anticoagulant activity on these PVAL coated membranes, these membranes were activated (through hydroxyl groups) by using CNBr as an activating agent, and then heparin was attached through the active sites, covalently. Details of this procedure have been described previously (*8*). Briefly, an activating solution containing CNBr (CNBr concentration: 0.0166 g/ml) was prepared and its pH adjusted to 11-12 with 2M NaOH. The PC membrane (surface area: 32 cm^2) was then added to the activation solution and incubated for 90 minutes. After completion of the activation step, the membrane was incubated with a heparin solution (heparin concentration: 3.33 mg/ml) at room temperature for 4 h.

Inner Membrane. Polyethersulphone (PES) membranes were prepared by a solvent casting technique (*9*). A 6% (w/v) solution of PES in a mixture of 2-methoxyethanol and DMF (25% + 75%, v/v) was cast on a glass plate. The polymer film (the thickness : 25 μm) was then formed by evaporation of the solvents in a conditioned room (relative humidity: 80%, and temperature: 20°C). The membranes were stored in vacuum desiccator until use.

As an alternative to the PES membranes, cellulose acetate (CA with 39.8% acetyl content) membranes (with a thickness of 25 μm) were also prepared by solvent casting method. The composition of the casting solution was 2% CA (w/v) in acetone. Film formation was achieved by evaporation of acetone at room temperature (20°C) in 80% relative humidity environment.

PC and its glow-discharge modified forms were also tested as the inner membrane.

Enzyme Layer. The enzyme layer was prepared by crosslinking of glucose oxidase (GOD) with glutaraldehyde, in the presence excess bovine serum albumin (BSA). In a typical procedure, 30 mg of GOD was mixed with a 200 mg BSA in 1 ml of buffer solution. A 6 μl volume of this mixture was then rapidly mixed with 3 μl of glutaraldehyde (5%, v/v) in a Gilson pipette tip and placed between the PC and PES membranes. This sandwich was then compressed between two glass slides and held under hand pressure for 5 minutes. The glass slides were pried apart and the membrane/enzyme laminate was left to dry at room temperature for a further 5 minutes, washed with buffer and then placed over a Rank oxygen electrode.

Performance of the Electrodes

Three different types of tests, i.e., selectivity, linearity and blood-compatibility were investigated to ascertain the performance of the electrodes.

Selectivity Tests. In these tests, a Rank electrode was used with a single membrane layer, instead of the sandwich type of laminate. The PES, PC and its modified forms, and CA membranes were used in these tests. Hydrogen peroxide (H_2O_2), catechol, uric acid, paracetamol and ascorbate were used as the analytes. Each analyte was used separately. The initial concentration of the analyte was 0.2 mM.

Linearity Tests. In these tests, sandwich type of laminates were prepared by using the PC membrane with a 0.05 μm pore size as the inner membrane. The GOD layer was prepared by the method described above. As the outer membrane, the PC membranes with different pore sizes (i.e., 0.01, 0.03, 0.05 μm), the glow-discharge treated PC membrane (PC/HMDS), the polyvinylalcohol coated PC membranes (PC/PVAL) and the heparin attached PC/PVAL membranes (PC/PVAL/H) were tested. In these tests, the test solution was an isotonic buffer solution (pH: 7.4) containing 0.0528 mol/l Na_2HPO_4, 0.0156 mol/l NaH_2PO_2, 0.0051 mol/l NaCl and 0.00015 mol/l EDTA. In the tests, the enzyme electrode was first allowed to stabilize for 5-10 min in the buffer solution, before the first reading was recorded. Then, 5 µl of an aqueous solution of glucose was added to achieve a 2 mM glucose concentration in the medium. The response was taken. This procedure was repeated several times, without changing the solution but adding enough glucose solution to increase its concentration 2 mM each time, until the deviation from linearity was observed.

Blood-Compatibility Tests. Blood compatibility tests were carried out in three groups: (1) protein-membrane interactions; (2) blood-membrane interactions; and (3) blood cloting times on the membranes.

The protein-membrane interaction tests were carried out by using plasma and plasma proteins (i.e., human serum albumin, HSA, human serum fibrinogen, HSF and human immunoglobulin, HIgG) obtained from the Kızılay Blood Bank (Ankara, Turkey). These proteins were labelled with ^{99m}Tc-pertechnetate as described elsewhere (*10*). In a typical test, a 16 cm^2 of the membrane sample was placed into a pre-silanised petri-dish containing 5 ml of plasma with proper amounts of radiolabelled proteins. After a 30 minutes of contact time, the membrane was washed with isotonic solution, and the radioactivity of both on the membrane and the plasma sample from the petri-dish was measured by a gamma-scintillation counter (Berthold, BF 2500, Germany).

In the blood-membrane interaction studies, fresh blood samples taken from healthy human donors were used. 5 ml of human blood was exposed to a 32 cm^2 of membrane in a pre-silanised blood-sampling tube for 1 h, by continuously rotating the tubes. At the end of interaction period, the blood samples were homogenized and the blood cells were counted by using a coulter counter (Coulter TC 10, USA). Plasma of the blood samples were seperated by centrifuging at 3500 rpm for 10 minutes. The plasma samples were then analyzed by biochemical autoanalyser (Ciba-Corning, Express 550, USA) to determine the concentrations of the HSA, HSF and total protein.

In the blood-clotting time experiments, fresh blood samples taken from healthy human donors were used. The membranes were contacted with the non-heparinised blood, the extrinsic and intrinsic clotting times (PTZ and PTT, respectively) were determined by using a Coagulameter (Amelung KC-10).

Result and Discussion

Description of the Electrode. Sandwich type amperometric enzyme electrodes are amongst the most widely studied electrodes for diagnostic medicine (*11*). The working principle of a typical sandwich type glucose electrode is schematically described in Figure 3. The outer membrane mainly controls the substrate (i.e., glucose) diffusion and therefore provides the linearity. This membrane should also allow sufficient oxygen transfer which is vital for the enzymatic reaction. It should be noted that the blood-compatibility of the outer membrane is one of the most important considerations. If it is non-compatible, blood proteins and cells (mainly platelets) are accumulated at the blood-contacting side of the membrane, and may diminish very significantly the transport of the substrate and therefore cause the failure of the electrode (*12*).

In the middle of the sandwich laminate, an enzyme layer is located which reacts with the substrate by using oxygen and produces electrochemically detectable species (e.g., H_2O_2). Note that the enzyme is used in the immobilized (crosslinked) form in order to prevent the enzyme leakage and to increase the enzymatic stability.

To control the enzymatic conversion rate, the enzyme concentration in this layer is controlled by adding diluting agents. In the case of the glucose electrode constructed in this study, GOD molecules were co-crosslinked with BSA molecules in a proper ratio (i.e., 30/200) to control the enzymatic activity.

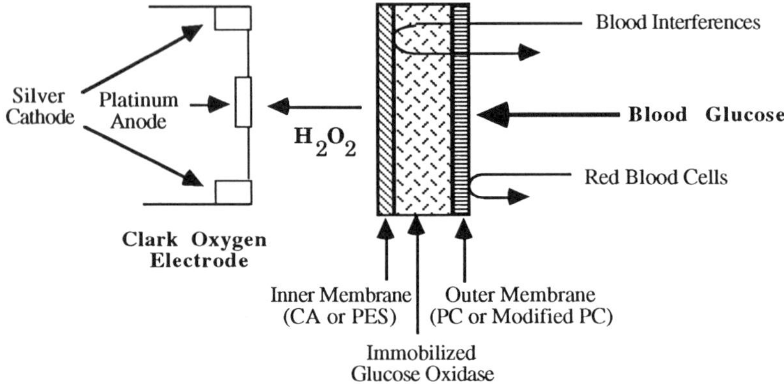

Figure 3. Configuration of a biosensor.

In the glucose electrode, the reaction product is hydrogen peroxide. H_2O_2 molecules pass through the inner membrane, reach to the electrode surface, and create the electrode response by oxidation. The most important property of this inner membrane is it's permselectivity. This membrane should only permit the H_2O_2 molecules, but not the interfering molecules (e.g., catechol, uric acid, paracetamol and ascorbate) which may be present in blood (*13*).

Selectivity Tests. As described above, in this group of studies the selectivity of some candidate membranes (PES, CA, PC and PC/HMDS) were studied. Note once again that each membrane was tested as a single layer, not a whole sandwich. Electrode responses in nanoampers were measured for each electroactive analytes (i.e., hydrogen peroxide, catechol, uric acid, paracetamol and ascorbate) with an initial concentration of 0.2 mM. Figure 4 shows permeabilities of some selected membranes tested in this study. Notice that the H_2O_2 permeabilities of all the membranes presented here were approximately equal except PC (pore size: 0.05 μm). However, the permeabilities of the other electroactive compounds were significantly different.

The original polycarbonate (PC) membrane allowed transport all of the analytes mentioned above. The glow-discharge treatment diminished the respective permeabilities but considerable permeations of the electroactive compounds were still observed. Therefore, it can be concluded that, the PC based membranes cannot be used as the inner selective membrane due to the strong interference of the potential electroactive compounds which would exist in the direct blood applications.

Cellulose acetate (CA) membranes with different pore structures were prepared by solvent casting and tested here as potential inner membranes. Permeabilities of different analytes through a typical CA membrane were given in Figure 4. Note that this membrane give the best performance of all CA membranes tested in this study, with respect to interferences. As shown here, the H_2O_2 permeability was high, it was not permeable for most of the electroactive compounds tested here, except catechol ions. It was not possible to prevent the passage of ascorbate ions through any of the CA based membranes that we have prepared. Therefore, we decided not to consider the CA based membranes as the inner membrane of our sandwich laminates.

Very promising results were obtained with the polyethersulphone (PES) membranes. It is clearly observed from Figure 4 that, the PES membrane prepared by us following the method described above fulfilled the all requirements from a permselective membrane. Namely, the PES membrane allowed the H_2O_2 transport but not to the other interfering compounds (i.e., electroactive analytes). Therefore, we decided to use this membrane as the inner membrane for the construction of our sandwich type of glucose electrode.

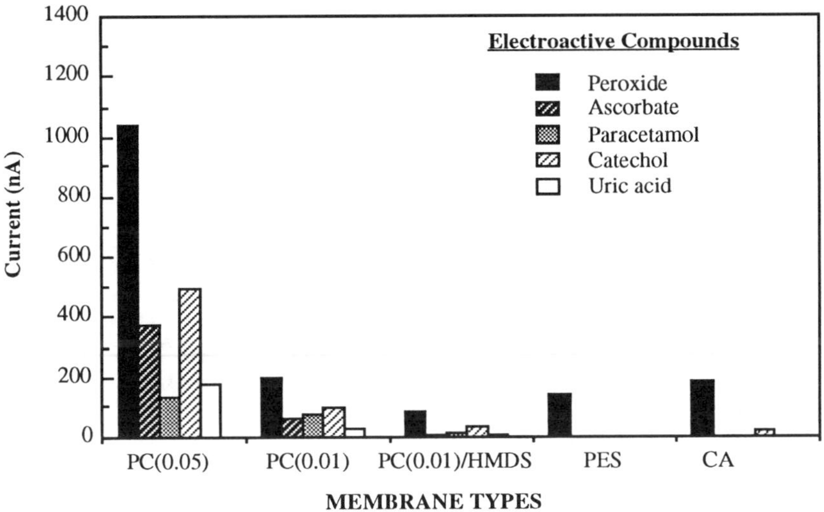

Figure 4. Permeabilities of some selected membranes.

Linearity Tests. As mentioned before, one of the most important parameters of a sandwich type of enzyme electrode is its linearity. For the glucose measurements in the blood samples (e.g., measurement of glucose level in the blood of the diabetic patients) this linearity level is expected to be extended up to about 40 mM of glucose (*14*). To extend the linearity range of our sandwich type glucose electrode was one of the major goals of this study.

We prepared sandwich type of laminates by using different type of PC based membranes as the outer membrane. The inner membrane was the PES membrane which was described in the previous section. The enzyme layer prepared by using 6 μl solution of GOD with BSA and 3 μl of glutaraldehyde (5%). The results of the studies related to the optimization of the composition of the enzyme layer were given elsewhere in detail (*15*). We used these laminates with the Rank electrode described above, and measured the electrode response of the glucose in buffer solutions (5 ml) containing different amounts of glucose (up to 100 mM).

Two groups of studies were performed. In the first group, the PC membranes with three different pore sizes, namely 0.01, 0.03 and 0.05 μm were tested. Note that these PC membranes were prepared by a tract-etching method, which leads pores as straight cylindrical tubes (turtousity is about one) with approximately equal diameters as shown in Figure 5. Therefore, one expects significant difference in the transport properties due to sharp cutoffs. The effect of pore size of the PC membranes on the linearity of the glucose electrode is given in Figure 6. It is clearly shown that the linearity increased as the pore size decreases. Therefore, we concluded that this type of tract-etched PC membranes containing pores with narrower diameter would be more suitable for the extension of the linearity range. However, it should be noted that there is no tract-etched PC membrane having pores with a diameter smaller then 0.01 μm commercially available.

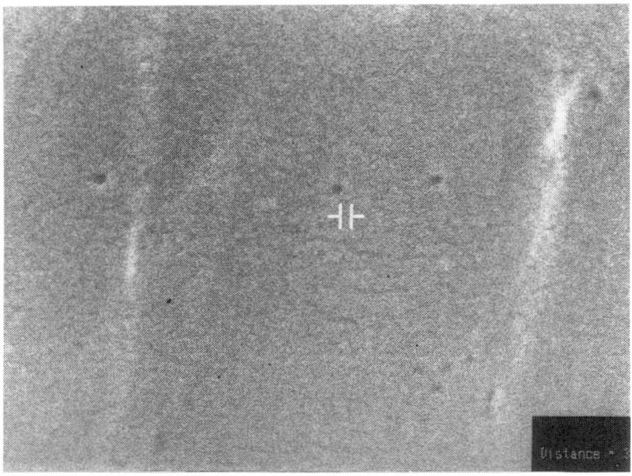

Figure 5. SEM micrograph of plain PC membrane with 0.03 μm pore size (Mag: X 60K).

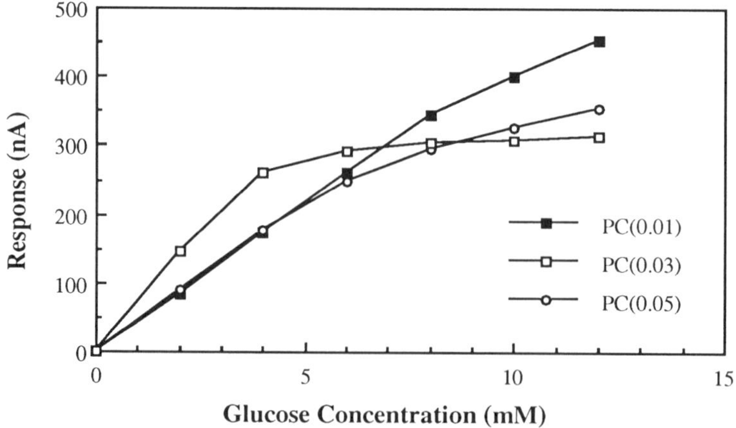

Figure 6. The effect of pore size on the electrode linearity.

In order to increase the linearity range of the commercially available PC membranes by decreasing the pore size, we attempted to use a novel technique, which is so-called "glow-discharge treatment". Note that by this technique, it is possible to change the surface properties of a membrane without significantly changing the bulk structure (*16*). We deposited hexamethlydisiloxane (HMDS) on the PC membranes with different pore sizes by plasma polymerization in a glow-discharge reactor. We changed the glow-discharge conditions (i.e., discharge power and period, monomer flow rate and pressure in the reactor) in order to optimize the treatment. Details of these studies were given elsewhere (*17*). Here, we presented the data related to the HMDS glow-discharge treated membrane prepared at optimum conditions which are as follows: treatment time: 20 min, monomer flow rate: 30 ml/min, pressure: 10^{-3}–10^{-4} mbar, discharge power: 10 W).

Our previous experience lead us to select HMDS as the monomer for the treatment. We have coated several nonblood-compatible materials by plasma polymerization of this monomer in a glow-discharge system to create blood-compatible surfaces for diverse biomedical applications (*18-21*). We have shown that this treatment significantly increase the blood-compatibility without affecting the substrate material properties.

Figure 7 illustrates the effect of glow-discharge modification on the glucose electrode linearity. As shown here the glow-discharge treatment significantly increased the linearity. For instance, the linearity range of the original PC membrane with a diameter of 0.01 µm was 10 mM of glucose, while the linearity range was extended to 35 mM of glucose when this PC membrane was treated with HMDS (i.e., the PC/HMDS membrane) in the glow-discharge reactor. The scanning electron micrograph of the PC/HMDS membrane is given in Figure 8. As seen here, at the glow-discharge conditions applied here, there was an observable size reduction of the pores from 50 nm to 34.2 nm, which may cause the extension of the linearity range of the electrode due to the changes in the mass transfer properties of the membrane after treatment.

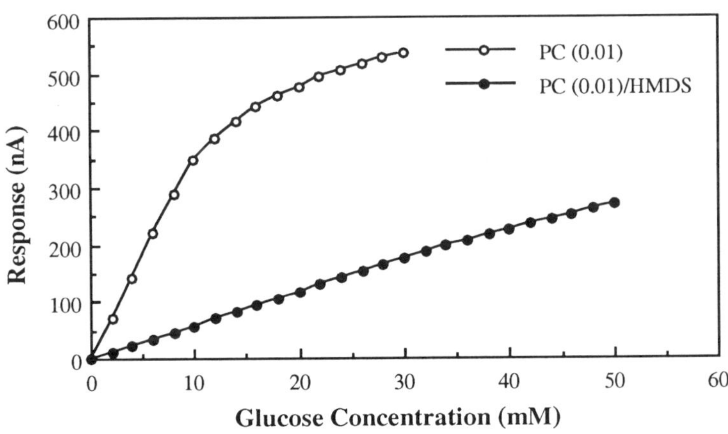

Figure 7. The effect of glow discharge modification of surfaces on the electrode linearity.

In this group of experiments, linearities of two more membranes were also tested. These membranes were the polyvinylalcohol coated PC (PC/PVAL) and its heparin attached form (PC/PVAL/H). As it will be discussed in the coming section, these modifications were carried out to increase both the linearity and the blood-compatibility of the PC membranes. Figure 9 shows the linearity test results of these membranes. As seen here, the PVAL coating (both chemically crosslinked and crosslinked with γ-irradiation) extended the linearity range of the original PC membrane from 10 mM to about 30 mM of glucose. The heparin attachment did not change the linearity further. The effect of the PVAL coating may be explained by considering the change in the mass transfer properties of the original PC membrane.

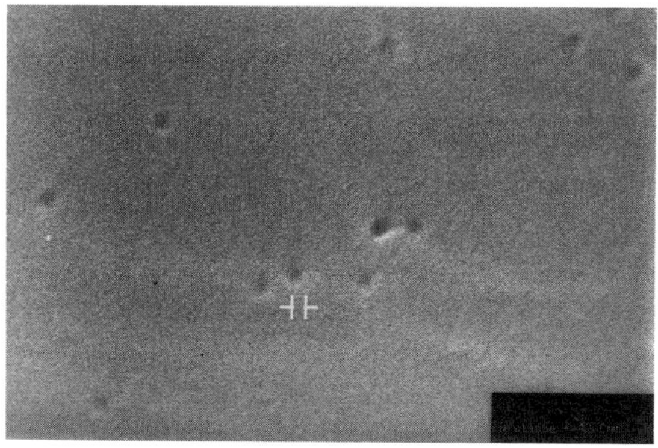

Figure 8. SEM micrograph of glow-discharge treated PC membrane with a pore size of 0.05 μm (Mag: X 60K).

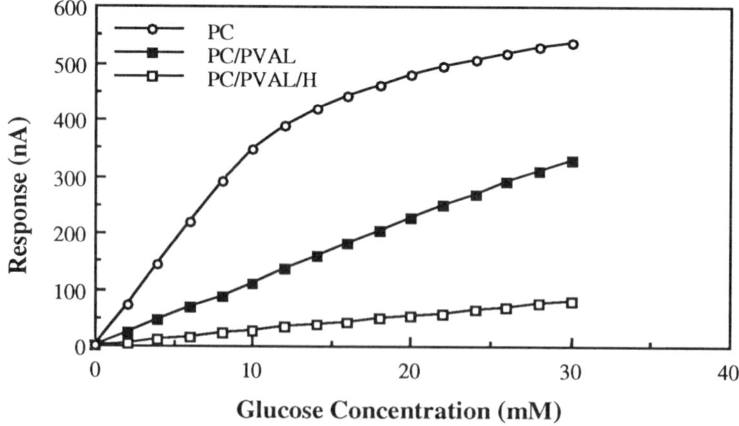

Figure 9. The effect of chemical modification of surfaces on the electrode linearity.

Blood Compatibility Tests. The aim of this study is to develop an enzyme electrode for the measurement of glucose levels mainly in vivo. In addition to developing a system that is linear over the physiological range another goal was to develop a system that is biocompatible. For this purpose, we conducted the following standard blood-compatibility tests.

In the first group of experiments, three main blood proteins, i.e., albumin (HSA), fibrinogen (HSF) and globulin (HIgG) were radiolabelled with ^{99m}Tc-pertechnetate. These radiolabelled proteins were added within 5 ml of blood plasma samples obtained the Kızılay Blood Bank (Ankara, Turkey). The potential membranes tested in this study as the outer membrane were then incubated with the plasma samples. After 1 h of incubation period the radioactivity levels both in the plasma phase and on the membrane surfaces were counted. Figure 10 shows the protein adsorption on different membrane surfaces. Note that the values on the vertical axis in the figure are relative values in which adsorptions of each protein on the original PC membrane were assumed to be 100. As seen here, adsorption of blood plasma proteins on the PVAL coated PC membrane were considerably lower than the PC surfaces. These results are very similar to the literature stating that PVAL is a very inert surface (*22*). But, it should be noted that inert surfaces are not always blood-compatible, they may indirectly cause other incompatibility problems, such as platelet activation and agglomeration, complement activation, etc. (*23*). There were significantly higher level of protein adsorption on the HMDS glow-discharge treated surface. However, it should also be noted that high protein adsorption does not lead always non-compatibility. The type of the protein layer and its time depended behaviour should be considered (*24*).

In the second group of experiments, membranes were incubated directly with blood in heparinized tubes. After 1h incubation period, blood cells were counted by means of a coulter counter. The blood cell counts were given Table I. Note that the tube containing no membrane was called as "control". All of the membranes caused drop in the blood cell counts (especially in platelet counts), however,these changes may not be considered as significant. In addition, the differences between the membranes were also not very noticeable.

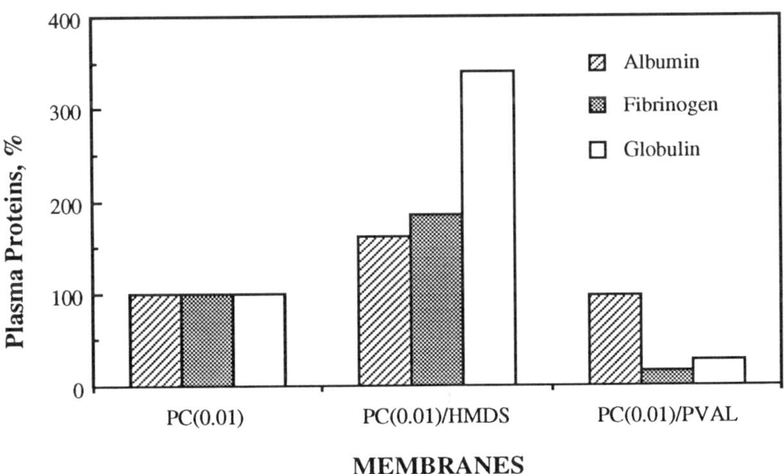

Figure 10. Protein adsorption on different membrane surfaces.

Table I. The Change in the Amount of Blood Cells After 1 h of Contact with Different Membranes in Total Human Blood

Membrane	Leucocyte (10^9/l)	Erytrocyte (10^{12}/l)	Thrombocyte (10^9/l)
Control	6.4	5.14	165
PC	5.7	4.89	147
PC/HMDS	5.6	4.92	139
PC/PVAL	5.6	4.81	142
PC/PVAL/H	5.9	4.84	130

In the third group of experiments, the extrinsic and intrinsic blood clotting times were determined. Table II summarizes these data. All of the membranes tested here did only slightly changed the extrinsic blood clotting times (PTZ) and the intrinsic blood clotting times (PTT). The PTT value for the PC membrane was little higher than the other (better surface), but this is again not very significant.

Table II. The Extrinsic and Intrinsic Blood Clotting Times for Various Membrane Surfaces

Membrane	PTZ (sec)	PTT (sec)
Control	14	45
PC	16	54
PC/HMDS	15	50
PC/PVAL	15	51
PC/PVAL/H	16	50

Conclusion

In this study, we aimed to develop a sandwich type of glucose electrode to measure the glucose level in blood. The results of the selectivity tests showed that the PES membrane permits only H_2O_2 but not the other electroactive analytes which may interfere the electrode response in blood applications. The linearity and blood-compatibility of the outer membrane have been considered as two important parameters in blood contacting applications of enzyme electrodes. This study showed that the linearity of the original system utilizing untreated polycarbonate membranes could be extended by either glow discharge treatment with HMDS or chemical coating by PVAL. These modifications also caused different extents of adsorption of blood plasma proteins. However, if we consider the insignificant changes in the levels of the blood cells and also in the clotting times, we can say that the respective modifications do not improve the blood compatibility of the original PC membranes, which we were not expecting. If we consider both the linearity and blood-compatibility at the same time, we may concluded as the HMDS glow-discharge treated or the PVAL-heparin coated membranes should be used as the outer membrane of the sandwich type of glucose electrode described in this study.

Acknowledgement

Dr. Mehmet Mutlu gratefully acknowledges financial support from International Atomic Energy Agency (IAEA) (Grant No: C6/TUR/8813). The authors wishes to thank Abbas Yousufi Rad for his technical assistance.

Literature Cited

1. *Biosensors: Fundamentals and Applications*; Turner, A.P.F.; Karube, I. and Wilson, G. S., Eds.; Oxford University Press, Oxford, 1987.
2. Matthews, D. R.; Holman, R. R.; Bown, E.; Steemson, J.; Watson, A.; Hughes, S. and Scott, D. *Lancet* **1987**,*vol 1*, 778.
3. Vadgama, P.; Alberti, K.G.M.M. In *Recent advances in Clinical Biochemistry* ; Price, C. P. and Alberti , K.G.M.M. Eds.; Churchill Livingstone, 1985, Vol. 3.
4. McDonnell, M. B. and Vadgama, P. M. *Membranes: Separation Principles and Sensing; Selective Electrode Review* ; 1989; Vol. 11, 17-67.
5. Vadgama, P. In *Biosensors in vitro and in vivo; Clinical Biochemistry Nearer the Patient II* ; Marks, V. and Alberti, K.G.M.M. Eds.; Bailliere Tindall, 1986.
6. Tang, L.X.; Koochaki, Z.B.; Vadgama, P. *Analytica Chimica Acta,* **1990**,*vol 232*, 357-365.
7. Boomgard, V.D.; King, T.A.; Tadros, F.; Tang, T.; Vincent, B. *J. Colloid Interface Sci.,* **1978**,*vol 66*, 68.
8. Kohn, J.; Wilchek, M. *App. Biochem. Biotech.* **1983**, *vol 8*, 227.
9. Battersby C. M. and Vadgama, P. *Analytica Chimica Acta* **1986**, *vol 183*, 59-66.
10. Atkins, H.L.; Richards, P. *United States Atomic Energy Committee Publication,* **1990**, *BNL-11831,* 1-19.
11. Vadgama, P.; Desai M. A. In *Biosensor Principles and Applications*; Blum, L. J. and Coulet, P. R. Eds.; Marcel Dekker, Inc., 1991, 303-338.
12. Woodward, S. C. *Diabetes Care* 1982, *vol 5*, 278-281.
13. *The Merck Manual of Diagnosis and Therapy* , Holvey, D. N. and Talbott, J. H., Eds.; Merck & Co., Inc., 1972, 1826-1830.
14. *Samson Wright's Applied Physiology*, Keele, C. A. and Neil E., Eds.; Oxford University Press, 1961, 461-473.
15. Mutlu, M.; Mutlu, S.; Vadgama, P. In *1991 IChemE Research Event*; Kenny, C.N. Darton, R. Davidson, P.J. Seaton, N.A. and Stitt E.H. Eds.; Institute of Chemical Engineers, Rugby, UK, 1991.
16. Pişkin, E. *J. Biomater. Sci. Polymer Edn.* **1992**, *vol 4*, 45-60.
17. Mutlu. S., *Ph. D. Dissertation*, Hacettepe University, Ankara, 1993.
18. Mutlu, M.; Mutlu, S.; Rosenberg, M.F.; Kane, J.; Jones, M.N.; Vadgama, P. *J. Materials Chemistry,* **1991**,*vol 1(3),* 447-450.
19. Mutlu, M.; Ercan, M.T.; Pişkin, E. *Clinical Materials,* **1989**,*vol 4*, 61-76.
20. Pişkin, E.; Evren, V. *Life Support Systems,* **1986**,*vol 4*, 2-17.
21. Özdural, A.; Hameed, J.; Bölük, M.Y.; Pişkin, E. *ASAIO J,* **1980**, *vol 3,* 116-119.
22. Brinkman, E.; van der Does, L. and Bantjes, A. *Biomaterials* **1991**, *vol 12,* 63-70.
23. Bruck, S.D. *J. Biomat. Med. Dev. Artif. Org.,* **1973**,*vol 1(1),* 79-98.
24. Andrade, J. D.; Hlady, V. *J. Biomater. Sci. Polymer Edn.* **1991**, *vol 2*, 161-172.

RECEIVED September 17, 1993

Chapter 7

Reproducible Electrodeposition Technique for Immobilizing Glucose Oxidase

and a Differentially Permeable Outer-Membrane Material for Use on a Miniature Implantable Glucose Sensor

K. W. Johnson, D. J. Allen, J. J. Mastrototaro, R. J. Morff, and R. S. Nevin

Medical Devices and Diagnostics Division, Eli Lilly and Company, Lilly Corporate Center, Indianapolis, IN 46285

Two novel technologies have been developed to address problems associated with the fabrication of a miniature electroenzymatic glucose sensor for implantation in the subcutaneous tissue of humans with diabetes. An electrodeposition technique was developed to electrically attract glucose oxidase and albumin onto the surface of the working electrode of the glucose sensor. The resulting enzyme/albumin layer was chemically crosslinked with glutaraldehyde. This technique facilitated the simultaneous deposition of enzyme onto numerous miniature sensors. A polyethylene glycol (PEG)/polyurethane copolymer was developed for use as the outer membrane on the sensor. This membrane was designed and optimized to provide differential permeability of oxygen relative to glucose in order to avoid the oxygen deficit encountered in physiologic tissues. The membrane was also found to be highly biocompatible.

An implantable glucose sensor has been recognized for many years as the critical component necessary for optimal control of blood glucose concentrations in diabetic patients. A sensor that would yield continuous readings of blood glucose levels so that a person with diabetes could take the appropriate corrective steps during the initial onset of hyper- or hypoglycemia would be a valuable addition to diabetes treatment. In addition, incorporating such a device into a closed-loop system with a microprocessor and an insulin infusion pump could provide automatic regulation of the patient's blood glucose.

A majority of the glucose sensors intended for *in vivo* implantation currently being developed are electroenzymatic (*1-4*). In these sensors, glucose oxidase catalyzes the reaction of glucose and oxygen, on an equal-molar basis, to form hydrogen peroxide and gluconic acid. Either the amount of hydrogen peroxide produced or the oxygen consumed by the enzymatic reaction is measured electrochemically. We chose to quantify hydrogen peroxide by electrochemical oxidation at the working electrode of a three-electrode system poised at +0.6 V relative to a Ag/AgCl reference electrode. The current generated by the oxidation of the hydrogen peroxide is linearly related to the glucose concentration, provided the enzymatic reaction takes place in an excess of oxygen relative to glucose.

0097–6156/94/0556–0084$08.00/0

In subcutaneous tissue, the concentration of oxygen can be 100 to 1000 times lower than the concentration of glucose, making oxygen the rate limiting substrate. Membranes with differential permeabilities to oxygen and glucose have been incorporated into the sensor to solve this “oxygen deficit” problem (*5,6*). We have developed a polyethylene glycol(PEG)/polyurethane copolymer for use as an outer membrane. Be varying the ratio of the membrane components, we optimized the differential permeabilities of oxygen and glucose in order to minimize any variation in sensor output due to changes in the oxygen concentration. The membrane created an oxygen-rich environment in the enzyme layer by restricting glucose and allowing oxygen molecules unaltered diffusion, while the sensor output remained proportional to variation in the glucose concentration.

A variety of techniques have been utilized for the immobilization of enzymes for use in biosensors *(5-12)*. Unfortunately, very few of these techniques are adaptable to the miniature electrochemical biosensors currently being fabricated using semiconductor techniques. The semiconductor techniques allow a sensor with a surface area of 1.0 μm^2 to be possible. For the fabrication of enzyme sensors, it is desirable to accurately deposit and immobilize the enzyme only on the surface of the working electrode, which is used to monitor a product of the enzymatic reaction. Enzyme deposited on other areas of the sensor would be reacting to produce a product that could not be detected, which is useless and should be minimized due to the expense of the enzyme. For the development of these techniques, it is important to appreciate that the enzymes are very fragile molecules in that their activity is easily degraded upon exposure to most organic chemicals, temperatures above approximately 50° C, and extreme pH (high or low). These considerations limit the arsenal of available techniques for enzyme deposition.

Kimura et al. *(13)* developed a method for enzyme deposition using lift-off techniques with photoresist. Heineman and others have evaluated the potential of gamma irradiation as a means of immobilizing an enzyme in a polymer matrix on an electrode surface *(14)*. This method can also be used with chemicals that polymerize upon exposure to ultraviolet light. Shinohara et al. have synthesized enzyme-containing membrane films by electrochemical polymerization of pyrrole or aniline in the presence of enzyme *(15)*. Platinized platinum, sometimes referred to as "platinum black", was employed as the supporting matrix for enzyme deposition by Ikariyama et al. *(16)*. The enzyme was incorporated into these pores by immersing the platinum blacked electrode into a phosphate buffered solution (pH = 7.0) containing the enzyme. The enzyme molecules adsorbed to the platinum black surface and could be further anchored to the sensor by covalent crosslinking with glutaraldehyde. This group also modified their procedure by adding the enzyme to the hexachloroplatinate solution and simultaneously electrodepositing platinum black and glucose oxidase *(17)*.

A technique has been developed for reproducibly depositing an enzyme or other biomolecule onto the surface of a conductor. Specifically, glucose oxidase has been electrodeposited onto the surface of an electrode previously electroplated with platinum black. This deposition takes place in an aqueous solution (pH = 7.4) without the assistance of an electropolymerizable matrix material such as polypyrrole. The deposition process does not require a time-consuming alignment step that is characteristic of some deposition techniques utilizing photoimageable matrix materials; yet it results in the enzyme precisely deposited only on the working electrode. The thickness of the resulting enzyme layer can be varied from Angstroms to microns by varying the deposition parameters. With proper fixturing, potentially hundreds of sensors can be simultaneously coated with enzyme.

Fabrication

Figure 1 illustrates the various layers and chemical reactions associated with the sensor developed in our laboratories. We chose to mass-produce this type of sensor by developing a fabrication scheme that utilized thin/thick film cleanroom processing techniques similar to those used in the integrated circuit industry. The sensors were fabricated in batches of 112, although this number can be easily increased and the process automated. This approach was intended to address the problem responsible for the absence of an implantable glucose sensor from the market, namely the inability to mass-produce a reliable, reproducible, and economical disposable sensor.

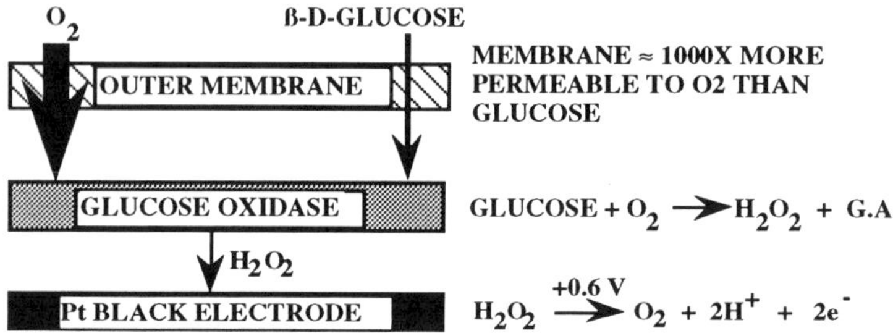

Figure 1. Block schematic of a hydrogen peroxide-based electroenzymatic glucose sensor with a differentially permeable outer membrane layer.

The actual steps involved in the fabrication of the sensors have been described elsewhere (*18,19*). The amperometric sensor onto which the glucose oxidase and outer membrane are deposited was fabricated using thin/thick film processing techniques. This process resulted in a three-electrode sensor with a 0.1 mm^2 platinum-black working electrode. The sensor also contained a platinum-blacked counter electrode and a Ag/AgCl reference electrode. The preparation, deposition, and characterization of the outer membrane and glucose oxidase layers will be discussed here.

Enzyme Deposition Methods

Enzyme Deposition. Glucose oxidase (type X-S), bovine albumin (fraction V), and glutaraldehyde were obtained from Sigma Chemical Company and used as supplied. A silicone defoaming agent (B emulsion) was from Dow Corning. A nonionic surfactant (Tergitol NP-10) from Union Carbide Corporation was also incorporated.

A 5% by weight solution of glucose oxidase was prepared utilizing a phosphate buffered saline solution (pH 7.4) as the solvent. While dissolving the glucose oxidase by stirring, it was important to prevent foaming of the solution which could denature the glucose oxidase. The electrodeposition cell was filled with this solution and the electrode onto which the enzyme was to be deposited,and a counter electrode were submerged inside the cell. Electrical connections were made to a galvanostat and current was applied such that a constant density of 5 mA/cm2 resulted at the electrode surface. The current flow was oriented in a direction to result in a positive charge on the electrode surface to attract the negatively charged glucose oxidase molecules. The current was applied for 2 minutes. The actual voltage at the working electrode varied with the impedance. The electrode was removed from the enzyme solution and submerged in deionized, distilled water with no agitation for 5 seconds to remove residual glucose oxidase. Following this, the electrode was submerged in a solution containing 2.5% by volume glutaraldehyde in phosphate buffered saline (pH 7.4) for 30 minutes to covalently crosslink the glucose oxidase molecules and form a water-insoluble layer. The glutaraldehyde crosslinking prevented the glucose oxidase from going back into solution. The electrode was again submerged in deionized, distilled water for 5 seconds and allowed to dry in air for 30 minutes. The enzymatic component of the biosensor was functional at this point.

Biomolecule Codeposition. Another important aspect of this method was the capability to simultaneously electrodeposit two or more different biomolecules providing their isoelectric points are all either higher or lower than the pH of the deposition solution. A solution was prepared containing 5% by weight glucose oxidase and 5% by weight albumin (bovine) in a phosphate buffered saline solution (pH 7.4). The glucose oxidase and albumin had isoelectric point values of 4.3 and 4.7 respectively, resulting in both molecules possessing a negative charge in the pH 7.4 solution. A current density of 5.0 mA/cm2 was applied to the electrode for 2 minutes while submerged in the glucose oxidase/albumin solution. The molecules were crosslinked in the glutaraldehyde solution as described above, rinsed, and dried.

Additives. The non-ionic surfactant and anti-foaming agent were added to the solutions at a concentration of 1 drop per 100 mL of solution to enhance the deposition with no loss of enzyme activity in the resulting enzyme layer.

Outer Membrane Methods

Polymer Synthesis. Polyethylene glycol (PEG) 600, 1000, and 1500 were obtained from Aldrich Chemical Company and dried by azeotroping with toluene. Molecular weights for the polyethylene glycols were determined from hydroxyl numbers. Diethylene glycol (DEG) from Fisher Scientific was purified by vacuum distillation over metallic sodium. Hexamethylene diisocyanate (HMDI) from Aldrich Chemical Company was vacuum distilled. Dicyclohexylmethane-4,4'-diisocyanate (DCDI) (Desmodur W) from Mobay Chemical Company was used as received. Dibutyltin bis-(2-ethylhexanoate) from Kodak was stored over phosphorus pentoxide. N,N-dimethylformamide (DMF) from EM Science and 4-methyl-2-pentanone from Aldrich Chemical Company were dried over 3A molecular sieves.

For a typical bulk polymerization 4.80 g of PEG 600, 4.24 g of DEG, and 8.07 g of HMDI were placed in a 100 mL flask and heated to 50°C while purging continuously with nitrogen. 10 mg of dibutyltin bis-(2-ethylhexanoate) was dissolved in 7 mL of 4-methyl-2-pentanone and added to the warm reaction mixture. The reaction quickly became exothermic, with the temperature rising to 100°C within a few minutes. When the reaction cooled to 90°C, heating was resumed to maintain that temperature for another 60 min. The polymer was dissolved in 200 mL of 90°C DMF. When the polymer solution had cooled to room temperature, it was poured into 2 L of vigorously stirred deionized water to precipitate the polymer. The precipitated polymer was torn into small pieces and soaked in deionized water for 24 hours, with several water changes. The polymer was dried in vacuo at 50°C.

For a typical solution polymerization 14.40 g of PEG 600, 12.73 g of DEG, and 24.22 g of HMDI were dissolved in 250 mL of DMF in a 1000 mL flask. The solution was heated to 50°C as the flask was continuously purged with nitrogen. 30 mg of dibutyltin bis-(2-ethylhexanoate) was dissolved in 25 mL of 4-methyl-2-pentanone and added to the flask. There was a slightly exothermic reaction, with the temperature rising to 55°C. The reaction was heated to 75°C for 120 min. and then at 90°C for another 120 min. When the reaction solution cooled to room temperature, it was diluted with 100 mL of DMF then poured into 5 L of vigorously stirred water to precipitate the polymer. The precipitated polymer was soaked and dried as above.

Determination of Molecular Weights. Peak molecular weights were determined by gel permeation chromatography using a Waters GPC I equipped with two Waters Ultrastyragel linear columns, a Waters R401 differential refractometer, and Waters 730 Data Module. Determinations were done at 25°C. Samples were dissolved in chloroform at a concentration of 0.25% (w/v) and an injection of 150 µL was used. The mobile phase was chloroform at 1.0 mL/min. Molecular weights were determined by comparison of peak retention times with retention times for a series of nine polystyrene standards run using the same conditions.

Determination of Water Pickup. 0.5 g of polymer was dissolved in 10 mL of warm chloroform. The solution was poured into an aluminum container ~5 cm in diameter and left for the solvent to evaporate at ambient conditions. The film was removed from the container and dried to constant weight at 50°C in vacuo. The dried film was weighed, immersed in deionized water for 24 hours, removed, blotted dry, and weighed immediately. The percent water pickup for the polymer was calculated from the formula

$$\%\ \text{pickup} = (W_w - W_d) \times 100$$

where W_w is the weight of the hydrated film and W_d is the weight of the dry film.

Diffusion Studies. Diffusion membranes were prepared by spreading a chloroform solution of the polymer on a flat, level glass plate with a 2 inch wide Gardner adjustable knife (Gardner Labs). The membranes were air dried overnight. The dry membrane was hydrated with deionized water for 30 min. then removed from the glass plate and supported on a Mylar sheet. The thickness of the hydrated film was measured with a micrometer while on the support. Membranes ~10 µm thick were used for the diffusion studies.

Oxygen diffusion studies were done at 37°C with standard cells (Crown Glass Company) having a combined volume of 100 mL and an area of 7.0 cm^2 for the opening between the cells. The hydrated membrane was tightly sandwiched between the cell halves using rubber gaskets. Both halves of the cell were filled with PBS. The donor half was saturated with air while the receptor half was saturated with nitrogen. A calibrated oxygen sensor (Microelectrodes, Inc.) was placed in the receptor half, and measurements were taken at 5 min. intervals until equilibrium was reached.

Glucose diffusion studies were done at 37°C with standard cells (Crown Glass Company) having a combined volume of 7.0 mL and an area of 0.64 cm^2 for the opening between the cells. The membrane was placed as above. The donor half of the cell was filled with PBS containing glucose at a concentration of 300 mg/dL, and the receptor half was filled with PBS alone. Glucose concentrations in both halves were measured at appropriate intervals using a Cooper Assist Clinical Analyzer.

Enhancement of Polymer Biocompatibility. Films from about 0.5 g. of polymer were cast as for the water pickup determinations. The dry films were placed in 100 ml. of deionized water and left at ambient temperature. The water was changed every 24 hours for a total of 72 hours. The films were dried in vacuo at 50°C. Then the films were placed in 100 mL of 10% (v/v) dichloromethane/hexanes (both AR - Mallinkrodt) and left at ambient temperature. The dichloromethane/hexanes was change every 24 hours for a total of 72 hours. The films were dried in vacuo at 50°C.

***In vitro* Characterization of Sensor Performance.** The *in vitro* performance of the sensors was evaluated by submersion in phosphate-buffered saline solutions (Sigma Chemical Company), pH 7.4, with the temperature maintained at 37° C. The glucose concentration is adjusted by spiking with a concentrated (10,000 mg/dL) dextrose solution that had set at room temperature for 24 hours prior to use to assure complete mutarotation of the glucose. The oxygen effect on sensor output was evaluated by placing the sensors in a solution having a fixed glucose concentration, and the oxygen content was varied from 1 to 21% (PO_2 from 7 to 150 mmHg) by bubbling with gases of known oxygen concentration; the normal oxygen concentration of the subcutaneous tissue is approximately 3-4%. A potential of +0.6 V (vs. Ag/AgCl) was applied to the sensors using a potentiostat fabricated in our laboratories.

Most of the *in vitro* data was collected using a computer-controlled calibration system which varied the glucose and oxygen concentrations automatically (*20*).

Results and Discussion

Enzyme Deposition. The conditions outlined above normally resulted in a glucose oxidase layer approximately 1.5 µm thick when dry as measured by a stylus profilometer (Dektak). The amount of enzyme deposited could be changed by varying the current density during the electrodeposition step. Figure 2 illustrates the resulting current from electrodes with identical surface area coated using the Ikariyama technique (*16*) and the method described here utilizing two different current levels. For this test, the sensors did not have an outer membrane on top of the enzyme layer, so the current from the sensor was a direct indication of the amount of glucose oxidase immobilized on the surface of the electrode. From this data, it was evident that the application of current to the electrode enhances the attraction and subsequent attachment of glucose oxidase to the surface of the working electrode compared to simple adsorption. In addition, the amount of glucose oxidase deposited on the electrode was in proportion to the current density. Care was taken when selecting the current density so the corresponding voltage did not climb to a level where deleterious gas bubbles were generated, resulting in bubbles entrapped in the biomolecule layer.

The codeposition of glucose oxidase and albumin resulted in a layer approximately 5 µm thick. This combination of biomolecules resulted in sensors which had a more stable long-term output in a 100 mg/dL glucose solution than glucose oxidase alone as illustrated in Figures 3 and 4. The sensors having glucose oxidase and albumin immobilized together did not exhibit the downward drift over time seen with the sensors comprised of glucose oxidase only. This long-term stability over a 72 hour period was thought to be absolutely necessary for an implantable biosensor. This enhanced stability may have been due to the formation of a membrane layer which was more tightly crosslinked upon exposure to glutaraldehyde which did not allow glucose oxidase to leach out. In addition, the albumin may have provided an environment that was more hospitable to the glucose oxidase in terms of hydrophobicity, localized pH, etc.

The deposition of the enzyme was very site-specific as shown in the scanning electron micrograph, Figure 5. This photo illustrates the enzyme layer deposited on top of the electrode only in the region where the top insulation layer was removed. This deposition of the enzyme only on the working electrode minimized the amount of glucose oxidase wasted and thus minimized the production costs.

Due to the electrical nature of this process, large numbers of sensors (28) were coated with biomolecules in a simultaneous, parallel manner. In addition, the

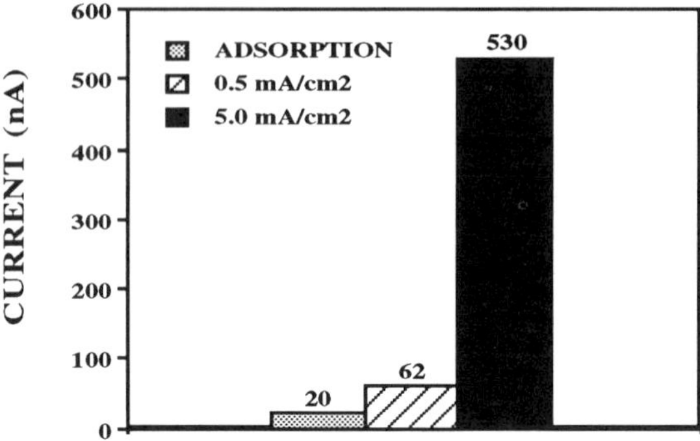

Figure 2. Comparison of current output for three sensors in a 100 mg/dL glucose solution *in vitro*. The sensors were identical in electrode area and method of fabrication, except for the enzyme deposition. One of the sensors was coated using the Ikariyama adsorption technique *(16)* and the other two were coated using the method described here utilizing two different current values. The sensors were not covered with an outer membrane. (Reproduced with permission from ref. 19. Copyright 1991 Elsevier Sequoia.)

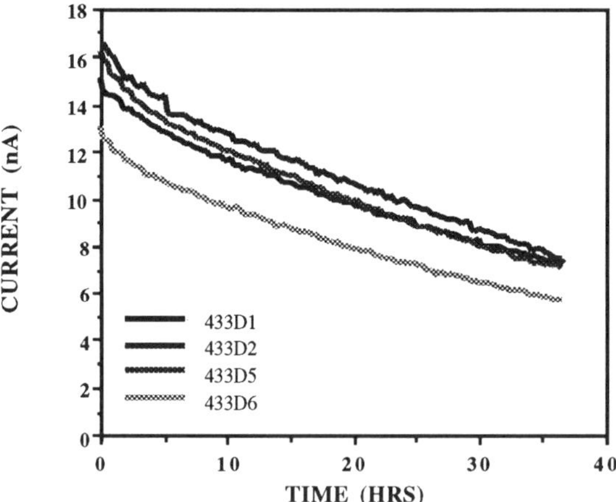

Figure 3. Current output vs. time for a group of four sensors in a 100 mg/dL glucose solution *in vitro*. The working electrode had glucose oxidase only electrodeposited on its surface, followed by crosslinking in glutaraldehyde. The sensors were covered with an outer membrane. (Reproduced with permission from ref. 19. Copyright 1991 Elsevier Sequoia.)

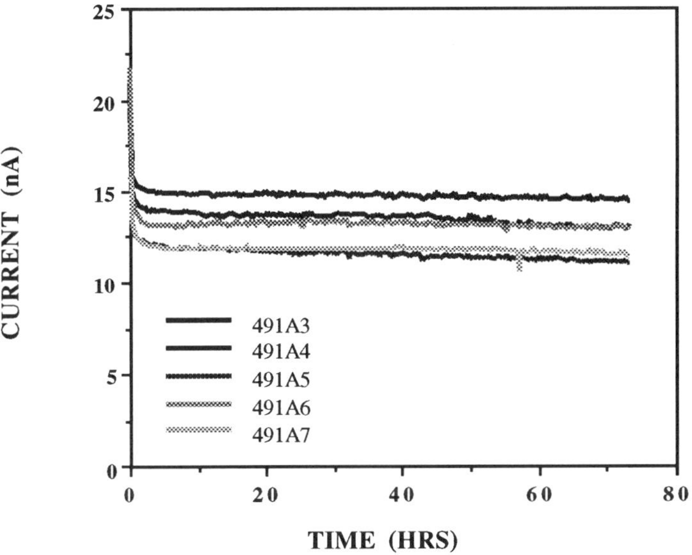

Figure 4. Current output vs. time for a group of four sensors in a 100 mg/dL glucose solution *in vitro*. The working electrode had glucose oxidse and albumin codeposited on its surface, followed by crosslinking in glutaraldehyde. The sensors were covered with an outer membrane. (Reproduced with permission from ref. 19. Copyright 1991 Elsevier Sequoia.)

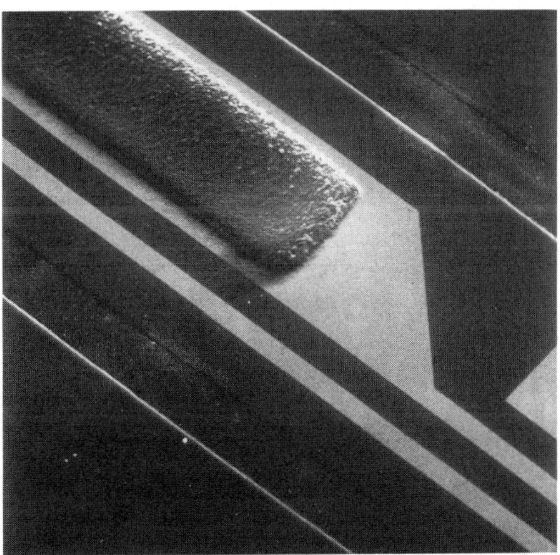

Figure 5. SEM photograph (200x) of a portion of the working electrode that had glucose oxidase and albumin electrodeposited and crosslinked onto the exposed surface. The visible metal layer surrounding the layer of glucose oxidase and albumin is the conductor underlying the insulation layer. This sensor was not covered with an outer membrane.

electrodeposition process would allow the coating of conductors with uneven topographies or geometrical surfaces such as the inside of metal tubes for use in flow-through processes or HPLC columns.

Polymer Synthesis and Characterization. Table I lists a representative sample of the polymers prepared. The molar ratios of the monomers used in the polymerization are listed in parenthesis behind the monomer. Polymerization reactions proceeded without incident. The dissolution of the bulk polymers in DMF required a few hours since the DMF was added to the flask in 25 mL portions.

Table I. Polymer Composition (Molar Ratio of Monomers) & Characterization

Polymer	Diisocyanate	Glycol	PEG	Mol. Wt.	%Water
1	HMDI (6)	DEG (5)	600 (1)	90,000	24.5
2	HMDI (7)	DEG (6)	1500 (1)	87,000	56.0
3	HMDI (9)	DEG (8)	1000 (1)	50,000	21.8
4	HMDI (13)	DEG (12)	600 (1)	92,000	9.4
5	HMDI (13)	DEG (12)	1000 (1)	36,000	15.0
6	DCDI (15)	DEG (14)	1500 (1)	125,000	13.4

Molecular weight data are also given in Table I. Generally the molecular weight of the polymers was higher than 85,000, except for the polymers using PEG 1000. The actual molecular weight of these polymers may have been lower due to a slight error in the hydroxyl number or incomplete drying.

Table I shows the water pickup of each polymer. As expected the water pickup was affected by both the molecular weight and amount of PEG used. An increase in either the amount or molecular weight of the PEG also increased the water pickup. The PEG amount and molecular weight were not the only factors effecting the water pickup. The water pickup was also effected by the nature of the hydrophobic portion of the polymer. Making this region more hydrophobic would also decrease the water pickup.

Polymers with water pickup greater than 56% were prepared but were not included here because they were not suitable for use on the sensor due to poor adhesion, unacceptable swelling, poor diffusion characteristics, etc. Additional polymers were prepared using ethylene glycol or 1,4-butanediol instead of diethylene glycol, however; those polymers were not soluble in the appropriate solvents for film casting by the methods listed above. Changing the film casting solvent and method were also found to change the water pickup and diffusion characteristics of a polymer membrane. The molecular weights of these polymers could not be determined on our equipment as the polymer precipitated from chloroform as the solution cooled to 25°C.

Relative Diffusion Coefficients. The sensor was intended to operate in the subcutaneous tissue. In this environment the concentration of glucose can exceed the concentration of oxygen. If this happens, oxygen becomes the limiting factor, and the sensor output is no longer indicative of the glucose concentration. A membrane that can regulate the amount of glucose that gets to the sensor without affecting the oxygen permeability was necessary.

Membranes of the polymers were evaluated as to their relative diffusion coefficients for glucose and oxygen. Relative diffusion coefficients were calculated using the formula

$$D = \frac{\text{slope}\,(V_1 + V_2)\,h}{-A}$$

where V_1 and V_2 were the volume (cm^3) of the diffusion cell halves, h was the membrane thickness (cm), A was the area (cm^2) of the opening between the diffusion cell halves, and the slope was determined from a plot of $\ln(C_d - C_r)$ vs time (sec) where C_d and C_r were the concentration in the donor and receptor halves of the diffusion cell. The relative diffusion coefficients calculated for the polymers are shown in Table II. The coefficients are considered relative since the system was not calibrated with know materials. However, the relative diffusion coefficients can still be used to evaluate the potential utility of the different polymers.

The oxygen diffusion coefficients varied only by a factor of approximately 2. The glucose diffusion coefficients varied over a much greater range, with one polymer being impermeable to glucose. The actual diffusion coefficients were not as important as the ratio of the coefficients. The ratios of $D_{oxygen}/D_{glucose}$ are given in Table II. These ratios varied by a factor of almost 200. Polymer 1 had a ratio that made it a good candidate as the outer membrane for the sensor.

Enhancement of Polymer Biocompatibility. The polymers as initially prepared were all found to be cytotoxic, most likely due to residual catalyst in the polymer. The extraction scheme was designed to remove water and organic extractable materials. After following the extraction scheme, all of the polymers passed the cytotoxicity screen.

Figure 6 illustrates that the output from the sensor was unaffected by varying the oxygen concentration from 20.9% down to 1% (P_{O2} from 150 to 7 mmHg). This study has also been performed in solutions with glucose concentrations as high as 400 mg/dL. The lack of change in the output is indicative of the outer membrane successfully overcoming the oxygen deficit problem.

The remainder of the *in vitro* performance data for the sensor will be summarized here as it has been published previously (*20*). The current output from the sensor upon exposure to unstirred, air-saturated solutions with varying glucose concentrations from 0 to 400 mg/dL at 37° C was linear (R^2>0.98) and 10 mg/dL incremental changes to the glucose concentration of a test solution were easily resolved. The background current was determined to be 1.3 nA +0.3 (mean + 1 SD). The 90% response time of the sensor was determined to be approximately 90 seconds for both increases and decreases in the glucose concentration. Finally, the long-term stability of the sensor's output was stable, drifting less than 10%, during more than 72 hours of continuous operation. The percentage yield of properly functioning sensors was greater than 95%.

Conclusions

Outer Membrane. We have shown that the hydrophilic polyurethanes have a differential permeability to glucose and oxygen. While the relative diffusion of oxygen remained stable over a wide range of water pickup values, the glucose diffusion varied over a much larger range. This allowed a polymer with an appropriate ratio of oxygen diffusion to glucose diffusion to be selected for use as the outer membrane on the sensor.

We have developed a method to determine relative diffusion coefficients for oxygen and glucose so that the diffusion coefficients can be used to screen materials for use as an outer membrane on the sensor. Thus the polymers ded not have to be screened on an actual sensor. This accelerates the selection process and also reduced the expense of the evaluation.

Table II. Relative Diffusion Coefficients

Polymer	D_{Oxygen} (X 10^{-6}) [a]	$D_{Glucose}$ (X 10^{-8}) [b]	Ratio
1	8.83±1.65	0.23± .01	3790
2	6.93±1.40	7.60± .58	20
3	4.59± .64	1.81± .39	254
4	3.87± .51	Imperm.	-------
5	5.72± .91	3.85 [c]	149
6	4.83±1.02	4.75± .19	101

[a] n=6, [b] n=3, [c] n=1

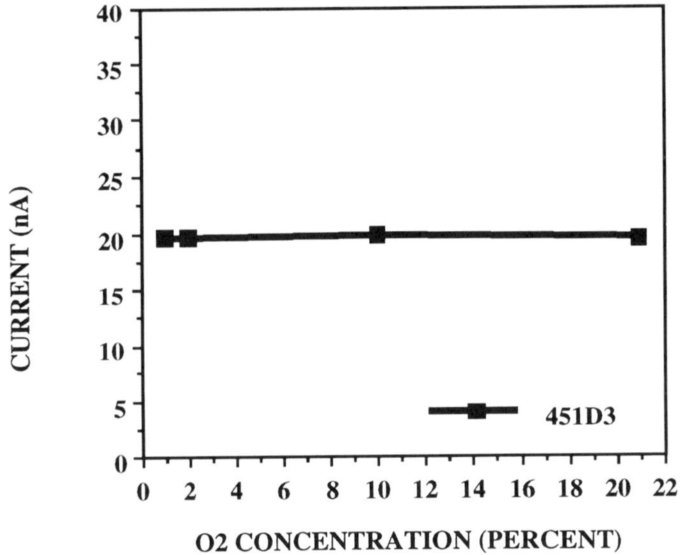

Figure 6. Current output vs. oxygen concentration for a typical sensor in a 100 mg/dL glucose solution *in vitro*. The sensor was covered with a membrane cast from Polymer 1.

We have illustrated that the selected hydrophilic polyurethane can be used as an outer membrane on a small subcutaneous glucose sensor. The membrane had good adhesion, did not change its properties over a 72 hour period of implantation, and passed appropriate biocompatibility tests.

Enzyme Electrodeposition. A method has also been developed for reproducibly electrodepositing an enzyme or other biomolecule onto the surface of conductors, regardless of the conductors' size and topography. This deposition took place from an aqueous solution whose pH was matched to the hospitable range and the isolectric point for the biomolecule. The process also allowed the simultaneous deposition of two or more biomolecule proteins.

This deposition technique did not require a supporting polymer matrix or optical alignment step characteristic of other techniques. The long term activity of the enzyme was stable for at least 72 hours at physiologic temperatures in a sensor configuration when coimmobilized with albumin.

The electrodeposition process allowed large numbers of sensors to be coated simultaneously due to the electrical nature. Yet the enzyme was precisely deposited only on the surface of the conductor exposed to the solution and the flow of current. Finally, the thickness of the enzyme layer was controlled by varying the electrical deposition parameters.

***In Vivo* Sensor Performance.** The ultimate test for the completed glucose sensor, fabricated utilizing the methods described here, was *in vivo* function. Tests performed in rabbits and human volunteers have shown an excellent correlation between glucose levels indicated by the subcutaneous sensor and commercial glucose analyzers (*20*).

Acknowledgments

The technical assistance of Charles Andrew, Brad Noffke, Nancy Bryan-Poole, William McMahan, Nan Cox, and Jennifer Simon was greatedly appreciated.

Literature Cited

1. Shichiri, M.; Kawamaori, R.; Hakui, N.; Yamasaki, Y.; Abe, H., *Diabetes* **1984**, 33, pp. 1200-2.
2. Bindra, D. S.; Zhang, Y.; Wilson, G. S.; Sternberg, R.; Thevenot, D. R.; Moatti, D.; Reach, G., *Biosensors & Bioelectronics* **1991**, 63, pp. 1692-6.
3. Bruckel, J.; Zier, H.; Kerner, W.; Pfeiffer, E. F., *Horm. Metab. Res.* **1990**, 22, pp. 382-384.
4. Koudelka, M.; Rohner-Jeanrenaud, F.; Terrettaz, J.; Bobbioni-Harsch, E.; DeRooij, N. F.; Jeanrenaud, B., *Biosensors & Bioelectronics* **1991**, 6, pp. 31-6.
5. Gough, D. A.; Lucisano, J. Y.; Tse, P. H. S., *Anal. Chem.* **1985**, 57, pp. 2351-7.
6. Clark, C. C., Jr.; Spokane, R. b.; Sudan, R.; Stroup, T. L., *Trans. Amer. Soc. Artif. Intern. Organs* **1987**, 33, pp. 323-8.
7. Updike, S. J.; Hicks, G. P., *Nature* **1967**, 214, pp. 986-8.
8. Romette, J. L.; Froment, B.; Thomas, D., *Clin. Chim. Acta.* **1979**, 95, pp. 249-53.
9. Claremont, D. J.; Sambrook, I. E.; Penton, C.; Pickup, J. C., *Diabetologia* **1986,** 29 pp. 817-21.
10. Fischer, U.; Abel, P., *Trans. Am. Soc. Artif. Intern. Organs* **1982**, 28, pp. 245-8.
11. Shichiri, M.; Kawamori, R.; Goriya, Y.; Yamasaki, Y.; Nomura, M.; Hakui, N.; Abe, H., *Diabetologia* **1983**, 24, pp. 179-84.
12. Gorton , L.; Jonsson, G., *J. Mol . Catalysis* **1986**, 38, pp. 157-9.
13. Kimura, J.; Ito, N.; Toshihide, K.; Kikuchi, M.; Arai, T.; Negishi, N.; Tomita, N., *J. Electrochem. Soc.* **1989**, 136-6, pp. 1744-7 .
14. Heineman, W. R.; Hajizadeh, K.; Tieman, R. S.; Rauen, K. L.; Coury, L. A. Jr., Proc. Third Intl. Meeting Chem. Sensors, Cleveland, OH, U.S.A., Sept. 24-26, 1990, pp. 369-372.
15. Shinohara, H.; Chiba, T.; Aizawa, M., *Sensors and Actuators* **1988**, 13, pp. 79-86.
16. Ikariyama, Y.; Yamauchi, S.;Yukiashi, T.; Ushioda, H., *Anal. Lett.* **1987**, 20(9), pp. 1407-1416.
17. Ikariyama, Y.; Yamauchi, S.;Yukiashi, T.; Ushioda, H., *J. Electrochem. Soc.* **1989**, 136(3), pp. 702-706.
18. Mastrototaro, J. J.; Johnson, K. W.; Morff, R. J.; Lipson, D.; Andrew, C. C.; Allen, D. J., *Sensors and Actuators B* **1991**, 5, pp. 139-144.
19. Johnson, K. W., *Sensors and Actuators B* **1991**, 5, pp. 85-89.
20. Johnson, K. W.;Mastrototaro, J. J.; Howey, D. C.; Brunelle, R. L.; Burden-Brady, P. L.; Bryan, N. A.; Andrew, C. C.; Rowe, H. M.; Allen, D. J.; Noffke, B. W.; McMahan, W. C.; Morff, R. J.; Lipson, D.; Nevin, R. S., *Biosensors & Bioelectronics* **1992**, 7, pp. 709-714.

RECEIVED August 3, 1993

Chapter 8

Thin-Layer Flow-Through Enzyme Immunosensor Based on Polycaprolactam Net

Lizhi Yin, Keli Xing, Hong Du, Hui Miao, Xi Wang, and Yuanming Li

Department of Transduction Technology, Tianjin Medical College, Tianjin 300070, People's Republic of China

An enzyme immunosensor composed of an oxygen electrode and a sandwich structure membrane is described. Antibodies to hepatitis B surface antigen (anti–HBsAg) is used as a model antibody, and polycaprolactam net as a support for preparing the antibody–bound membrane. Anti–HBsAg is covalently immobilized on polycaprolactam net with dimethyl sulphate, 1,6–hexanediamine and glutaraldehyde. The resulting antibody–bound membrane exhibits high mechanical strength, high binding capacity of antibody, small steric hindrance and good permeability for gas or liquid, and is extremely suited to construct an enzyme immunosensor with a gas sensing electrode. A heterogeneous sandwich enzyme immunoassay with glucose oxidase–labeled antibody is used in this work as the model assay. The sandwich structure membrane formed by this method is fixed onto the polypropylene membrane of the oxygen electrode. The depression of the O_2 concentration due to the enzyme reaction is measured by the oxygen electrode in the thin–layer flow cell. The detection limit for HBsAg is 5 ng / ml.

Immunoassay is a highly specific and sensitive approach for the determination of physiologically important trace substances. The use of immunoassay in diagnostic medicine is well documented and a number of routine clinical methods have been established for the detection of a wide variety of common antigen, hapten, and antibody. The use of enzyme labels in stead of radioisotope labels for the measurement of antigens, haptens, and antibodies has greatly stimulated the development of enzyme immunoassay (EIA). EIA can match RIA (radioimmunoassay) in terms of sensitivity and specificity; It has advantages of speed, convenience, and reduced cost, and has become generally accepted as an alternative

0097–6156/94/0556–0096$08.00/0

and even a substitute for other immunoassays (1). Enzyme labels in immunoassay provide the chemical amplification which allows the detection of very low concentration of analyte (2). Modern electroanalytical methods have the wide dynamic range and very low detection limits. Therefore, the enzyme immunosensors have been developed to take the advantage of the specificity of antibody, the enzyme amplification, and the ease with which small amounts of the enzyme–generated product can be detected electrochemically (3). A number of enzyme immunoassays with electrochemical detection have been demonstrated. For example, the assays for IgG (4,5), human chorionic gonadotropin (6), and α–fetoprotein (7), based on electrode covered with an antibody–bound membrane or the membrane obtained by the sandwich EIA procedure have been developed. Electrochemical enzyme immunoassays for rabbit IgG (8,9) and factor VIII–related antigen (10) using passive adsorption of its antibody to polystyrenecuvettes or tubes have been reported. Using an electropolymerized polytyramine modified electrode, human IgG has been detected with GOD–labeled anti–IgG via the detection of H_2O_2 (11). The Clark oxygen electrode is one of the best chemical transducers. It does not need an external reference electrode, the entire sensor assembly can be immersed in the test solution as a single unit, and its platinum electrode is not deteriorated by proteins and other materials. When it is used as a chemical transducer to construct a flow enzyme immunosensor, it is desirable that the support used as antibody should be characterized by good permeability for gas and liquid, high mechanical strength and a lot of activable groups. Polycaprolactam net is a good membranous material accorded with these demands. On this net anti–HBsAg is covalently immobilized through a bifunctional spacer molecule using a cross–linking agent, and the resulting antibody–bound membranes exhibit high mechanical strength, high binding capacity of antibody, small steric hindrance, and good permeability for gas and liquid. The sandwich enzyme immunoassay is running on the membrane. It involves the initial binding of the antigen to the membrane–bound antibody and the use of the enzyme–labeled antibody to determine the antigen. In the sandwich–type–enzyme immunoassay, the antigen is sandwiched between the enzyme–labeled antibody and the membrane–bound antibody. The membrane of the ″Ab–Ag–Ab″ structure formed on the surface of the antibody–bound membrane as a result of sandwich EIA procedure, is named ″sandwich structure membrane″ in this paper. The sandwich structure membrane is fixed onto the polypropylene membrane of the oxygen electrode. When a glucose standard solution is introduced into the thin–layer flow cell at a given flow rate to contact the electrode, the depression of O_2 concentration, due to the enzyme reaction, is determined by the oxygen electrode. The signal of such an electrode is proportional to the O_2 depression and to the antigen concentration. The purpose of this study was to demonstrate the feasibility of the use of polycaprolactam net for preparing antibody–bound membrane and constructing a flow enzyme immunosensor with the Clark oxygen electrode. Here, the enzyme immunosensor reported is composed of a sandwich structure membrane and an oxygen electrode. HBsAg was chosen as a model antigen and glucose oxidase (GOD) as an enzyme label. Glucose oxidase catalyzes the reaction of glucose with oxygen to form hydrogen peroxide and gluconolactone. Polycaprolactam net was used as a support. Due to good permeability of the sandwich structure membrane, the response time of the sensor matches the electrode response time for the substance related to the enzyme reaction without sandwich structure membrane. We also studied the sensor optimization and analytical characteristics.

Materials and Methods

Apparatus. The apparatus assembly consisted of an amperometric device, a recorder, and a pump. A linear Y / t chart recorder was used to record the FIA signals, and the sample propulsion was made with a peristaltic pump. The connecting tube was of PTFE or polyethylene. The O_2 depression, due to the enzyme reaction, was monitored with an oxygen electrode flow–through cell system designed for FIA (Fig.1), which was laboratory–built and consisted of an oxygen electrode and a flow–through cell.

Materials. Glucose oxidase (100u / mg) was obtained from Toyobo Co. Ltd (Osaka, Japan). Purified hepatitis B surface antigen (HBsAg subtype adr) was purchased from Institute of Biologicals (Beijin, China), and anti–HBsAg (the IgG fraction from goat anti–HBsAg serum ,which has a titre of 1.6×10^5 in PHA test and 2×10^6 in RIA test) from antihepatitis Institute (Tianjin, China). Glutaraldehyde was 50% aqueous solution (Sigma). The other reagents used were of analytical grade,from China. Polycaprolactam net was from screen mesh Co. (Shanghai, China) with the following characteristics : thickness 85 μm and mesh 102 threads / cm.

Substrate Solution. Standard glucose solution was prepared from 1.0 g glucose in 100 ml of 0.02 M phosphate buffer (PB), pH 7.0 . Phosphate–buffered saline (PBS) was from 0.15 M NaCl containing 0.01 M phosphate buffer, pH7.4. PBS–T was from PBS containing 0.05% Tween 20.

Diluting and Washing Solution. The egg white of six chicken eggs was stirred for about 1 h in 500 ml PBS with 0.1% NaN_3. The suspension was then centrifuged at 10,000 g for 15 min to remove debris. The supernatant was diluted 3% (v / v) in PBS–T with 0.1% NaN_3 (PBS–T–EWB), to which glycine was added until a final concentration of 0.1 M was reached (0.1 M G–PBS–T–EWB).

Blocking Solution. Glycine was added to PBS–T–EWB until a final concentration of 1 M was reached (1 M G–PBS–T–EWB). All buffer solution were prepared from deionized H_2O of at least $10^6 \Omega$ resistivity.

Preparation of the GOD–labeled Anti–HBsAg. The conjugate of anti–HBsAg–GOD was prepared by a modification of the method reported by Tsuji et al. (11). 10 mg of GOD were dissolved in one ml of phosphate buffer (0.1 M, pH 6.8) with 0.4% glutaraldehyde and the mixed solution was incubated at room temperature for 18– 24 h. After incubation, this activated GOD solution was dialyzed at 4℃ against 0.15 M NaCl (200 volumes, 24 h, four changes). One ml of 5 mg / ml anti–HBsAg was added to the activated GOD solution and to this mixture 0.2 ml of 1 M Na_2CO_3 / $NaHCO_3$buffer, pH 9.5, was added. The solution was placed in an airtight vessel, stirred gently for 2–3 h at room temperature and then left at 4℃ overnight. After incubation at 4℃, the solution was dialyzed against 0.01 M phosphate–buffered saline, pH 7.4 (200 volumes, 24 h, four changes). Sodium azide (0.1% final concentration) was added to the conjugate, and it was stored at 4℃ for one week prior to use. It is important that the conjugate never be frozen. The conjugate was diluted with the diluting solution to adjust the GOD activity at 6 u / ml for use.

Preparation of the Antibody–bound Membrane. Antibody was immobilized on polycaprolactam net by a modification of the method devised by Horbby and Morris (12). Small disks measuring 6 mm in diameter were cut out of the polycaprolactam net and fixed around a hydrophobic plastic rod with polypropylene thread for ease handling. The rod was immersed in dimethyl sulphate–acetone (1:4, v / v) solution, moving back and forth at room temperature for 10 min. The rod was then transferred to a vessel containing anhydrous methanol for washing the membranes twice or until there was no white precipitate. The methylated polycaprolactam disks were removed from the rod, and placed in 0.5 M 1,6–hexanediamine solution, adjusted to pH ca.9 with hydrochloric acid, and stirred for 2 h at room temperature.The 1,6–hexanediamine acts as a spacer between the polycaprolactam structure and the antibody molecule. After thorough rinsing with 0.1 M sodium chloride solution, the small disks with the spacer were placed in a solution of 12%(v / v) glutaraldehyde in 0.1 M borate buffer, pH 8.5, and stirred for 45 min at room temperature. After further washing they were immersed in the antibody solution in 0.1 M phosphate buffer, pH 7.0, 300 μl / piece, for 2 h and then overnight at 4℃. The disks were washed thoroughly in PBS–T and placed in 1 M G–PBS–T–EWB overnight at room temperature for blocking unreacted groups. The disks after washing with PBS–T were stored in 0.1 M G–PBS–T–EWB at 4℃. The resulting disk is named antibody–bound membrane.

Enzyme Immunosensor and Assay Procedure. A piece of antibody–bound membrane was incubated in 200 μl of a solution of HBsAg diluted with the diluting solution, with gentle shaking for 2 h at room temperature. The membrane was taken out, washed with the washing solution and incubated in 200 μl of solution containing 1.2 u GOD activity of the GOD–labeled anti–HBsAg, for 40 min at room temperature with gentle shaking. After this incubation, this membrane was washed several times with the washing solution. The membrane thus obtained became the sandwich structure membrane. The enzyme immunosensor was constructed by fixing this membrane onto the polypropylene membrane of the oxygen electrode by means of a cap with screws fitted with packing (Fig.1). The working electrode was set at −600 mv vs. an Ag–AgCl reference electrode (0.5 M potassium chloride). Then 0.02 M PB, pH 7.0, was pumped into the flow–through cell at a flow–rate of 0.5 ml / min. After a baseline of the recorder was stabilized, a standard glucose solution in place of 0.02 M PB was pumped at the same flow–rate. If the antibody–bound membrane binds the GOD–labeled anti–HBsAg, the reaction of glucose with O_2 catalyzed by GOD on the sandwich structure membrane reduces the O_2 concentration in the thin–layer flow–through cell, thus depressing the oxygen electrode current. This depression is proportional to the GOD activity retained on the sandwich structure membrane and subsequently to HBsAg concentration.

Results and Discussion

O–methylation of Polycaprolactam Net and Attachment of Anti–HBsAg onto It. Polycaprolactam net is soluble in dimethyl sulphate, and hence its O–methylation is difficult for preparing antibody–bound membrane. Several organic solvents, such as acetone, methanol, ethanol, dichloromethane trichloromethane, 1,4–dioxane etc, have been tried in our study, and we found that acetone and 1,4–dioxane can be used to dilute dimethyl sulphate for O–mathylation. In the present test, actone was used as a solvent, since it does not dissolve polycaprolactam net. In a dimethyl sulphate solution diluted with acetone, the dissolved rate of polycaprolactam decreases with an increase in acetone concentration. If the O–methylation time of about 10 min is taken, 1:4 (v / v) dimethyl sulphate–acetone solution is preferred. The structure of the antibody–bound membrane, obtained by attaching antibody onto polycaprolactam net by means of 1,6–hexanediamine and glutaraldehyde, may be as follows:

$$\begin{array}{l} -\mathrm{C{-}NH(CH_2)_6\ N{=}CH(CH_2)_3\ CH{=}N{-}antibody} \\ \ \ \| \\ -\mathrm{N^{+}H} \end{array}$$

It is preferable to couple antibodies to polycaprolactam structures after the insertion of a spacer in the support. In this way, the antibody is attached to the support by relatively long flexible arms, which reduce the possibility of steric hindrance to benefit the immobilization of antibodies, the reaction of antibody / antigen, and the interaction between the labeled enzyme and its substrate.

Effect of Incubation Time on the Sensor Response. we determined the time required for the two antigen / antibody reactions to reach eqilibrium by using 1 μg / ml HBsAg standard and 6 u / ml GOD activity of the GOD–labeled antibody at the varying incubation time for each reaction, one at a time, The result shows that the membrane–bound antibody / antigen reaction required 2 h to obtain the maximal response of the sensor. The reaction between the antigen bound to the antibody–bound membrane and the labeled antibody , 40 min, reaction time was considered sufficient.

Effect of pH on the Sensor Response. The sandwich structure membrane prepared under the above conditions, was used, and the pH dependence of the sensor response was studied at room temperature at a flow–rate of 0.5 ml / min, over the range pH 6.0 – 8.0. The different pH solutions used are the phosphate solutions adjusted with sodium hydroxide. The optimum pH was found to be 7.0 (Fig.2). This value is in accord with the optimum pH determined with GOD membrane, which is obtained by the immobilization of GOD on polycaprolactam net with dimethyl sulphate, 1,6–hexanediamine and glutaraldehyde.

Flow–rate dependence. Effect of flow–rate at pH 7.0 over the range 0.26–1.26 ml / min was studied, using the sandwich structure membrane prepared under the above conditions. The resulting sensor response vs. flow–rate profile (Fig.3) indicates that the sensor response decreased with increasing flow–rate. The reason for this phenomenon may be that the flow–rate is too large to conduct the enzyme catalyzed glucose reaction with O_2 . A lower flow–rate can increases the sensor response but prolongs the detecting time. As a compromise, a flow–rate of 0.5 ml / min was employed in this study.

Antibody Concentration Dependence. Using a HBsAg concentration of 1 μg / ml and a GOD activity of the labeled antibody of 6 u / ml, the sandwich structure membrane was prepared. Under the above optimum conditions, the effect of different antibody concentrations on the sensor response (Fig.4) was investigated. It is observed from Fig.4 that 300 μl of 150 μg / ml anti–HBsAg cannot make all the activated groups to react. In order to obtain a better precision, we used 300 μl of 100 μg / ml anti–HBsAg to prepare the antibody–bound membrane. The resulting membrane does not exhibit steric hindered effect.

Elimination of Nonspecific Binding. Another problem present in the sensor is a large background current observed in the absence of antigen, which is from nonspecific binding of the GOD–labeled antibody.

Nonspecific binding of proteins has been decribed in detail. For nonspecific binding of proteins, there are two common modes. The first occurs from the interaction of the labeled antibody with some species other than the antigen. The second type of nonspecific binding is from the adsorption of the labeled antibody or antigen to the solid support (2). Using polycaprolactam net as support, in addition to the two modes described above, an electrostatic interaction between the surface and the protein is also present. This is because the support of the antibody–bound membrane itself carries positive charges (see structure). To eliminate the nonspecific bindings, we used the PBS–T containing egg white and glycine

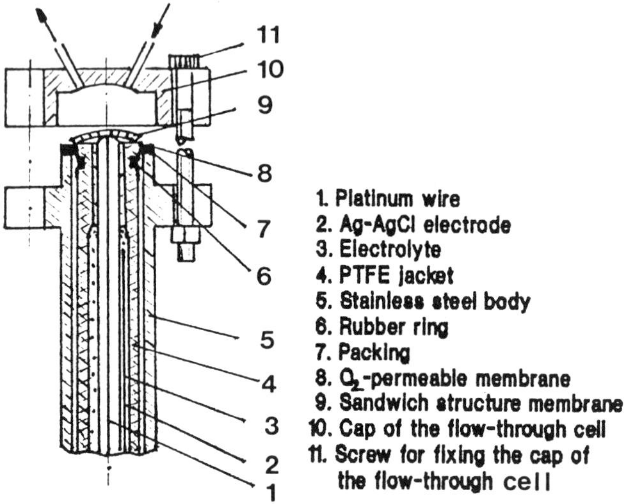

Figure 1. Diagram of the immunosensor electrode and flow-through cell.

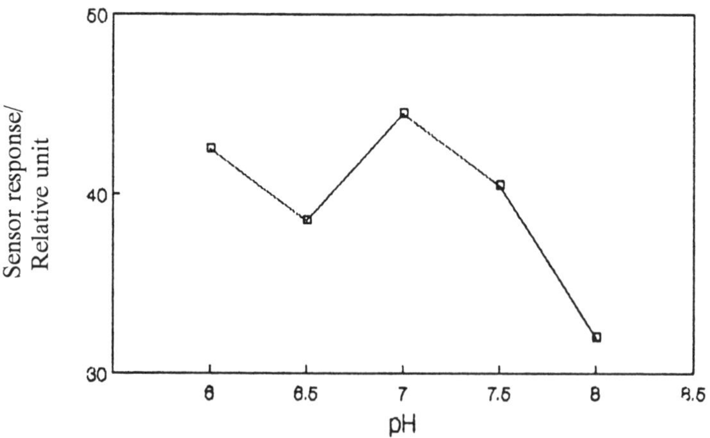

Figure 2. Effect of pH on the sensor response.

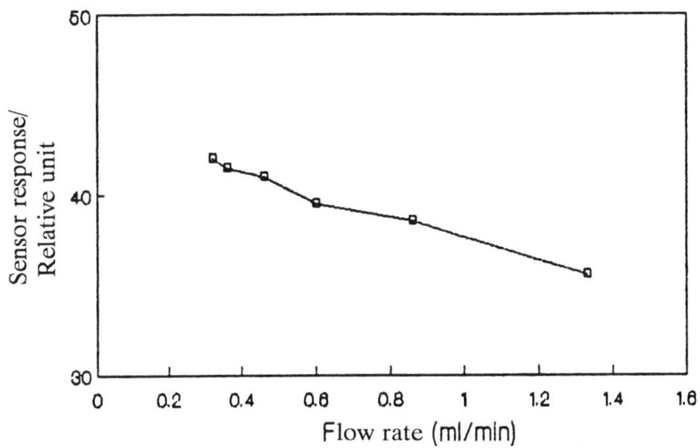

Figure 3. Effect of flow rate on the sensor response.

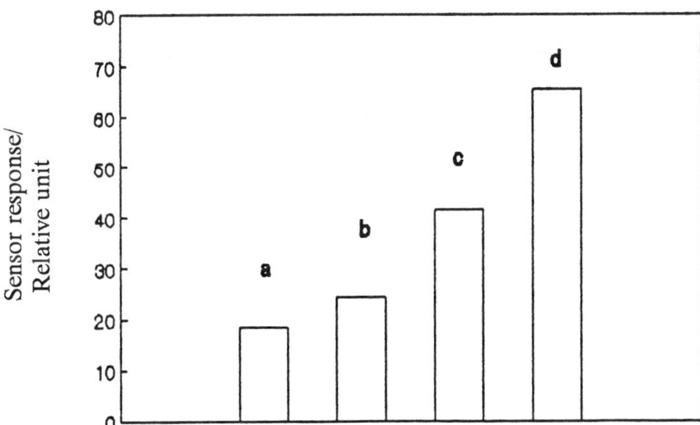

Figure 4. Dependence of the sensor response on anti-HBsAg concentration. Anti-HBsAg concentration: (μg/ml): 25 (a), 50 (b), 100 (c), and 150 (d).

for diluting the antigen or labeled antibody solution, and washing the membranes in the assay protocol (Fig 5,6). Glycine in neutral or alkaline solution carries a negative charge and may neutralize the positive charges on the membrane.

Calibration Curve of the Sensor. The sensor was calibrated at the optimized conditions described above with HBsAg standard over the range 2.8 ng/ml–28 μg/ml. Fig.7A illustrates typical chart recorder output and Fig.7B shows the corresponding calibration curve. The response time was < 25s, and the wash time < 35s. The lowest detectable concentration was 5 ng/ml.

Repeated sampling of two HBsAg standards (0.5 and 1.0 μg/ml) gave mean currents of 18.5 and 41.5 relative units respectively, the corresponding standard deviation was 1.0 and 2.5 relative units (n = 5).

Feasibility of Constructing the Enzyme Immunosensor for the Determination of Antigen. In the conventional immunoassay, to prevent nonspecific binding of immunoreagents, it is necessary to add inhibitors, which is usually protein solution containing a non–ionic detergent, such as Tween 20. The inhibitors, however, cause desorption of adsorbed antigen and adsorbed antibody. To diminish this desorption, covalent attachment of antibodies to insoluble supports can be done for use in immunoassays. A number of solid phase supports have been used for preparing insoluble enzymes. Among the better ones has been polycaprolactam which resembles polypeptides in chemical structure and is rich in activable peptide linkages. Our procedure for immobilizing antibodies on polycaprolactam net is very simple and quite mild and a large percentage of the antibody activity remains after the immobilization process. The antibody–bound membrane prepared with the polycaprolactam net, has high mechanical strength, high binding capacity of antibody, small steric hindrance and good permeability for gas or liquid. Because of its porosity it readily reacted with antigen, labeled antibody and substrate, and easily washed, compared with the conventional EIA using polystyrene cuvette. In addition, the sensitivity obtained by the antibody–bound membrane is much more than the conventional EIA (Fig.8).

The principal advantages of the Clark oxygen electrode are that proteins and other materials which can foul solid eletrode do not contact the platinum and reference electrodes. The entire sensor assembly can be immersed in the test solution as a single unit without an external reference electrode. Hence the oxygen electrode is well suited to construct a flow–through enzyme immunosensor with such a porous antibody–bound membrane using GOD or catalase as a label. Since the enzyme immunosensor has an excellent response time, the enzyme activity on the sandwich structure membrane can be determined with flow injection mode, which makes the detecting procedure simple and convenient.

The preliminary experimental results show that the enzyme immunosensor is feasible for practical application.

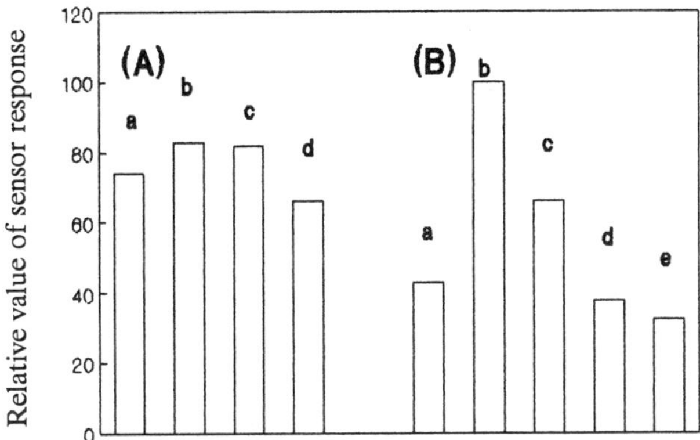

Figure 5. The sensor response dependence of the washing solution of different components and concentrations. A: 1.0 M glycine–PBS–T containing egg white (v/v), 1% (a), 3% (b), 5% (c), and 7% (d). B: 3% (v/v) egg white–PBS–T containing glycine, 0 M (a), 0.1 M (b), 0.35 M (c), 0.7 M (d), and 1.0 (e).

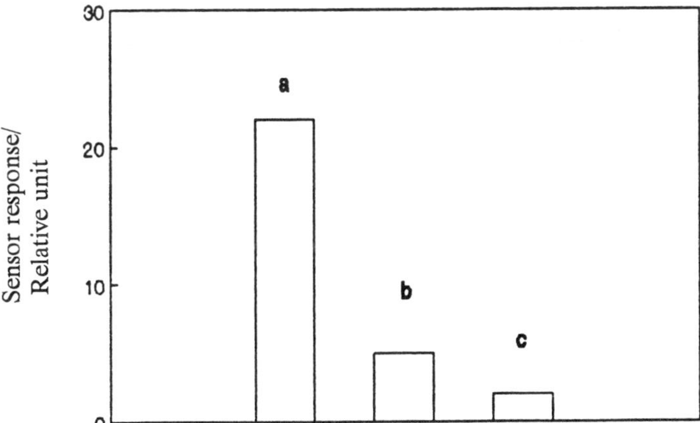

Figure 6. Different washing solution dependence of non-specific adsorption of the enzyme conjugate. a, PBS–T; b, 3% (v/v) egg white–PBS–T; c, 0.1 M glycine–3% (v/v) egg white–PBS–T.

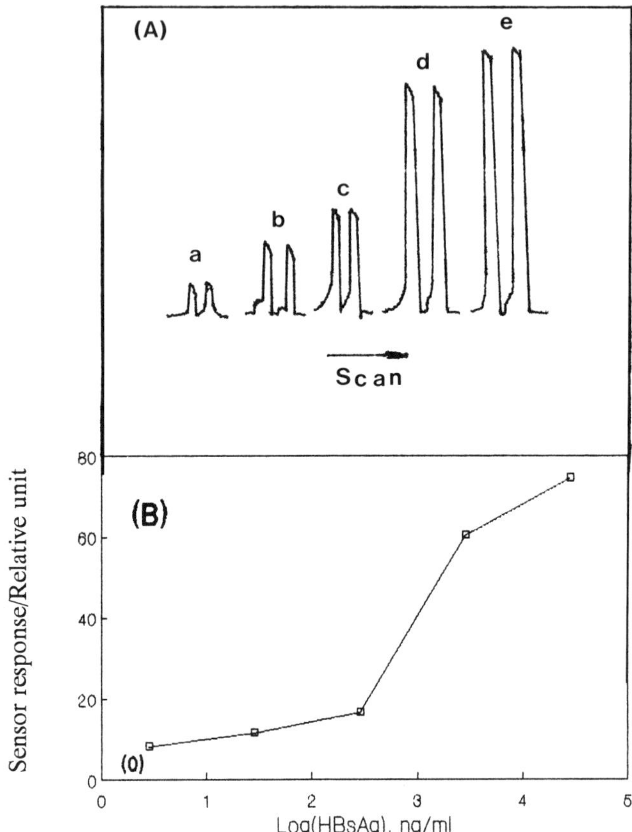

Figure 7. The typical chart recorder output (A) and calibration curve (B) for HBsAg. The brackets denote the 0 ng/ml standard HBsAg concentration (ng/ml): 2.8 (a), 28 (b), 280 (c), 2,800 (d), and 28,000 (e).

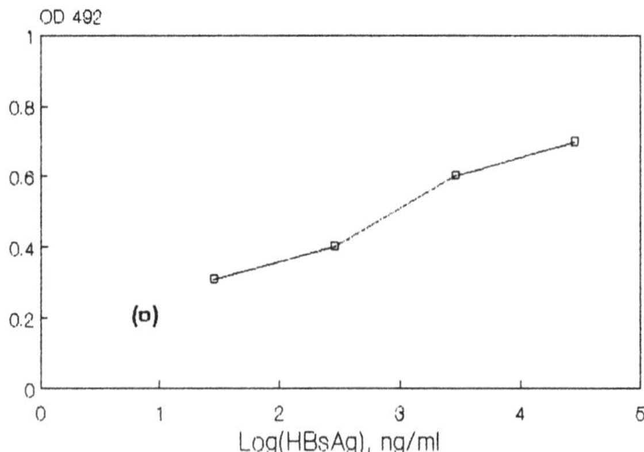

Figure 8. Calibration curve for HBsAg using the polystyrene cuvette coated by passive adsorption of anti-HBsAg and the conventional EIA. The brackets denote the 0 ng/ml standard HBsAg.

Acknowledgements

The authors acknowledge financial support from the Tianjin Committee of Science and Technology, China, and Professor J.F.Pan of Immunology Department, TMU, for her advice about immunology in this work.

References

1. Black, C.; Gould, B.J. Analyst 1984, 109. 533–547.
2. Jenkins, S.H.; Heineman, W.R.; Halsall, B. Anal. Biochem. 1988, 168, 292–299.
3. Stanley, C.J.; Cox,R.B.; Carblosi, M.W.; Turner, A.P.F. J. Immunol. Methods. 1988, 112, 153–161.
4. Aizawa, M.; Morioka, A.; Suzuki, S. J. Membr. Sci. 1978, 4, 221–228.
5. Yin, L.Z.; Li, Y.L.; Guo, S.Y.; Zhou, B.K. Chinese J. Immunol. 1987, 3, 60–63.
6. Aizawa, M.; Morioka, A.; Suzuki, S.; Nagamura, Y. Anal. Biochem. 1979, 94, 22–28.
7. Aizawa, M.; Morioka, A.; Suzuki, S. Anal. Chim. Acta. 1980, 115, 61–67.
8. Wehmeyer, K.R.; Heineman, W.R.; Halsall, H.B. Clin. Chem. 1985, 31, 1546–1549.
9. Wehmeyer, K.R.; Brian, H.; Heineman, W.R. Anal. Chem. 1986, 58, 135–139.
10. Renneberg, R.; SchoBler, W.; Scheller, F. Anal. Lett. 1983, 16, 1279–1289.
11. Tsuji, I.; Eguchi, H.; Yasukouchi, K.; Taniguchi, I. Biosensors & Bioelectronics, 1990, 5, 87–101.
12. Hornby, W.E.; Morris, D.L. In Immobilized Enzymes, Antibodies, and Peptides; Weetall, H.H., Ed.; Marcel Dekker, New York, 1975, pp 141–169.

RECEIVED July 26, 1993

Biosensor Polymers and Membranes

Chapter 9

Conducting Polymers and Their Application in Amperometric Biosensors

Wolfgang Schuhmann

Lehrstuhl für Allgemeine Chemie und Biochemie, Technische Universität München, Vöttingerstrasse 40, D–85350 Freising-Weihenstephan, Germany

The development of reagentless enzyme electrodes implies the covalent binding of enzymes, redox mediators and/or coenzymes on the sensor surface to prevent contamination of the sample by sensor components. Additionally, miniaturization and mass production of enzyme electrodes require techniques for biosensor assembling avoiding manual deposition procedures. The electrochemical deposition of conducting-polymer layers occurs at the surface of an electrode independent of its size and form. Subsequent to the functionalization of conducting-polymer films, electrode surfaces can be obtained which are suitable for the binding of enzymes, coenzymes or redox mediators. The morphology of the conducting polymer films can be adapted to exclude cooxidizable compounds from approaching the electrode surface, thus improving the selectivity of the amperometric enzyme electrodes. The deposition and functionalization of polypyrrole, polyazulen and poly(dithienylpyrrole) as well as the covalent binding of redox mediators, enzymes and mediator-modified enzymes will be described. Main emphasis is focused towards an improvement of the electron-transfer reactions between the immobilized enzymes and the modified electrode surfaces.

Most of the amperometric biosensors described until now are constructed either by crosslinking a suitable redox enzyme within a polymeric gel on the electrode surface or by assembling a preformed enzyme-containing membrane on top of the electrode. Although this approach has led to amperometric biosensors for substrates of most oxidases or enzyme sequences involving oxidases, besides the inherent thermal instability of proteins two major drawbacks have to be focused on.

First, due to the fact that the active site of redox enzymes is in general deeply buried within the protein shell, direct electron transfer between enzymes and electrode surfaces is rarely encountered. This is especially true for enzymes which are integrated within non-conducting polymer membranes in front of the electrode surface. Hence, electron transfer is usually performed according to a 'shuttle' mechanism involving free-diffusing electron-transfering redox species for example the natural electron acceptor O_2 or artificial redox mediators like ferrocene derivatives (*1*), osmium

0097–6156/94/0556–0110$08.00/0

complexes (*2*), quinones (*3*). Due to the necessity for the redox mediators to diffuse freely between the active sites of the enzymes and the electrode surface, these electrodes show a limited long-term stability as a consequence of the indispensable leaking of the mediator from the sensor surface (*4*). Additionally in the case of the natural redox couple O_2/H_2O_2, the sensor signal is dependent on the O_2 partial pressure, and a high operation potential has to be applied to the working electrode giving rise to possible interferences from cooxidizable compounds. The second drawback is related to the fabrication of these sensors. The physical assembling of an enzyme membrane and an electrode is extremely difficult to automate and thus in principal incompatible with microelectronic fabrication techniques. Additionally, the miniaturization as well as the integration of individual biosensors into a miniaturized sensor array is impossible with techniques which are mainly based on the manual deposition of a droplet of the membrane-forming mixture onto the electrode surface.

Consequently, the next generation of amperometric enzyme electrodes has to be based on immobilization techniques which are compatible with microelectronic mass-production processes and easy to miniaturize. Additionally, the integration of all necessary sensor components on the surface of the electrode has to prevent the leaking of enzymes and mediators simultaneously improving the electron-transfer pathway from the active site of the enzyme to the electrode surface. In this communication, functionalized conducting polymers electrochemically deposited on the electrode surface, are investigated with respect to their possibilities for the modification of electrodes and the covalent attachment or entrapment of sensor components.

Possibilities for the Development of Reagentless Amperometric Enzyme Electrodes

A fundamental presupposition for the construction of reagentless amperometric enzyme electrodes is the design of a suitable electron-transfer pathway from the active site of the enzyme to the electrode surface. According to the Marcus theory (*5*), a redox mediator with a low reorganization energy after the electron transfer has to be able to penetrate into the active site of the enzyme to shorten the distance between the prosthetic group (e.g. $FAD/FADH_2$) and the mediator. Hence, the rate constant of the electron-transfer reaction can be increased. After this 'first' electron transfer the redox equivalents have to be transported to the electrode surface with a rate constant which is in the range of the turnover rate of the enzyme. In the shuttle mechanism mentioned above, this electron transport is connected to a mass transfer correlated to the diffusion of sufficiently soluble redox mediators. Theoretically, there is no difference if the mass transfer of a single mediator molecule is substituted by an electron hopping between adjacent mediator molecules tremendously decreasing the necessary diffusion distance of the individual molecule. However, for the electron-exchange distance to be short and thus the rate constant to be high the mediator molecules have to be bound allowing a certain flexibility. Obviously, for the development of reagentless amperometric enzyme electrodes one has to assure that the distance between the deeply buried active site of the immobilized enzyme and the electrode surface can be divided into numerous electron-transfer reactions between immobilized but flexible redox mediators.

The first approach in this direction was made by Degani and Heller (*6,7*) who immobilized ferrocene derivatives near the active site of glucose oxidase after a partial separation of the subunits of the enzyme. In the absence of oxygen a direct electron transfer from the enzyme to an electrode was observed assuming a sequence of electron-transfer reactions from the $FADH_2$ via the enzyme-linked ferrocene moieties to the surface of the enzyme and finally to the electrode. This approach, however, although preventing the redox mediator to leak from the electrode surface is dependent on the diffusion of the ferrocene-modified enzyme itself. Recently, it has been shown that the mediator can be bound via long and flexible chains to the surface of the enzyme (*8*), and that the electron transfer occurs according to a 'wipe'-mechanism by swinging the mediator into the active site of the enzyme and then out to the electrode suface (Fig. 1). It could be demonstrated that the intramolecular electron-transfer mechanism is predominant with long spacer chains, while in the case of short spacer chains the mechanism is intermolecular with one mediator-modified enzyme acting as a high-molecular but free-diffusing redox mediator for a second one.

Another approach is based on the application of redox polymers, e.g. osmium complex-modified poly(vinyl pyridine) (*9-11*) or ferrocene-modified poly(siloxanes) (*12,13*), crosslinked together with an enzyme on the top of the electrode. The electron transfer from the active site of the polymer-entrapped enzyme to the electrode surface occurs to a first polymer-bound mediator which has sufficiently approached the prosthetic group to attain a fast rate constant for the electron-transfer reaction. From this first mediator the redox equivalents are transported along the polymer chains by means of electron hopping between adjacent polymer-linked mediator molecules (Fig. 2). Extremely fast amperometric enzyme electrodes have been obtained with significantly decreased dependence from the oxygen partial pressure. However, the redox polymer/enzyme/crosslinker mixture has to be applied either manually or by dip-coating procedures onto the electrode surface.

For the construction of reagentless enzyme electrodes one has to focus on a technique for the modification and functionalization of electrode and even micro-electrode surfaces to allow the covalent binding of the enzyme and the redox mediator taking into account the presuppositions for an effective and fast electron transfer between the enzyme and the electrode.

Functionalized Conducting Polymers for the Covalent Binding of Enzymes

Conducting polymers like polypyrrole, polyaniline, polythiophene can be obtained by electrochemical polymerization either potentiostatically, galvanostatically or by means of multi-sweep experiments (*14*). Well-adhering polymer films are formed on the anode surface, thus allowing the selective modification of ultra-micro electrodes. By measuring the charge transferred during the electrochemical polymerization process and by controlling parameters like temperature, monomer concentration, polymerization potential or current, as well as the concentration and nature of the supporting electrolyte, polymer films with defined thickness and morphology can be obtained. Hence, conducting polymer films show interesting properties concerning the decrease of the influence of interfering compounds due to their size-exclusion and ion-exchange characteristics (*15,16*).

For the integration of bioselective compounds and/or redox mediators in the conducting polymer, the preparation of functionalized conducting-polymer films is an indispensable prerequisite. Two possibilities for the modification or functionalization of conducting polymer layers are obvious (Fig. 3). On the one hand one can synthesize the functionalized monomers and subsequently polymerize them or copolymerize them together with the non-functionalized monomer to the corresponding functionalized polymers, or on the other hand an already-grown conducting-polymer film can be modified by means of heterogeneous reaction sequences.

The polymerization of functionalized monomers has the advantage to lead to a uniform distribution and a high concentration of the functional groups within the polymer film. However, the polymerization of substituted monomers often fails, especially when the introduced functional group show a substantial nucleophilicity and thus can react with the radical-cation intermediates during the electrochemical polymerization reaction. Additionally, the corresponding films show different morphology or stability compared with the unsubstituted one. In the case of polypyrrole, for example, the conductivity of the N-substituted polymers drops by a factor of up to 10^5.

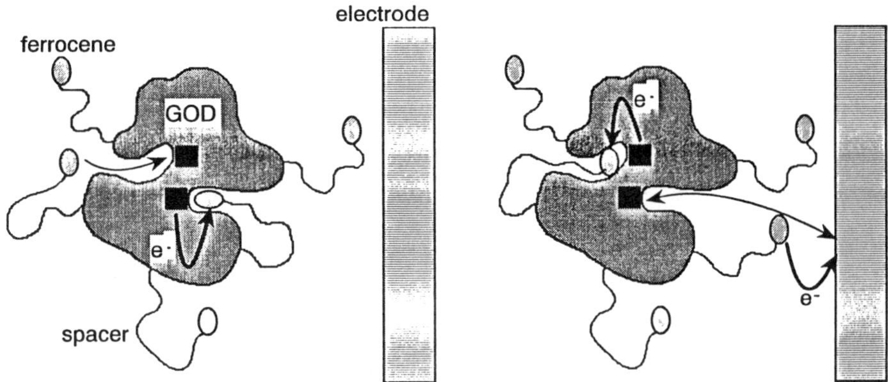

Figure 1. Glucose oxidase with redox mediators bound via flexible spacer chains to the enzyme surface. Schematical representation of the 'wipe' mechanism of electron-transfer from the enzyme's active site to the electrode surface.

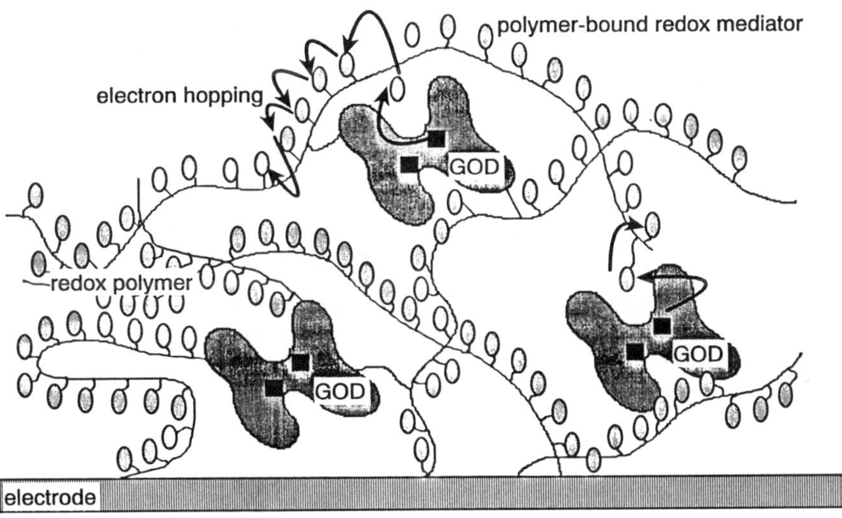

Figure 2. Electron transfer in a redox polymer via electron hopping between adjacent redox centers.

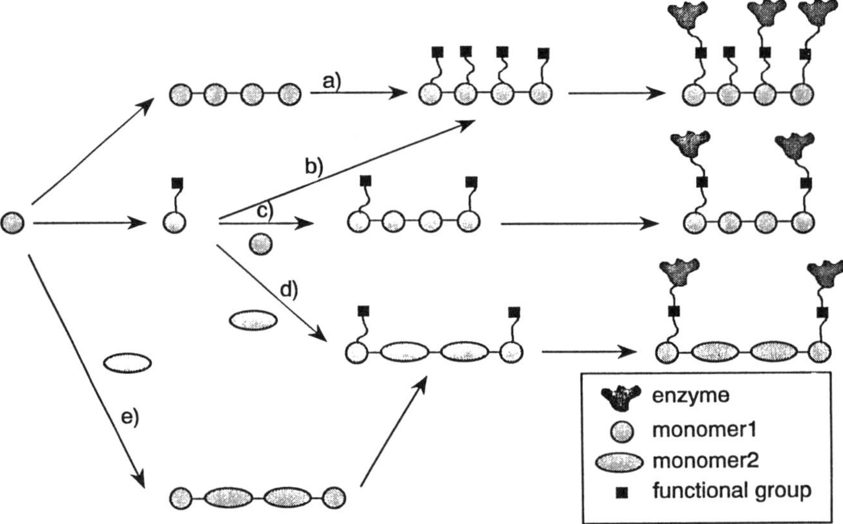

Figure 3. Possibilities for the preparation of functionalized conducting polymers and subsequent covalent binding of enzymes. a) Functionalization of already grown polymer films. b) Polymerization of functionalized monomers. c) Copolymerization of functionalized and non-functionalized monomers. d) Copolymerization of functionalized and non-functionalized monomers of different type. e) Copolymerization of monomers of different types and subsequent functionalization.

β-Substituted polypyrrole films in contrast show high conductivity, however, the synthesis of related monomers is complicated due to the high reactivity at the 2 and 2′-position of the heterocycle (*17*). In contradistinction, N-substituted pyrrole monomers (*18*), substituted azulen derivatives (*19*) or functionalized dithienylpyrroles (*20*) could be obtained and polymerized to the corresponding conducting-polymer films (Fig. 4). Another possible approach is the modification of already grown polymer films by immersing the polymer-modified electrodes into a suitable reagent solution. The introduction of β-amino-groups in polypyrrole has been used as a means for the functionalization of conducting polymer films (*21,22*).

These polymer films can be used as functionalized surfaces for the covalent immobilization of enzymes and/or redox mediators (Fig. 5). As the high-molecular weight enzyme molecules cannot penetrate into the polymer network they are exclusively bound at the outer surface of the polymer film. Thus, their active sites are easily attainable by the substrate allowing the use of the size-exclusion properties of conducting polymer films to decrease the influence of interfering compounds like ascorbic acid or acetaminophen (*15,23,24*). The covalent immobilization of enzymes to amino group-modified polypyrrole films lead to fast and stable biosensors. However, the functionalization of the already grown polymer leads to a non-uniform distribution of the functional groups at the polymer and thus to a suboptimal reproducibility in the production of these enzyme electrodes. To overcome this problem, glucose oxidase has been bound covalently at amino-functionalized polyazulen (*25*) and amino-functionalized dithienylpyrroles (*26*). A remarkably high enzyme activity could be immobilized at these polymer-modified electrode surfaces leading to a sufficiently large current density in dependence from the substrate concentration (Fig. 6). A good long-term stability up to 4 weeks of continuous operation and up to 5,000 injections of glucose could be observed with these electrodes proving the strong covalent fixation of the enzyme on the functionalized conducting-polymer film. However, until now a direct electrical wiring between the covalently bound enzyme and the electrode surface involving the conducting polymer chains could not be encountered (*27*). Thus, the covalent immobilization of enzymes on functionalized conducting-polymer surfaces can be advantageously used for the development of amperometric enzyme electrodes independently from the size and accessibility of the electrode surface to be modified. Additionally, the electrochemical deposition of conducting-polymer films can be simultaneously used - even on wafer-level - for the uniform and reproducible modification and functionalization of multi-electrode arrays.

Entrapment of Enzymes within Electrochemically-Grown Conducting Polymer Films

Another possibility for the immobilization of enzymes within conducting-polymer layers is the entrapment of the enzymes within the growing polymeric network during the electrochemical polymerization process (*28,29*). Obviously, the mass-transfer characteristics imply the diffusion of the substrate through the conducting-polymer network leading to a dramatic dependence of the signal from the film thickness, the morphology and porosity of the conducting-polymer film (*30*). However, although the entrapment of negatively-charged enzymes as counter anions for polycationic polypyrrole was proposed as a reason for the immobilization process (*31*), it is more likely that the entrapment is only due to a statistical enclosure of the enzymes present in the vicinity of the electrode surface.

Copolymerization of Pyrrole-Modified Glucose Oxidase and Pyrrole. To achieve a more defined entrapment of enzymes within polypyrrole films and simultaneously to increase the immobilized enzyme activity, the copolymerization of a pyrrole-modified enzyme together with pyrrole has been envisaged (*32*). Water-soluble β-

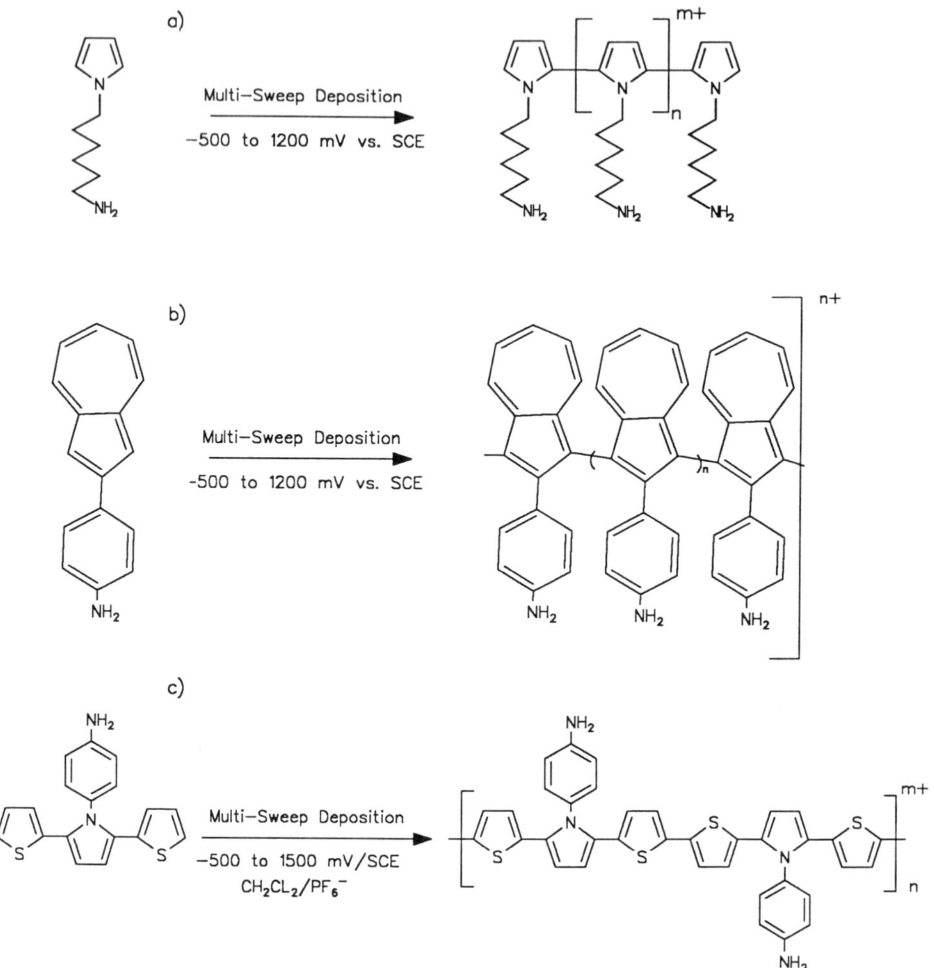

Figure 4. Electrochemical preparation of functionalized conducting-polymer films. a) Polymerization of N-substituted pyrrole derivatives. b) Polymerization of 4-(aminophenyl)azulen. c) Polymerization of N-(4-aminophenyl-2,2′-dithienyl)-pyrrole.

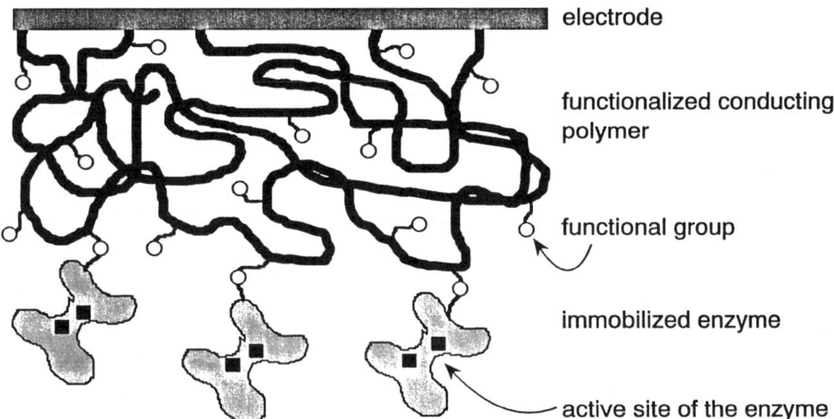

Figure 5. Schematic representation of an amperometric biosensor with the enzyme covalently bound to a functionalized conducting-polymer film.

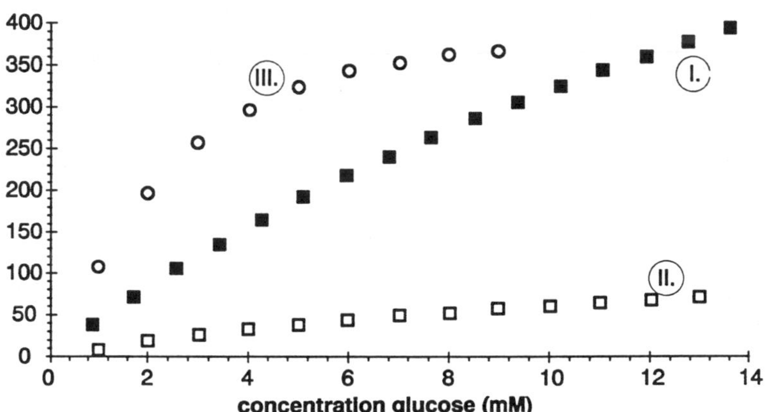

Figure 6. Calibration graphs for glucose obtained with amperometric enzyme electrodes with glucose oxidase covalently bound to different conducting polymers. I. β-amino(polypyrrole). II. poly-(4-aminophenyl)azulen. III. poly[N-(4-aminophenyl)-2,2′-dithienyl]pyrrole.

(aminoalkyl)pyrrole derivatives can be linked to IO_4^--oxidized glucose oxidase (*33*) with formation of Schiff bases which are subsequently reduced to the respective secondary amines by means of $NaBH_4$ (Fig. 7a). The modification of the enzyme with water-insoluble pyrrole derivatives can be performed in a phase-transfer reaction with individual enzyme molecules entrapped within the water pool of inverse micelles (*34*) (Fig. 7b). The pyrrole-modified enzyme can be copolymerized with pyrrole to "enzyme-loaded" conducting polymer films (Fig. 8a) showing a significantly increased immobilized enzyme activity. Calibration graphs for glucose obtained with a pyrrole-modified glucose oxidase/pyrrole copolymer electrode in comparison to a native glucose oxidase conventionally entrapped within an electrochemically grown polypyrrole film demonstrate the improved features of these electrodes (Fig. 8b). Independently, very recently a similar approach has been reported with short-chain N-(carboxyalkyl)pyrroles linked to glucose oxidase (*35*). A significant increase of the stability of the related polypyrrole/enzyme electrodes has been observed.

By optimizing the number of introduced pyrrole units at the enzyme surface, a homopolymerization of the pyrrole-modified glucose oxidase or at least a significant shift of the ratio of the enzyme-bound pyrrole units and non-modified pyrrole should be envisaged to even more increase the immobilized enzyme activity.

Entrapment of Mediator-Modified Enzymes within Conducting-Polymer Films. The copolymerization of pyrrole-modified enzymes and the covalent binding of enzymes to functionalized conducting-polymer films lead to integration of only one of the necessary sensor components, the biological recognition element. As we and several others could not observe a direct electron transfer from the conducting-polymer bound enzyme via the ramified polymer network to the electrode surface the described conducting-polymer based enzyme electrodes are still dependent on the shuttling of electrons with the aid of free-diffusing mediators.

How is it possible to integrate simultaneously the enzyme and the mediator on the electrode surface, and how can we achieve electron-transfer reactions with a fast rate constant using immobilized mediators? Obviously, it is necessary to combine the knowledge about electron-transfer mechanisms based on spacer-linked mediators mentioned above and the advantages from the electrochemical deposition of conducting-polymer layers as immobilization matrix for enzymes and/or redox mediators.

Ferrocene-modified glucose oxidase (*8*), entrapped within a polypyrrole film during the electrochemical polymerization reaction (Fig. 9a), was tested excluding oxygen in a constant potential measurement (*32*). The calibration graph for glucose (Fig. 9b) demonstrates that electron transfer from the ferrocene-modified enzyme to the electrode surface is possible via the sensor-integrated redox mediators. This is the first example of a reagentless oxidase electrode based on conducting polymers and mediator-modified enzymes.

Conclusion and Outlook

Knowledge of possible electron-transfer pathways play an important role in the development of amperometric biosensors with improved operation characteristics. Obviously, a compromise has to be found between tight immobilization of all sensor components on the electrode surface and the need to maintain a sufficient high flexibility to allow proper approach of redox centers involved in the electron-transfer chain to obtain a high rate constant of the overall reaction. These features are in principle accomplished with enzyme electrodes based on redox-sensitive hydrogels, however, the manual deposition of these hydrogels is not compatible with mass-production techniques. The electrochemical deposition of conducting-polymer layers, obtained with reproducible thickness and morphology, occurs exclusively on the electrode surface and can hence be used as a versatile tool for the immobilization of

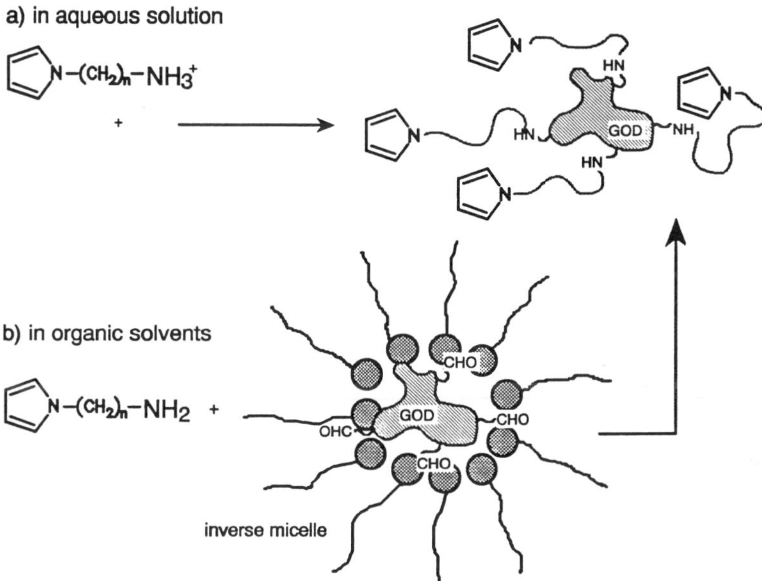

Figure 7. Synthesis of pyrrole-modified glucose oxidase. a) in aqueous solution b) in organic solvents.

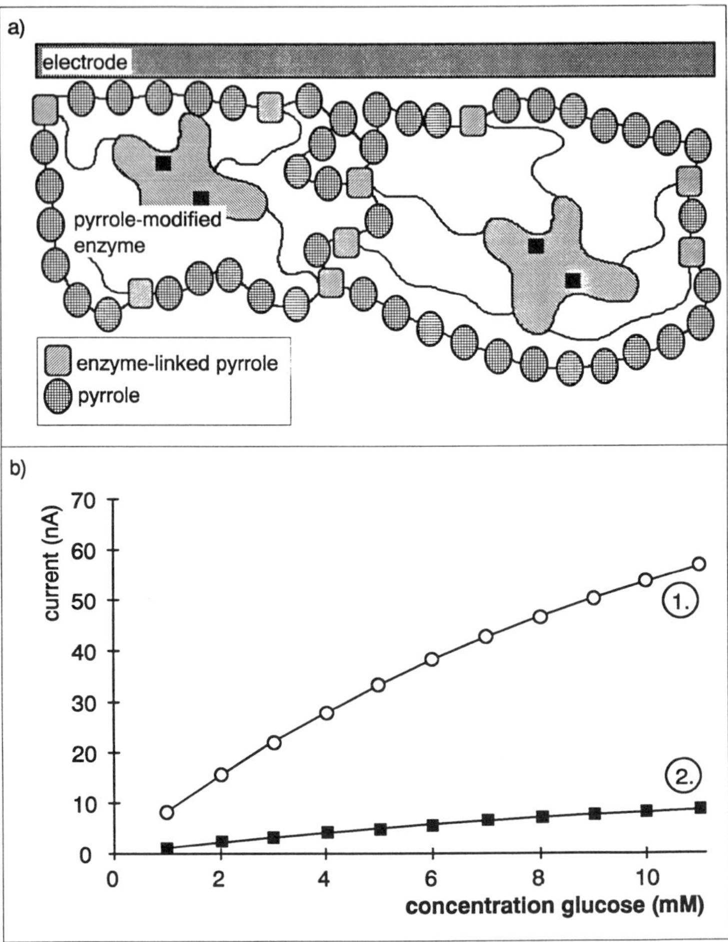

Figure 8. a) Schematical representation of an amperometric biosensor with a pyrrole-modified enzyme copolymerized with pyrrole. b) Comparison of calibration graphs for glucose obtained with electrodes based on 1) pyrrole-modified glucose oxidase copolymerized with pyrrole (deposition charge 184.2 μA*s) and 2) conventionally polypyrrole-entrapped native glucose oxidase (deposition charge 184.6 μA*s).

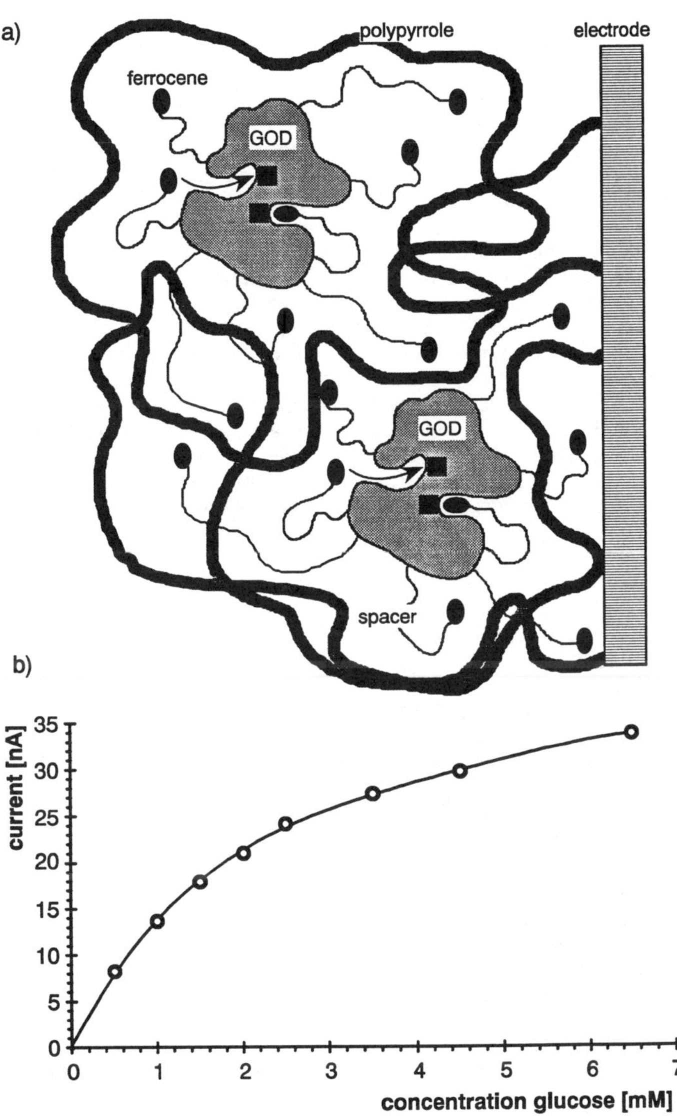

Figure 9. a) Entrapment of mediator-modified enzymes within a conducting-polymer film. b) Calibration graph for glucose obtained with a ferrocene-modified glucose oxidase entrapped within polypyrrole (350 mV; exclusion of O_2).

enzymes either covalently using functionalities on the polymer film or physically entrapped within the growing polymer film. A remarkable increase of the immobilized enzyme activity could be observed after copolymerization of pyrrole-modified glucose oxidase and pyrrole. As the conducting-polymer film itself does not participate in the electron transfer, mediator-modified enzymes entrapped within a polypyrrole layer has been used for the construction of a reagentless oxidase electrode.

Future work will be directed towards the entrapment of mediator-modified enzymes into mediator-modified polymer layers. As a final goal, reagentless amperometric enzyme electrodes with low working potential, decreased influence of the oxygen partial pressure and interfering compounds should be envisaged, simultaneously taking into account the need for mass production and miniaturization.

Acknowledgment. The author is greatful to Prof. Dr. H.-L. Schmidt for his support of the work. Without the excellent technical assistance of Johanna Huber and Heidi Wohlschläger most of the work would not have been done. Many discussions with Prof. Dr. A. Heller, Ling Ye, Dr. B. Gregg (University of Texas at Austin) have given me important insights into the problem of electron-transfer between enzymes and electrode surfaces. The author thanks Prof. Dr. J. Kulys (Lithunian Academy of Science, Vilnius) for his advice concerning the derivatization of enzymes using inverse micelles. The synthesis of the amino-substituted azulen by A. Mirlach and Prof. Dr. J. Daub (University of Regensburg, FRG) and of the functionalized di(thienyl)pyrroles by H. Röckel and Prof. Dr. R. Gleiter (University of Heidelberg, FRG) is gratefully acknowledged. The work was supported by the Bundesministerium für Forschung und Technologie, Projektträger BEO, FRG.

Literature Cited

1. Cass, A. G.; Davis, G.; Francis, G. D.; Hill, H. A. O.; Aston, W. J.; Higgins, I. J.; Plotkin, E. V.; Scott, L. D. L.; Turner, A. P. F. *Anal. Chem.* **1984,** *56* , 667-671.
2. Heller, A. *Acc. Chem. Res.* **1990,** *23* , 128-33.
3. Kulys, J. J.; Cenas, N. K. *Biochim. Biophys. Acta* **1983,** *744* , 57-63.
4. Schuhmann, W.; Wohlschläger, H.; Lammert, R.; Schmidt, H.-L.; Löffler, U.; Wiemhöfer, H.-D.; Göpel, W.; *Sens. Actuat.* **1990,** *B1,* 571-75.
5. Marcus, R. A.; Sutin, N. *Biochim. Biophys. Acta* **1985,** *811* , 265-322.
6. Degani, Y.; Heller, A. *J. Phys. Chem.* **1987,** *91* , 1285-1289.
7. Degani, Y.; Heller, A. *J. Am. Chem. Soc.* **1988,** *110* , 2615-2620.
8. Schuhmann, W.; Ohara, T. J.; Schmidt, H.-L.; Heller, A. *J. Am. Chem. Soc.* **1991,** *113,* 1394-97.
9. Gregg, B. A.; Heller, A. *Anal. Chem.* **1990,** *62* , 258-263.
10. Gregg, B. A.; Heller, A. *J. Phys. Chem.* **1991,** *95* , 5970-75.
11. Heller, A. *J. Phys. Chem.* **1992,** *96* , 3579-87.
12. Hale, P. D.; Inagaki, T.; Karan, H. I.; Okamoto, Y.; Skotheim, T. A. *J. Am. Chem. Soc.* **1989,** *111* , 3482-3484.
13. Hale, P. D.; Boguslavsky, L. I.; Skotheim, T. A.; Liu, L. F.; Lee, H. S.; Karan, H. I.; Lan, H. L.; Okamoto, Y. *ACS Symp. Ser.* **1992,** *487* , 111-24.
14. Imisides, M. D.; John, R.; Riley, P. J.; Wallace, G. G. *Electroanalysis* **1991,** *3* , 879-89.
15. Schuhmann, W.; *Sens. Actuat.* **1991,** *B4,* 41-49.
16. Wang, J.; Chen, S.-P.; Lin, M. S.; *J. Electroanal. Chem.* **1989,** *273,* 231-42.
17. Stefan, K.-P;. Schuhmann, W.; Parlar, H.; Korte, F.; *Chem. Ber.* **1989,** *122,* 169-74.
18. Schalkhammer, T.; Mann-Buxbaum, E.; Pittner, F.; Urban, G. *Sens. Actuat.* **1991,** *B4,* 273-81.

19. Daub, J.; Feuerer, M.; Mirlach, A.; Salbeck, J. *Synth. Met.* **1991**, *41-43*, 1551-55.
20. Röckel, H. Diplomarbeit Universität Heidelberg **1991**.
21. Schuhmann, W.; Lammert, R.; Uhe, B.; Schmidt, H.-L. *Sens. Actuat.* **1990**, *B1*, 537-41.
22. Schuhmann, W. *Synth. Met.* **1991**, *41*, 429-32
23. Hämmerle, M.; Schuhmann, W.; Schmidt, H.-L. In *GBF-Monographie*; Scheller, F.; Schmid, R.D.; eds.; VCH Weinheim, 1992, Vol. 17; pp. 111-114.
24. Cooper, J.; Hämmerle, M.; Schuhmann, W.; Schmidt, H.-L. *Biosens. Bioelectr.* **1993**, *8*, 65-74.
25. Schuhmann, W.; Huber, J.; Mirlach, A.; Daub, J. *Adv. Mater.* **1993,** *5*, 124-126.
26. Röckel, H.; Huber, J.; Gleiter, R.; Schuhmann, W., *Adv. Mater.*, submitted.
27. Bélanger, D.; Nadreau, J.; Fortier, G. *J. Electroanal. Chem.* **1989**, *274*, 143-55.
28. Foulds, N.C.; Lowe, C. R. *J. Chem. Soc. Faraday Trans.* **1986**, *82*, 1259-64.
29. Umana, M.; Waller, J. *Anal. Chem.* **1986**, *58* , 2979-2983.
30. Hämmerle, M.; Schuhmann, W.; Schmidt, H.-L. *Sens. Actuat* **1992**, *B6*, 106-112.
31. Foulds, N. C.; Lowe, C. R. *Anal. Chem.* **1988**, *60*, 2473-78.
32. Schuhmann, W. in "*Proceeedings of BIOELECTROANALYSIS,2*", 11.10.-15.10.1992, Matrafüred, Hungary, Akadémiai Kiadó, Budapest, **1993**, 113-138.
33. Schuhmann, W.; Kittsteiner-Eberle, R. *Biosen. Bioelectron.* **1991**, *6*, 263-273.
34. Kabanov, A. V.; Levashov, A. V.; Khrutskaya, M. M.; Kabanov, V. A. *Makromol. Chem.* **1990**, *191*, 2801-2814.
35. Wolowacz, S. E.; Yon Hin, B. F. Y.; Lowe, C. R. *Anal. Chem.* **1992**, *64*, 1541-45

RECEIVED January 24, 1994

Chapter 10

Poly(ether amine quinone)s as Electron-Transfer Relay Systems in Amperometric Glucose Sensors

Hsing Lin Lan[1,3], T. Kaku[1], Hiroko I. Karan[2], and Yoshiyuki Okamoto[1,4]

[1]Department of Chemistry and Polymer Research Institute, Polytechnic University, 6 Metrotech Center, Brooklyn, NY 11201
[2]Department of Physical Sciences and Computer Science, Medgar Evers College, City University of New York, Brooklyn, NY 11225

Various poly(etheramine quinone)s were readity prepared and tested for their efficiency as electron transfer relay systems in amperometric glucose biosensors. Cyclic voltammometry and constant applied potential measurements showed that poly(etheramine quinone) relay systems efficiently mediated electron transfer from reduced glucose oxidase to a conventional carbon paste electrode. Sensors containing these relay systems and glucose oxidase respond rapidly to low (<0.1 mM) glucose concentrations and reach steady state current responses in less than 1 min. Electron mediation efficiency depends on the number of methyl substituents on quinone. The degree of crosslinking in poly(etheramine quinone), and these systems are compared to monomeric quinones and quinone-modified polymers previously studied in our laboratory.

Amperometric glucose electrodes based on glucose oxidase undergo several chemical or electrochemical steps which produce a measurable current that is related to the glucose concentration. In the initial steps, glucose converts the oxidized flavin adenine dinucleotide (FAD) center of the enzyme into its reduced form ($FADH_2$). Because these redox centers are essentially electrically insulated within the enzyme molecule, direct electron transfer to the surface of a conventional electrode does not occur to any measurable degree. The most common method[1-4] of indirectly measuring the amount of glucose present relies on the natural enzymatic reaction:

[3]Current address: Institute of Biomedical Engineering, National Yang Ming Medical College, Taipei 11221, Taiwan
[4]Corresponding author

0097–6156/94/0556–0124$08.00/0

$$\text{glucose} + O_2 \xrightarrow{\text{glucose oxidase}} \delta\text{-gluconolactone} + H_2O_2$$

where oxygen is the electron acceptor for glucose oxidase. The oxygen is reduced by the $FADH_2$ to hydrogen peroxide, which may then diffuse out of the enzyme and be detected electrochemically. The working potential of such a device is quite high (H_2O_2 is oxidized at approximately +0.7V vs. the saturated calomel electrode, SCE), however, and the sensor is therefore highly sensitive to many common interfering electroactive species, such as uric acid and ascorbic acid; H_2O_2 is also known to have a detrimental effect on glucose oxidase activity. Alternatively, one could use the electrode to measure the change in oxygen concentration that occurs during the above reaction. In both of these measuring schemes, this type of sensor has the considerable disadvantage of being extremely sensitive to the ambient concentration of O_2.

In recent years, systems have been developed which use a non-physiological redox couple to shuttle electrons between the $FADH_2$ and the electrode by the following mechanism;

$$\text{glucose} + \text{GO(FAD)} \rightarrow \text{gluconolactone} + \text{GO(FADH}_2\text{)}$$

$$\text{GO(FADH}_2\text{)} + 2M_{ox} \rightarrow \text{GO(FAD)} + 2M_{red} + 2H^+$$

$$2M_{red} \rightarrow 2M_{ox} + 2e^-$$

In this scheme, GO(FAD) represents the oxidized form of glucose oxidase and GO($FADH_2$) refers to the reduced form. The mediating species M_{ox} /M_{red} is assume to be a one-electron couple. Sensors based on derivatives of the ferrocene/ferricinium redox couple[5-8] and on quinone derivatives[9-12], have recently been reported. In potential clinical applications, however, sensors based on electron-shuttling redox couples suffer from an inherent drawback: the soluble or partially soluble, mediating species can diffuse away from the electrode surface into the bulk solution, which would preclude the use of these devices as implantable probes in clinical applications, and restrict their use in long-term *in situ* measurements (e.g. fermentation monitoring). Most recently, these studies have been extended to include systems in which the mediating redox moieties are covalently attached to polymers such as polypyrrole[13-16], poly(vinylpyridine)[17-19], and polysiloxane[20-24]. Polymers containing *p*- and *o*-quinone group have been investigated by Kulys and his coworkers as electron transfer relay systems for redox enzymes and these systems were shown to effectively catalyzing the electroxidation of glucose[25,26].

Polysiloxanes containing quinone groups (Figure 1-A) were found to be efficient mediators of electron transfer from reduced glucose oxidase to a conventional carbon paste electrode[11,12]. Heller recently postulated that because these flexible polysiloxanes are somewhat low molecular weight polymers, they may act as diffusional mediators instead of polymeric electron transfer relay systems[27]. We have also synthesized an acrylonitrile-ethylene copolymers with corvalently attached *p*-hydroquinone (Figure 1-B). The molecular weight of the polymer was estimated as Mw = 15,000: Mn = 9,000[12]. The polymeric system was found to be effective electron transfer relay system in an amperometric glucose biosensor. They were , however, less efficient than those of polysiloxane systems. The difference may due to the flexibility polymeric backbone and the ease of electron transfer between quinone-hydroquinone moieties. The acrylonitrile-ethylene copolymer investigated was a large molecular system and cannot act as a diffusional mediator. Thus in the words of Heller, these systems serve to "electrical wire", facilitating a flow of electrons from the enzyme to the electrode.

Erhan and his coworkers, recently reported that polyetheramine-benzoquinone polymers can be readily prepared from hydroquinone and diamine. The polymer adheres to metal surface and resists most organic solvents after being cured chemically or thermally[28]. In the present paper, the synthesis of various polyetheramine-benzoquinone polymers and crosslinked polymeric systems are reported together with use of these polymers in amperometric glucose sensors. For comparison, monomeric quinones and previously synthesized quinone modified siloxane and other polymers[11-12] are also discussed as an electron relay system in these glucose sensors.

Experimental

Chemicals

Glucose oxidase (E.C.1.1.3.4, Type VII , from Aspergillus niger) was obtained from Sigma(St. Louis, Missouri). Graphite powder and paraffin oil were obtained from Fluka (Ronkonkoma, New York). Hydroquinone and methylhydroquinone were obtained from Aldrich (Milwaukee, Wisconsin). 2,5-Dimethylbenzoquinone was obtained from Tokyo Kasei Kogyo (Tokyo, Japan), and Jeffamine poly(oxypropylene)diamines were obtained from Texaco. Glucose (Sigma) solutions were prepared by dissolving appropriate amounts in 0.1M phosphate/0.1M KCl buffer (pH 7.0); the glucose was allowed to reach mutarotational equilibrium before use (ca. 24hr). All other chemicals were reagent grade and were used as received.

Polymer synthesis

The poly(etheramine quinone) polymers were prepared according to Scheme 1[28]. Hydroquinone [4.40g(0.04 mole)] or methylhydroquinone [4.96g(0.04 mole)] was suspended in 50mL methylene chloride in a round bottom flask equipped with a reflux condenser and a mechanical stirrer, and 11.5g calcium hypochlorite (65-70% of available chlorine) was added to the suspension with vigorous stirring, followed by 0.18g (2.5 mole% of hydroquinone) phase transfer catalyst, benzyltrimethylammonium chloride, with stirring continued under reflux. After about one hour all hydroquinone was oxidized to quinone. When all hydroquinone was oxidized, the reaction mixture was cooled to 20-30°C and 16g

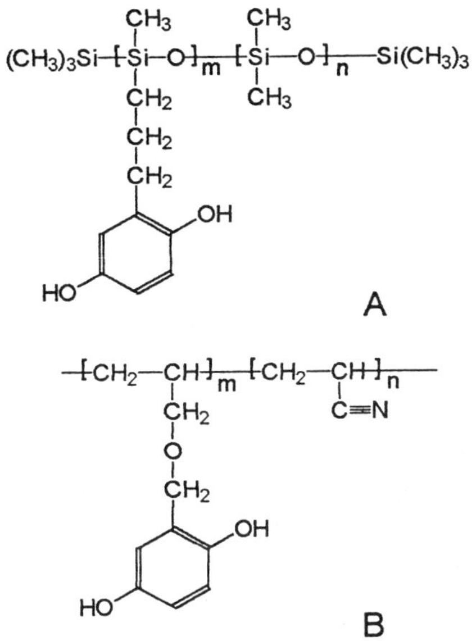

Figure 1 Schematic diagram of the polymers used in the quinone polymer/ glucose oxidase/ carbon paste electrodes: A. Hydroquinone modified polysiloxane; B. Hydroquinone modified poly(acrylonitrile/ethylene). The m:n ratio is approximately 1:2 for both systems.

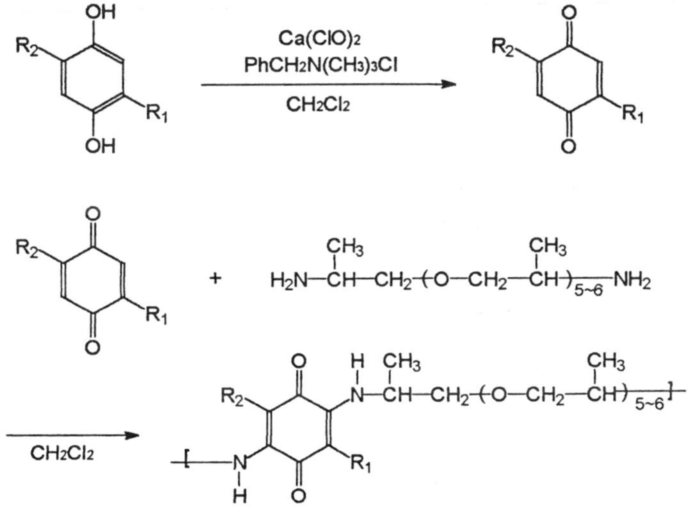

(a) $R_1 = R_2 = H$
(b) $R_1 = H$, $R_2 = CH_3$
(c) $R_1 = R_2 = CH_3$

Scheme 1.

Jeffamine D-400 (diamine with 5 to 6 units of propylene oxide) in 50 mL of methylene chloride was added slowly to prevent an increase of the reaction temperature. The reaction mixture was stirred vigorously for 1.5hr at room temperature. The reaction mixture was filtered and the filtrate of product was evaporated under vacuum to remove methylene chloride. The polymer obtained was dissolved in ethanol and precipitated into water. This step was repeated a couple of times and the polymer was dried in a vacuum at 35-40°C. For 2,5-dimethylbenzoquinone, which was dissolved in methylene chloride, the reaction was continued for 48hr at 35°C, then the product was purified by the same procedure as described in above. The polymers obtained were also crosslinked by heating in air. The polymers synthesized were characterized by measuring IR and NMR spectra.

Electrode Construction

The modified carbon paste for sensors was prepared by throughly mixing 100mg of graphite powder with a measured amount of either quinone or poly(etheramine quinone)s that were dissolved in THF or methylene chloride; the molar amount of the quinone moiety was the same for all electrodes (36μmol of quinone per gram of graphite powder). After evaporation of solvent, 10mg of glucose oxidase and 20μL of paraffin oil were added, and the resulting mixture was blended into a paste. The paste was packed into the cavity of a commercial plastic electrode (BAS, Lafayette, Indiana) which had previously been partially filled with unmodified carbon paste, leaving an approximately 2mm deep well at the base of the electrode(3.0mm o.d., 1.6mm i.d.,). The resulting surface area of the electrode was $0.02cm^2$. The electrodes were generally stored under dry conditions at 5°C. The electrode with crosslinked polymer was prepared as follows; the aqueous solution of non-crosslinked fresh quinone polymer was mixed with an aqueous solution of glucose oxidase and water was removed from the mixture under reduced pressure. The mixture was then heated at 40-45°C for 10-12 hr in air to form the crosslinked system. The resulting polymer was found to be in soluble in water and common organic solvents. The electrode was prepared by thoroughly mixing the crosslinked polymer-enzyme with graphite powder and a small amount (20μL) of paraffin oil.

Electrochemical Measurements

Cyclic voltammograms were performed using a Voltammetric Analyzer and the stationary potential measurements were performed using a Princeton Applied Research Polarographic Analyzer (Model 174A) and a Princeton Applied Research Potentiostat/Galvanostat (Model 363). All experiments were carried out in a conventional electrochemical cell containing 0.1M phosphate buffer with 0.1M KCl (pH 7.0) at 23 ±2°C. All experimental solutions were thoroughly deoxygenated by bubbling nitrogen through the solution for at least 10 min ; in the stationary potential experiments, a gentle flow of nitrogen was also used to facilitate stirring. In addition to the modified carbon paste working electrode , a Ag/AgCl reference electrode and a platinum wire auxiliary electrode were employed. In the constant potential experiments, the background current was allowed to decay a constant value before sample of stock glucose solutions were added to the buffer solution.

Results and Discussion

Cyclic voltammetry measurement

Figure 2 shows typical voltammetric results for carbon paste electrodes containing glucose oxidase and the poly(etheramine quinone)s as the electron relay system. With no glucose present, the voltammogram displays the typical anodic peaks corresponding to the oxidation of the polymer-bound quinone moieties in addition to the cathodic peaks corresponding to the reduction of benzoquinone. The sensors display an increase in the oxidation current upon addition of glucose that is compatible to that observed in sensors containing quinone-modified polysiloxane and quinone-modified acrylonitrile-ethylene copolymer[12]. The fact that the reduction current does not increase along with the oxidation current is indicative of the enzyme-dependent catalytic reduction of the benzoquinone moiety produced at oxidizing potential values. Comparison of the voltammograms with and without glucose indicates that poly(etheramine quinone)s can act as electron transfer relay systems between the FAD/$FADH_2$ centers of glucose oxidase and a carbon paste electrode. Voltammograms of carbon paste electrodes containing only glucose oxidase and no polymeric relay system do not display this catalytic behaviour.

The voltammetry in Figure 2 displays shifts in the quinone oxidation potentials upon addition of glucose, as do the large peak separations, Ep (for (a) $R_1=R_2=H$, 210mV, for (b) $R_1=H$, $R_2=CH_3$, 200mV and for (c) $R_1=R_2=CH_3$, 310mV)(Table 1), as observed in the case of quinone-modified polysiloxane and quinone-contained acrylonitrile-ethylene copolymer. As has been discussed by Miller and coworkers[29], who studied electrodes modified with quinoidal polymer films, the change in oxidation potential may be attributed to heterogeneous redox processes within the polymer, and to the possible formation of charge transfer complexes, such as quinhydrones, which could slow the propagation of the oxidation from the innermost sublayer to other polymer inner sublayer. The shifts of the oxidation and reduction potentials (E_{pa}, E_{pc}) toward the negative direction in both mononeric quinones and poly(etheramine quinone)s by introducing methyl group(s) on a quinone moiety are also observed (Table 1). The case of higher E_{pa} in (c) than in (b) may be explained by slower electron transfer through the polymer (c) than (b).

Constant potential measurement

The steady state response to glucose was measured at several applied potential values. The response to 21.3mM glucose for electrodes containing polymers (a), (b) and (c) as electron transfer mediators is shown in Figure 3. The anodic current increased with increasing positive applied potential and reached a limiting value at

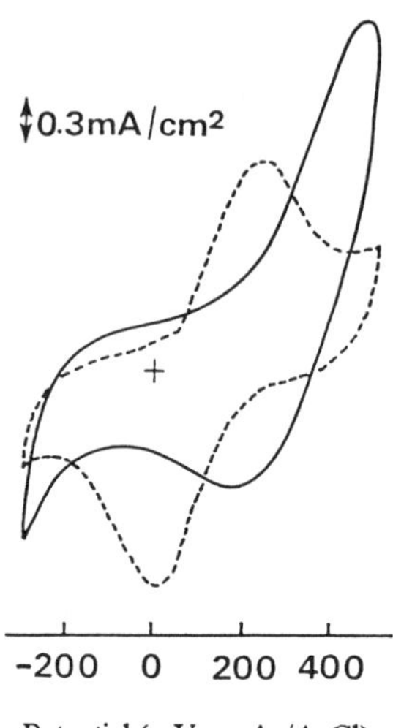

Figure 2(a) Cyclic voltammogram for the poly(etheramine quinone) polymer (a) / glucose oxidase / carbon paste electrode (scan rate: 5mV/s) in pH 7.0 phosphate buffer (with 0.1M KCl) solution with no glucose present (dashed line) and in the presense of 0.1M glucose (solid line).

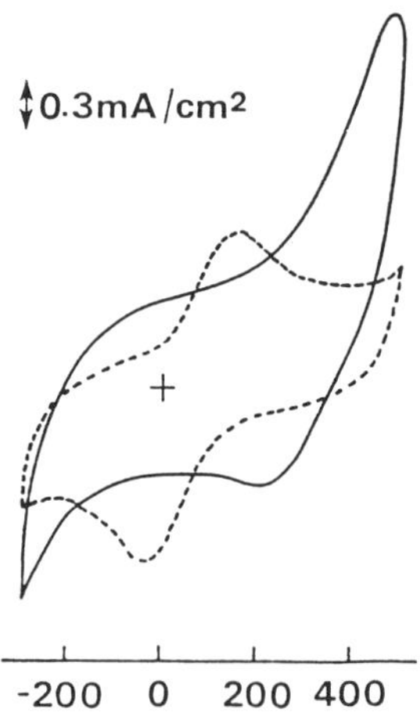

Figure 2(b) Cyclic voltammogram for the poly(etheramine quinone) polymer (b) / glucose oxidase / carbon paste electrode (scan rate: 5mV/s) in pH 7.0 phosphate buffer (with 0.1M KCl) solution with no glucose present (dashed line) and in the presense of 0.1M glucose (solid line).

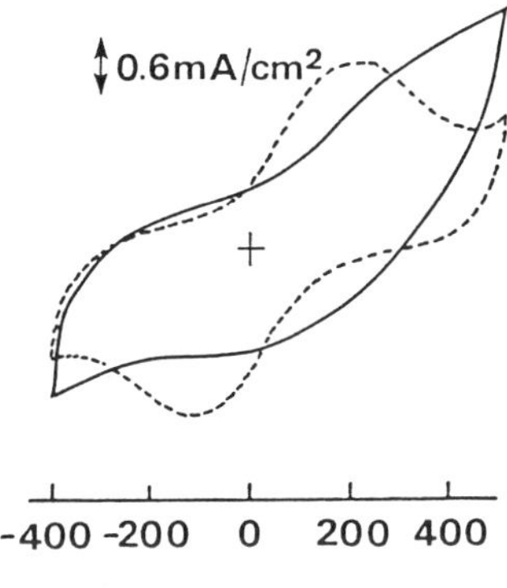

Figure 2(c) Cyclic voltammogram for the poly(etheramine quinone) polymer (c) / glucose oxidase / carbon paste electrode (scan rate: 5mV/s) in pH 7.0 phosphate buffer (with 0.1M KCl) solution with no glucose present (dashed line) and in the presense of 0.1M glucose (solid line)

Table 1 Summary of Cyclic Voltammograms

	Epa (mV)	Epc (mV)	Epa-Epc (mV)
Poly(etheramine quinone)	+220	+10	210
Poly(etheramine methylquinone)	+155	-45	200
Poly(etheramine dimethylquione)	+200	-110	310
1,4-Benzoquinone	+110	-30	140
Methylhydroquinone	+20	-130	150
2,5-Dimethylbenzoquinone	-20	-190	170

approximately +400mV vs. the Ag/AgCl electrode. These responses were very similar to those of hydroquinone-modified polysiloxane, except at more positive potentials(+300mV~ +400mV), poly(etheramine quinone)s (a) and (b) responded more efficiently to glucose than did the hydroquinone-modified polysiloxane[12]. The lower responses of sensors containing (c) could be due to the increasing hydrophobicity of the hydroquinone moieties because of the two methyl substituents, and may even due partly to the steric hindrance exhibited by those two methyl groups which prevent efficient interaction of polymeric mediators and glucose oxidase. Figure 4 shows a typical glucose calibration curves for sensors containing (a) and (b) indicating efficient electron mediation at an applied potential of +200mV vs. the Ag/AgCl electrode. Sensors containing these relay systems and glucose oxidase responded rapidly to low (< 0.1mM) glucose concentrations and reached steady state current responses in less than 1 min. (Figure 5). It should be noted that sensors based on the poly(etheramine quinone) relay system retained approximately 70% of the initial response after a month and then the value remained constant for another few months. One of the reasons for this initial instability of these sensors maybe doe to the gradual croslinking of poly(etheramine quinone) itself.

Hence, we prepared crosslinked poly(etheramine quinone) and costructed a glucose sensor. The electrode with the crosslinked polymer showed approximately 60% of the activity those the non-crosslinked polymer electrode. Although the efficiency of the sensors with the croslinked polymer was decreased, the activity of the sensor remained constant for over few months. This result supports that this crosslinked redox polymer system acts as an electron transfer relay system and not as a diffusioned mediator. Further investigation of the electron transfer properties of the crosslinked poly(etheramine quinone) are being carried out.

Apparent Michaelis-Menten constants

The linear response range of sensors was estimated from a Michaelis-Menten analysis of the glucose calibration curves in Figure 4. The apparent Michaelis-Menten constant K_M^{app} can be determined from the electrochemical Eadie-Hofstee form of the Michaelis-Menten equation[18].

$$j = j_{max} - K_M^{app} (j / C)$$

where j is the steady state current density, j_{max} is the maximum current density measured under conditions of enzyme saturation, and C is the glucose concentration. As shown in Figure 6 a plot of j vs. j / C produces a straight line, and K_M^{app} can be obtained from the slope and j_{max} from the y-intercept. The apparent Michaelis-Menten constant characterizes the enzyme electrode, not the enzyme itself, and provides a measure of the substrate concentration range over which the electrode response is approximately linear. For sensors containing (a) and (b), the K_M^{app} and j_{max} values from this analysis are 16.6mM, 106.7μA/cm^2, and 21 mM, 130μA/cm^2, respectively. These values correspond to mean values for three electodes. It is clear that the response of the sensors begins to deviate

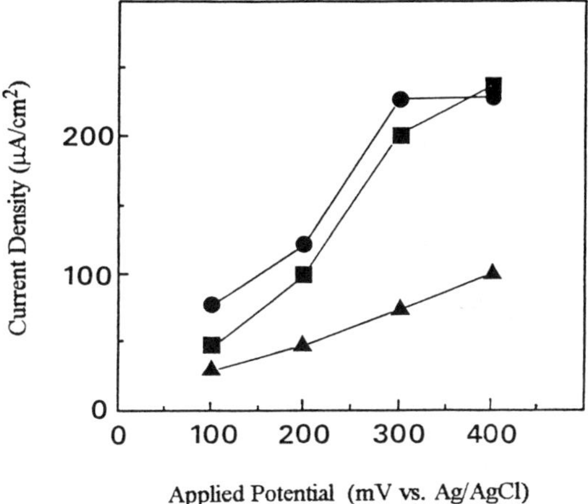

Figure 3 Steady state current responce to 21.3M glucose of the polyetheramine quinone polymers / glucose oxidase / carbon paste electrodes for several applied potential values. (■): polymer (a) ; (●): polymer (b) ; (▲): polymer (c).

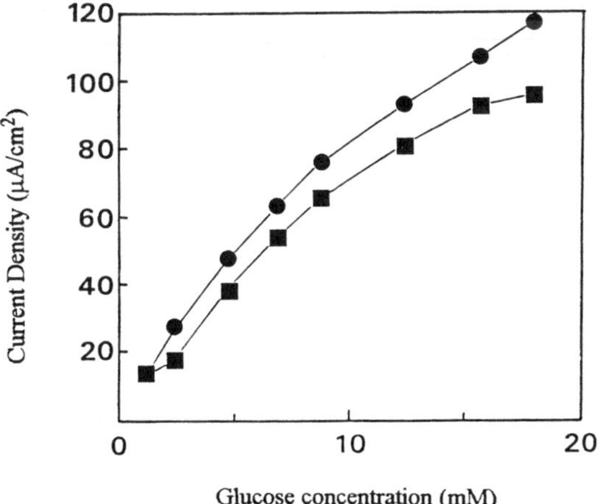

Figure 4 Glucose calibration curves for the polyetheramine quinone polymers / glucose oxidase / carbon paste electrodes at an applied potential of +200mV (vs. Ag/AgCl) ; (■) : polymer (a) ; (●) : polymer (b).

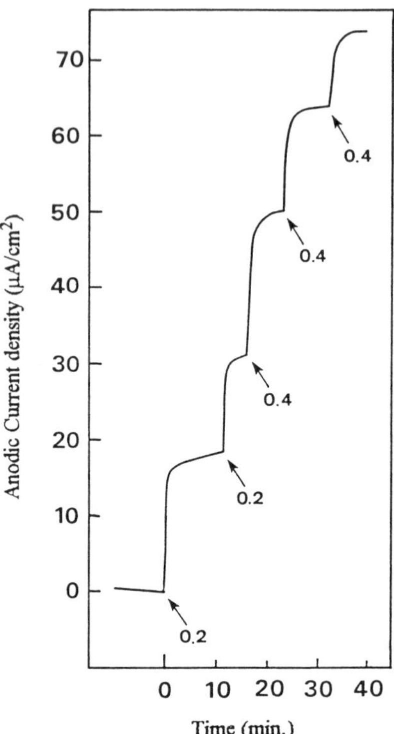

Figure 5 Typical response of the polyetheramine quinone polymers / glucose oxidase / carbon paste electrodes to addition of glucose at E = +200mV vs. SCE. Indicated under the response trace are the amounts (in ml) of 0.1M glucose injected into the test solution (initial volume : 10 ml).

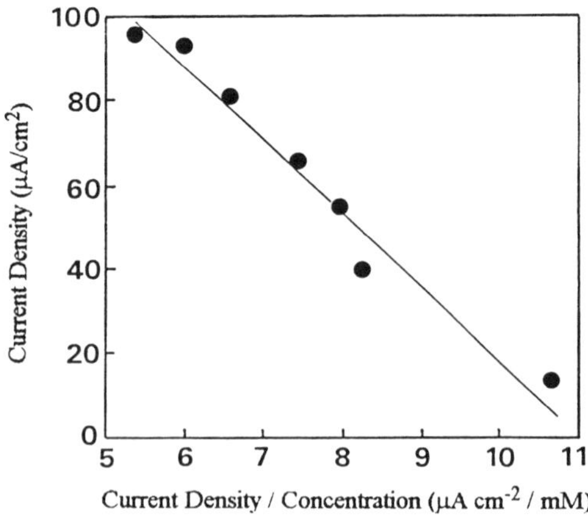

Figure 6(a) Typical electrochemical Eadie-Hofstee plot for the polyetheramine quinone polymer (a) / glucose oxidase / carbon paste electrode at E = +200mV (vs. Ag/AgCl).

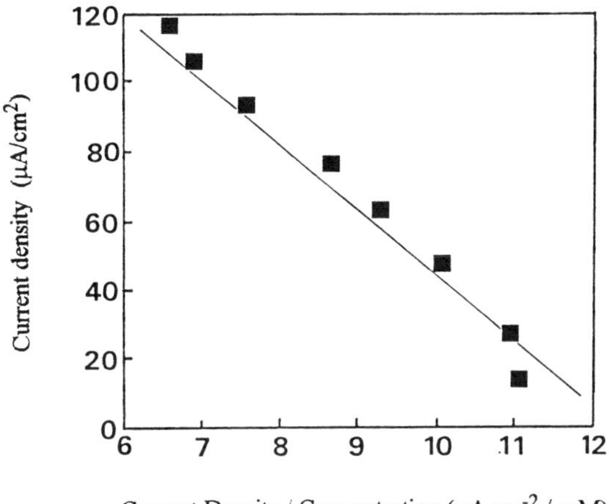

Figure 6(b) Typical electrochemical Eadie-Hofstee plot for the polyetheramine quinone polymer (b) / glucose oxidase / carbon paste electrode at E = +200mV (vs. Ag/AgCl).

from linearity even at glucose concentrations below 10mM (the response to glucose is expected to be strictly linear for concentrations less than or equal to ca. $0.1K_M^{app}$[23]. The linear range of these sensors can be increased substantially ($K_M^{app} > 200mM$) by using an additional polymer coating, such as a poly(ester sulfonic acid) cation exchanger, on the surface of sensor as was described previously[23]. As would be expected, such a polymer coating provides an additional resistance for diffusion of glucose to the active enzyme layer, and the overall kinetics would be controlled by this diffusion process. These coatings can also exclude anionic interferents and prevent electrode fouling by biological agents. In addition, sensor lifetime can be increased as glucose oxidase would not freely diffuse away from the electrode[23].

References

1. Clark, L.C. *Biosensors: Fundamentals and Applications,* Turner, A.P.F., Karube, I., and Wilson, C.S.,eds, Oxford University Press, New York (1987); Chapter 1.
2. Clark, L.C. and Lyons, C., *Ann. N.Y. Acad. Sci.*, **1962**, *102*, 29
3. Joensson, G. and Gorton, L., *Analyt. Letts*., **1987**, *20*, 839
4. Heider, G.H., Sass, S.V., Huang, K.-M., Yacynych, A.M., Wieck, H.J., *Analyt. Chem.*, **1990**, *62*, 1106
5. Cass, A.E.G., Davis, G., Francis, G.D., Hill, H.A.O., Aston, W.J., Higgins, I.J., Plotkin, E.V., Scott, L.D.L. and Turner, A.P.F., *Analyt. Chem.*, **1984**, *56*, 667
6. Lange, M.A.and Chambers, J.Q., *Analyt.Chem.Acta.*, **1985**, *175*, 89

7. Iwakura, C., Kajiya, Y. and Yoneyama, H., *J.Chem. Soc., Chem. Comm.*, **1988**, 1019
8. Joensson, G., Gorton, L.and Petterson, L., *Electroanalysis,* **1989**, *1,* 49
9. Ikeda, T., Shibata, T and Senda, S., *J. Electroanalyt. Chem.*, **1989**, *261,* 351
10. Kulys, J.J and Cenas, N.K., *Biochim. Biophys. Acta,* **1983**, *744,* 57
11. Inagaki. T., Lee, H. S., Hale, P. D., Skotheim, T. A., and Okamoto, Y., *Macromolecules*, **1989**, *22,* 4641
12. Karan, H.I., Hale, P.D., Lan, H.L., Lee, H.S., Liu, L.F., Skotheim, T.A., and Okamoto, Y., *Poly. Ad. Tech.*, **1991**, *2,* 229
13. Belanger, D., Nadrean, D. and Fortier, J., *J. Electroanal. Chem.*, **1989**, *274,* 143
14. Tamiya, G.and Karube, I., *Sens. Actuators,* **1989**, *18,* 297
15. Yabuki, S., Shinohara, H.,and Aizawa, M., *J. Chem. Soc., Chem. Comm.*, **1989**, 945
16. Trojanavicz, M., Matuszewski, W, and Podsiadla, *Biosensors Bioelectron.*, **1990**, 5, 149
17. Degani, Y. and Heller, A., *J. Am. Chem. Soc.,* **1989**, *111*, 2357
18. Gregg, B.A. and Heller, A., *Analyt.Chem.,* **1990**, *62,* 258
19. Pishko, M.V., Katakis, I., Lindquist, S.-E., Ye, L., Gregg, B.A., and Heller. A., *Angew.Chem.,Ed. Engl.,* **1990**, *29,* 82
20. Hale, P.D., Inagaki, T., Karan, H.I., Okamoto, Y. and Skotheim, T.A.. *J. Am. Chem. Soc.* **1989**, *111,* 3482
21. Inagaki, T., Lee, H.S., Skotheim, T.A. and Okamoto, Y., *J. Chem. Soc., Chem. Comm.,* **1989**, 1181
22. Hale, P.D., Inagaki, T., Lee, H.S., Karan, H.I., Okamoto, Y. and Skotheim, T.A., *Analyt. Chim. Acta,* **1990**, *228,* 31
23. Gorton, L., Karan, H.I., Hale, P.D., Inagaki, T., Okamoto, Y. and Skotheim, T.A., *Analyt. Chim. Acta,* **1990**, *228,* 23
24. Hale, P.D., Inagaki, T., Lee, H.S., Skotheim, T.A., Karan, H.I., and Okamoto, Y., *Biosensor Technology: Fundamentals and Applications,* Buck, R.P., Hatfield, W.E., Umana, M. and Bowden, E.F., eds., Marcel Dekker, New York , 1990 ; Chapter 14.
25. Cenas, N.K., Poius, A.K., and Kulys, J.J., *Bioelectrochem. and Bioenergetics,* **1984**, *12,* 583
26 Kulys, J.J., Biosensors 2,3, 1986
27. Heller, A., *J. Phys. Chem.,* **1992**, *96,* 3579
28. Nithianandam, V.S. and Erhan, S., *Polymer,* **1991**, *32(6),* 1146
29. Fukui, M., Kitani, A., Degrand, C., and Miller, L. L., *J. Am. Chem. Soc.,* **1982**, *28,* 104

RECEIVED February 15, 1994

Chapter 11

Flow-Injection Analysis of Some Anionic Species

Automated, Miniaturized, Dual Working Electrode Potentiostat Using a Poly(3-methylthiophene)-Modified Electrode

Elmo A. Blubaugh, George Russell[1], Merril Racke[2], Dwight Blubaugh, T. H. Ridgway, and H. B. Mark Jr.

Department of Chemistry, University of Cincinnati, Cincinnati, OH 45221–0172

The work presented in this manuscript is composed of two parts. The first part is concerned with the design and fabrication of a computer controlled dual working electrode potentiostat (DWEP). This automated andminiaturized DWEP was designed and built with utilization of a new technology in the field of electronics, specifically, surface mount electronic components (passive and active). This potentiostat design, also allows for the automated zeroing of current at each working electrodes, with the same nanoampere to microampere range, and a range of sensitivities from nanoampere to microampere.

The second part describes the application of this automated potentiostat as an electrochemical detector for various electro-inactive anionic species; nitrate (NO_3^-), nitrite (NO_2^-), sulfate (SO_4^{2-}), chloride (Cl^-), etc. (*1, 2*). The electrochemical detection involves the use of a conducting poly(3-methylthiophene) (P3MT) polymer modified platinum electrode in a flow injection analysis (FIA) mode.

The analytically important criterion of detection limits were determined to be 10^{-6} to 10^{-7} molar in anion concentration. The dynamic range was five to six orders in magnitude.

A new area within the field of organic electrochemistry was initiated by the first successful electropolymerization by Kern et.al. (*3*) in the 1950's. However, it was not until 1973 with the discovery of the metallic properties of $(SN)_x$ by Green et.al., that the area of conducting polymers started to evolve. Interest in the area of conducting polymers dramatically increased in the late seventies with publications by MacDiarmid (*4*) and Diaz (*5*). These publications centered on the ability of electrochemically synthesized polymers to exhibit relatively high electronic conductivity and undergo electrochemical redox transitions. Since the seventies, the publications using conducting polymers as electrodes has flourished. Also, conducting polymers is one of the more diverse and interdisciplinary fields found within science at the present time. Some of the possible applications for conducting polymers include electrocatalysis (*6*), electrochemical and electrochromic displays (*7*), thin layers for solar cells (*8*), rechargeable battery electrodes (*9*), corrosion protection and the extraction of

[1]Current address: Sabine River Laboratory, DuPont Chemical Company, P.O. Box 1089 (FM–1006), Orange, TX 77631–1089

[2]Current address: National Forensic Chemistry Center, U.S. Food and Drug Administration, 1141 Central Parkway, Cincinnati, OH 45202

0097–6156/94/0556–0137$08.00/0

ionic species (*10*). The ability of these thin polymeric layers to modify the behavior and response of the electrode from that behavior associated with the substrate electrode to the chemical and physical behavior associated with the modifying polymer layer is the focus of intense study. In this particular study the ability of conducting polymers to extract or become doped by ionic materials is the basis for this study.

Polythiophene and its derivatives can be polymerized by chemical or electrochemical techniques. In this study, the electrochemical method was utilized.The mechanism is a cationic radical polymerization (*11*). The polymerization pathway can be summarized in the following steps: 1) oxidation of the monomer to form a radical cation, 2) dimerization of the radical cations, 3) loss of proton to yield a neutral dimer, 4) oxidation of dimer to form a radical cation, 5) reaction of dimer radical cation with another radical cation, 6) repeat of the this study, are 3-methylthiophene, tetrabutylammonium tetrafluoroborate (TBATFB), as the supporting electrolyte. The organic solvent was acetonitrile. The resulting polymer was the first conducting polymer family found to be stable in air and water in both their doped or undoped state.

The doping/undoping mechanism for poly(3-methylthiophene) is shown in Figure 1. The oxidation or doping of the P3MT involves the transfer of an electron with the gain of an anion to form the charge compensated, or doped P3MT. Since Q_{OX} is approximately equal to Q_{RED}, this doping/undoping process is reversible (*12*). The changes from doped to undoped polymer are accompanied by color changes from dark blue to red.There is also a difference in the conductivity for these films between the doped or undoped state. The conductivity of a doped film approaches that of the metallic state, while the conductivity found for the undoped polymer approaches an insulator.

One of the first reported uses of a conducting polymer modified electrode for the detection of anions by flow injection analysis (FIA) utilized poly(pyrrole) (*13*). Low detection limits were achieved. However, poly(pyrrole) is irreversibly oxidized in the presence of oxygen at potentials of around +1.0 volts. Poly(3-methylthiophene) does not exhibit this behavior. Therefore, poly(3-methylthiophene appeared to be a good candidate as an electrochemical detector for anionic species by FIA with electrochemical detection at an poly(3-methylthiophene) modified electrode. Also P3MT does not need to be undoped (cycled between the negative and positive detection potentials) between successive injections as many other polymer systems must.

One of the objectives for this study was the design and fabrication of a physically small and highly automated DWEP. The eventual role that this potentiostat will play involves the remote sampling and monitoring of hazardous waste streams. Both the potentiostat hardware and the FIA sampling hardware lend themselves readily to computer control and multiple sample collection and evaluation, (*14,15*).

The approach taken in the design and manufacture for the potentiostat, which must have a small physical size, involved the utilization of a relatively new technology in the field of electronics, namely Surface Mount Technology (SMT). Not only does this technology allow for a smaller physical size, but also greater reliability. Both of these advantages are of a prime importance for a remote monitoring situation. Surface mounted electronic components do not have leads for connection to the printed circuit board (and hence no need for plated-thru holes), but rather are attached to the printed circuit board by soldering pads on only one side of the printed circuit board. The smaller physical size, which results from utilizing surface mounted components is the result of two factors: 1) surface mounted components are approximately one-third to one-fourth the size of analogous conventional thru-board components, 2) fewer plated thru-board holes, hence a smaller required space for placement of the component. There is a possible disadvantage associated with using SMT components. This disadvantage involves the very small size for the SMT components and the resulting difficulty in; 1) the assembly of these printed circuit boards and 2) the difficulty in the reworking or repair of these printed circuit boards without the aid of specialized and expensive equipment.

Experimental

Chemicals and Solvents. HPLC grade acetonitrile (Fisher Scientific Co.) was the solvent used for the electropolymerization. The monomer, 3-methylthiophene (3MT) (Aldrich Chemical Co., Milwaukee, WI), was distilled prior to use. The sodium salts of fluoride, chloride, bromide, iodide, nitrate and nitrite were purchased from Aldrich Chemical Company. Tetrabutyl ammonium tetrafluoroborate (TBATFB) was also purchased from Aldrich Chemical Company and used as received.

Aqueous solutions were prepared by dilution from stock solution or by dissolution of a weighed amount into distilled water. Distilled water was obtained from a NanoPure II system purchased from Barnstead (Newton, PA). The specific conductivity of the distilled water was approximately 10^{-18} S/cm.

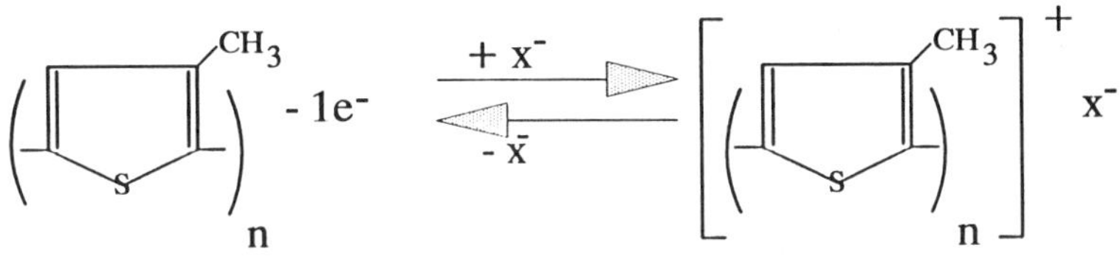

CH_3 - 1e$^-$ + x$^-$ - x$^-$ CH_3 + x$^-$ S n S n

Figure 1: Doping/Undoping Mechanism for Poly(3-methylthiophene)

Electrochemical Techniques. Cyclic Voltammetry (CV) was performed on most of the poly(3-methylthiophene) anion sensor electrodes. Cyclic voltammetry provided the electrochemical characterization and evaluation of the post treatment of the modified electrodes. The potential of the platinum electrodes (Model MF-1012, BAS Inc.), with or without modifying film of poly(3-methylthiophene) was controlled relative to an Ag/AgCl reference electrode (Bioanalytical Systems Inc., West Lafayette, IN). The auxiliary electrode was a platinum flag electrode or in the case of the FIA experiments a stainless steel block electrode (BAS, West Lafayette, IN).

Several different instruments were utilized to perform the cyclic voltammetry experiments. The model CV-IB (or CV-IA), Cyclic Voltammetry Unit, BAS-100 or BAS-100A Electrochemical Analyzer (BAS, West Lafayette, IN) were used as the instruments to provide potential control. The voltammograms were recorded using the HIPLOT DMP-40 Digital Series Plotter (Houston Instruments, Houston, TX). Cyclic voltammograms were also obtained with the automated DWEP, which is described elsewhere in this manuscript. Data storage and manipulation were accomplished via a PC-286 clone, in conjunction with an in-house developed data acquisition and software package (*16*).

Polymer films were formed by applying a potential of +1.600 volts to a platinum working electrode for 10 to 160 seconds. This variation of time gave polymer films, which had a thickness that varied in a direct proportion. Post-treatment of these films was performed by scanning the potential of the modified working electrode, which was immersed in a monomer free supporting electrolyte solution, between the potential limits of -200 to +1000 millivolts.

The electrosynthesis of the poly(3-methylthiophene) and any non-cyclic post-treatments were performed using a PAR Model-173 potentiostat/galvanostat equipped with a PAR Model-176 Current to Voltage Converter, which was connected to a PAR Model-179 Digital Coulometer (Princeton Applied Research, Princeton, NJ).

Electrodes and Electrochemical Cells. The majority of the work for this study used a dual platinum working electrode available from BAS, Inc. (Model MF-1012). All platinum electrodes were polished with metallurgical papers, followed by polishing with fine grade polishing paper, using a water and alumina suspension. After polishing, the electrode was then rinsed with distilled water and sonicated with methanol. The electrode was rinsed with the same solvent that was used in the electropolymerization experiment and used immediately.

The electrochemical cell used for the electropolymerization is of a unique design and has been described elsewhere (*17*) The electrochemical cell used in the FIA/anion analysis was a commercial flow-thru electrochemical cell available from BAS.

Automated Dual Working Electrode Potentiostat Design

Our design philosophy was based on several criteria. These criteria included, small physical size for the printed circuit board, computer control of both working electrodes independently of one another, a range of current sensitivities, which included nanoamperes through microamperes and automated compensation of background current in the same nanoampere to microampere range. We were also interested in the evaluation of a relatively new technology in the field of electronics, Surface Mount Technology (SMT).

The hardware design began with the capture of the hardware circuit schematic with the aid of a CAD software package called Schema II (Omation, Richardson, TX). The schematic capture software allows for identification, of each component and the interconnections between components. The schematic for the DWEP is presented in Figure 2. After the successful capture and verification for the DWEP schematic, the net list file containing the components and the interconnections used in the design was transferred to a printed circuit board artwork CAD program called Tango II (ACCEL Technology, San Diego, CA)

For the DWEP printed circuit board, a final physical size of 3 by 4 inches was achieved. In addition, to the small physical size, the routed printed circuit board was also composed of four layers of conductive copper. After post router verification of the board artwork design, five four-layered prototype printed circuit boards were manufactured and electrically and thermally tested by the fabrication shop (Sovereign Electronics, Youngstown, OH) prior to shipping. The total cost for these five prototype printed circuit boards was approximately $1800.

Surface mount components, such as resistors, capacitors and electrolytic capacitors were purchased from a commercial source (Digi-Key, Thief River Falls, MN). The operational amplifiers, OPA121 and INA105 were purchased from Burr-Brown, Tucson, AZ. Conventional soldering and desoldering techniques are applicable in the final step of populating and testing the printed circuit board with both active and passive surface mount components.

Flow Injection Analysis (FIA) with Poly(3-methylthiophene) Modified Electrodes

Figure 3 shows a schematic of the FIA system and details the thin-layer transducer cell, which contains the poly(3-methylthiophene) anion sensor electrode(s). Electrochemical detection in the amperometric mode was employed for the FIA/EC (flow injection analysis with electrochemical detection) studies. A typical potential used was +1.00 volts vs. Ag/AgCl reference electrode. The FIA-EC system consisted of an Altex 100A double reciprocation pump and a Rheodyne injector (Cotai, CA) with either a 20 or 50 microliter sample loop. Constant potentials were applied using the herein described DWEP, which is software controlled. The electronically transduced signals are converted from analog to digital by use of A/D and D/A converter circuits. The option exists to automatically save the data on the computer or print the results as the experiment progresses.

Results and Discussion

Hardware Design. The block diagram for the dual working electrode potentiostat is shown in Figure 4. We will use this diagram to explain in detail features associated with this hardware design. The two independently controlled working electrode voltages, E_1 and E_2, signals are DAC voltages which have a full scale voltage range of +/-10 volts. Each voltage signal is reduced by a factor of minus one-third (x-1/3). This attenuation is needed, in order to, bring these input voltages into a range of values (-/+3.33 volts), which are more commonly needed for solution electrochemistry. The attenuated voltage signal, E_1, is then sent to the control (or auxiliary) amplifier and is summed with the voltage signal from the reference electrode follower. This control amplifier (or feedback control loop) arrangement is of conventional design (*18*). However, the voltage signal, E_0, is a voltage that is amplified by a factor of three times the voltage level found at the output of the voltage follower amplifier.

This arrangement gives, E_0 voltage values, which are in the +/-10 volt range. The voltage range for E_0 and the input voltages, E_1 and E_2, use the full resolution available from the +/-10 volt bipolar voltage range, which was configured for the 12 bit DAC and ADC converter circuits. Working electrode, W_1, is connected to a "virtual ground", current to voltage amplifier. The voltage signal out, E_{W1}, is proportional to the current flowing to or from electrode W_1 multiplied with the feedback resistance, R_{F1} used. There are eight ranges of feedback resistances, R_{F1}, therefore eight ranges of current sensitivities, from nanoamperes to microamperes. The voltage signal,E_S, is a 12 bit DAC voltage which can be applied to one of eight resistances, R_{cmp1}. The resulting current, I_{cmp1}, is the background compensating current which is summed at the summing point node for this current to voltage converter amplifier.

The current measuring and potential control circuitry for working electrode, W_2, is somewhat different than the circuitry previously described for working electrode, W_1. The attenuated input voltages, E_1 and E_2 are subtracted with a differential amplifier, to give an output voltage which is proportional to (E_1-E_2). This voltage signal is directed to three separate amplifiers. One signal path involves subtracting a voltage, E_c, from the voltage (E_1-E_2). The resulting output signal is applied across one of eight compensation resistances, R_{cmp2}, which provides a background compensation current at the summing point node for the second working electrode, W_2. The second signal path involves directing the voltage signal (E_1-E_2) to the non-inverting inputs for two operational amplifier. This circuitry arrangement allows us to subtract the voltage (E_1-E_2) from the composite voltage output from the current to voltage converter for the second working electrode, W_2. The voltage, E_{W2}, at the subtracter circuit is a voltage, that is proportional to the current flowing from (or to) the working electrode, W_2.

The schematic for the DWEP is shown in Figure 2. There are several interesting design features which were incorporated into our hardware design. The inclusion of these features was dictated by the dual requirements of small physical size and superb reliability. The first feature was the exclusive use of the operational amplifier INA105, for each subtracter circuit. In addition to this amplifier being available as an SMT component, the four resistors required for a subtracter circuit are integrated into the op-amp package. These resistors are pre-trimmed at the factory, thereby assuring a total error of less than +/-0.015%. With the resistors integrated into the op-amp package, there is also an inherent space savings.

Another feature, which was included in the design to enhance space savings and reliability was the use of matched resistor contained in a network package (e.g. R_1 and R_2). Using the resistors in the resistor network package, for both the input and feedback resistors, gives a circuit whose temperature sensitivity is nearly zero, due to nearly equivalent temperature coefficient values for each resistor in the package.

The resistance values for the compensation resistances and the feedback resistances, for each current to voltage converter range in magnitude from 10^6 to 10^9 ohms. These resistance values are quite large and as a result problems with stability may become evident due to inductive coupling (*19*). We decided to avoid this potential problem by utilizing a Thevenin equivalent circuit. The Thevenin equivalent circuit requires three resistors of much lower resistance values. This approach mitigates the inductive coupling problem, thereby improving the stability. There are two disadvantages associated with the Thevenin circuit. The first disadvantage is the additional cost, in terms, of space required for three resistors versus one resistor. The second disadvantage is that due to the inherent limitation in the range of resistance values, which can be achieved using the Thevenin circuit. This range of resistance values is finite, and has a value of about 1000/1 (*19*). For a lower resistance value of 10^6 ohms, the equivalent circuit will allow an upper resistance value of about 10^9 ohms.

Electrical testing of an assembled prototype board involved the integration of this hardware, with the data acquisition hardware and the software driver package. A dual electrode dummy cell "test fixture" was constructed to aid in the electrical testing of this integrated system. This dummy cell was designed to make available eight discrete levels of resistance and capacitance values. The DWEP package could be tested for gain-frequency stability, signal to noise ratio as a function of the amount of current (10^{-6} to 10^{-9} amperes)

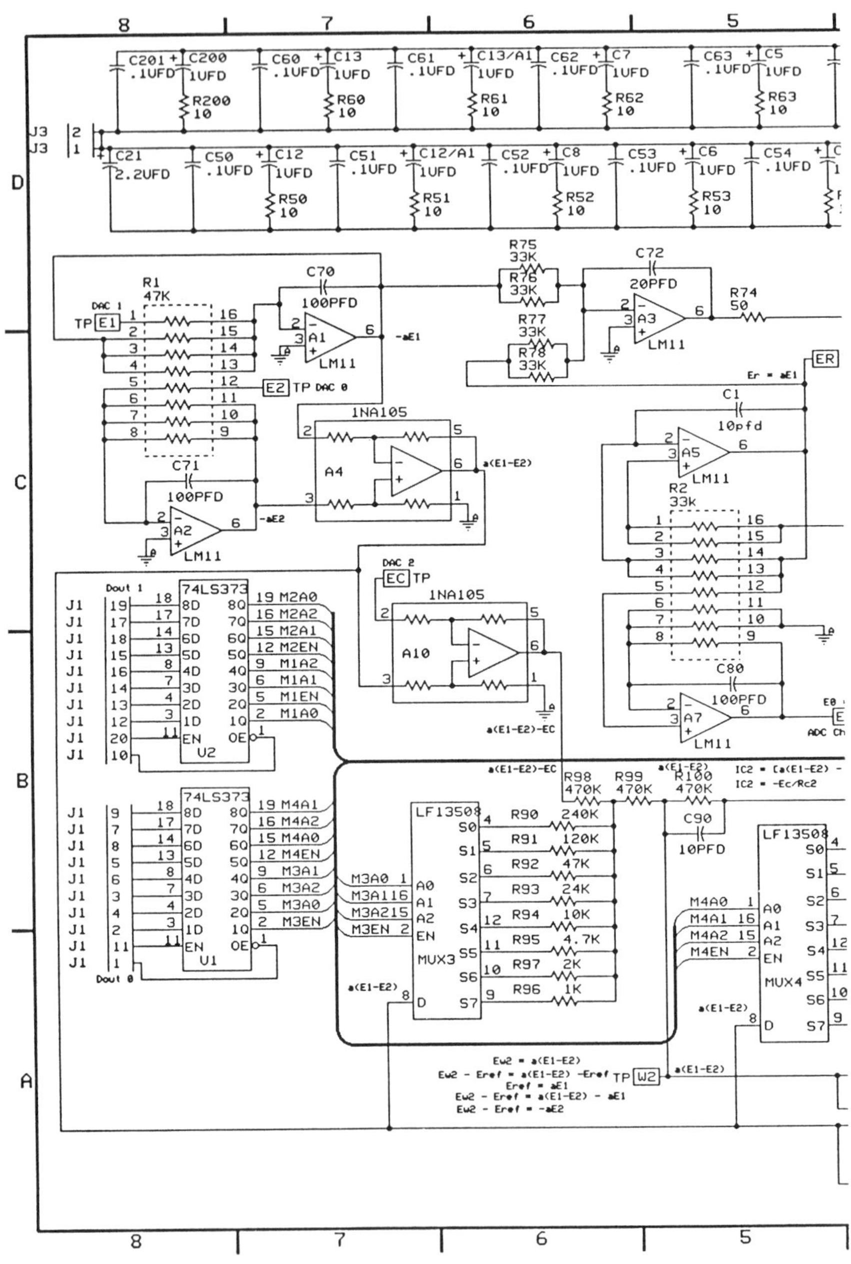

Figure 2: Schematic Diagram for the Dual

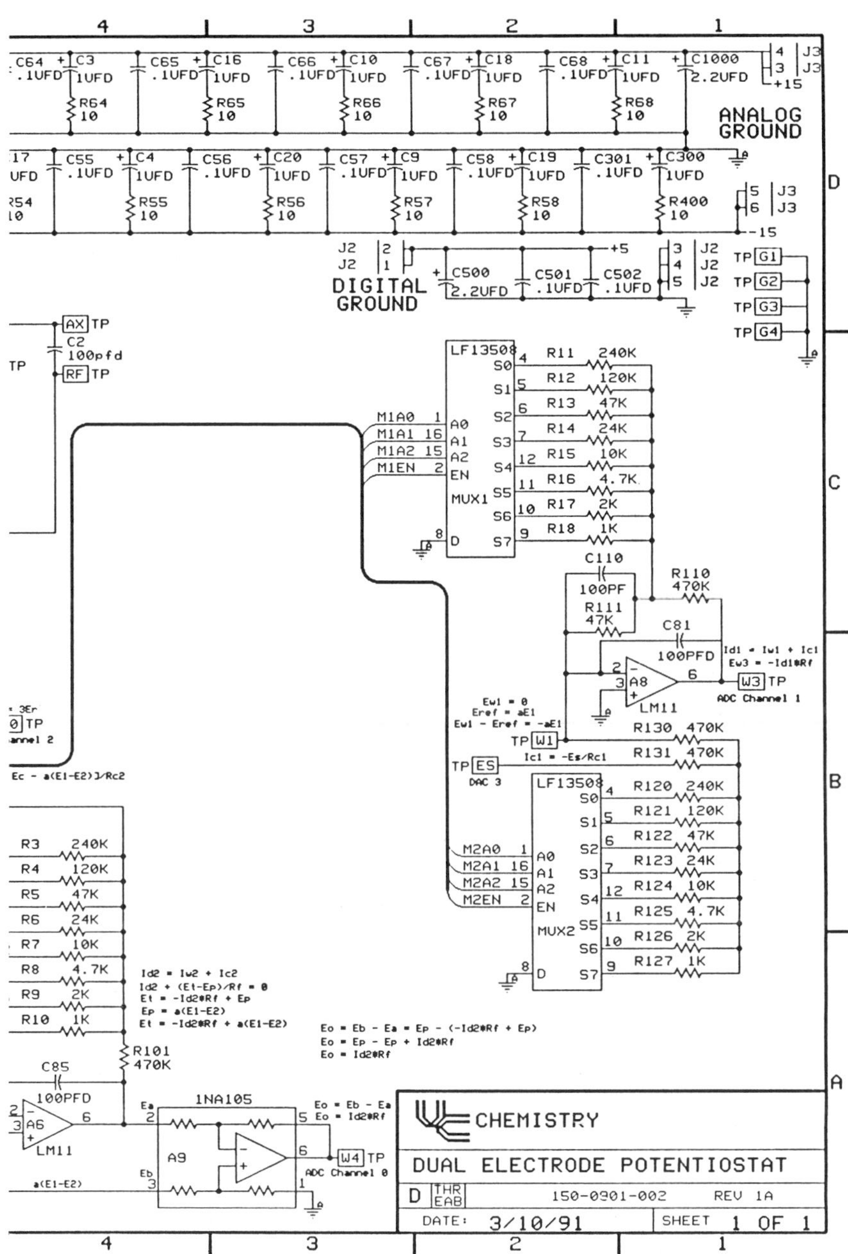

Working Electrode Potentiostat (DWEP)

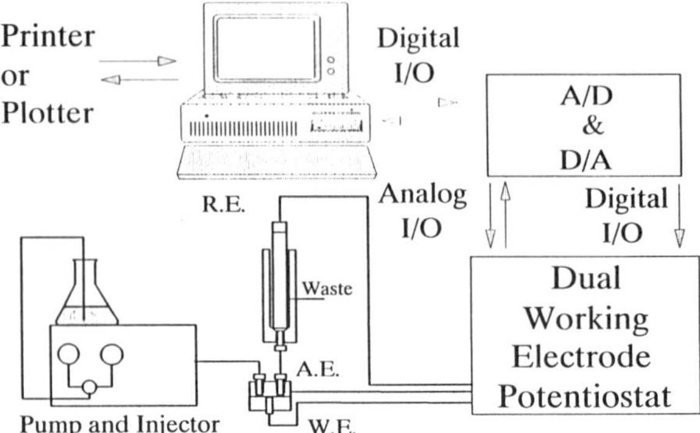

Figure 3: Block Diagram of the DWEP-Flow Injection Analysis System

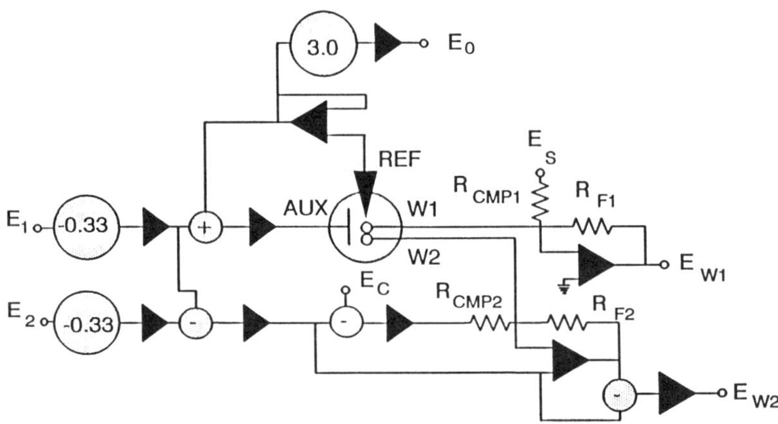

Figure 4: Block Diagram of the Dual Working Electrode Potentiostat

being measured. In general, over the entire range of current signal magnitude, the DWEP demonstrated a signal to noise ratio of better than 5 to 1, for the frequency range of 100 hertz to 2.5 kilohertz. These figures of merit are true for both electrodes. At frequencies greater than 2.5 kilohertz, the signal to noise ratio progressively decreased. When voltage pulses of time width less than 100 microseconds were applied to either (or both) electrodes severe noise spikes appeared on the current-time response. These noise spikes are most probably, the result of "cross-talk" between the two working electrodes. This problem of "cross-talk" has been observed in previous work involving multi-working electrodes (*20*). Further work needs to be done to determine the frequency range that this cross-talk occurs over and if any electronic countermeasures may be taken to mitigate this problem.

The automated compensation of background current is an intimate blend of hardware and software. This compensation is based on a closed feedback loop, where the output for each current to voltage converter is measured. If nonzero, then a calculated voltage is applied to the compensation resistance, resulting in a compensation current being applied at the summing point node for each current to voltage converter. The desired result of course is a zeroed output. With the background current fully compensated, the potentiostat behaved in a stable fashion. Before each experimental run, the output of each current to voltage converter was measured and if still zero, then the experiment would continue. If it was not zero, then the appropriate changes in compensation voltage and/or compensation resistance are made to rezero these outputs. These experiment to experiment incremental changes in the background compensation current were small (typically 20-50 picoamperes) and seemingly random. It is tempting to speculate, that the ultimate lower detection limit for our hardware/software package is approximately 100 picoamperes based on the observed random incremental change in background current, and a S/N ratio of 5.

Software Design. The DWEP is controlled by a custom graphical user interface (GUI) driven software package. This software package operates using multiple windows and screens to allow the user to choose an electrochemical experiment and set the various experimental parameters. By adopting a mouse driven, icon based format, the computer control software has been developed with the aim of simplifying the operator/machine interaction. With this GUI developed the computer display takes the form of a front panel of an instrument with icon-based thumbwheels, slide and pushbutton switches, as well as, annunciator, select, and data display. The program has a built-in notebook which the user can easily access to input annotations for the experiment performed. Also included is a screen where one can input the cell parameters. All information is automatically saved to the hard disk avoiding the common problem of lost or misplaced hardcopies and/or lost experimental conditions. The cell parameters, notebook input, etc. are stored together with the collected experimental data. Many of the higher level functions such as data analysis, graphing, etc. are facilitated by another program which can combine and analyze multiple experiment runs (i.e. scan rate studies).

The GUI experiment control package relies on a low level software package entitled Blubaugh Fisher Displays (BFD) (*16*). BFD is a software package which allows simplified hardware control from high level languages (e.g. BASIC, etc.). BFD acts as a software shell which provides an environment for real-time experiments that can be controlled and monitored by a higher level program. BFD provides four services which are essential to instrument software control: 1) data acquisition via analog to digital conversion subroutines, 2) experiment control via digital outputs and digital to analog conversion subroutines, 3) data manipulation subroutines (i.e. Fast Fourier Transform), and 4) graphic presentation of results to the user with both video graphics and hardcopy (plotters and printers). BFD is an assembly language package that provides the speed needed for data acquisition and manipulation that a higher level language cannot guarantee.

With BFD, the computer becomes excellent not only for data logging, but also for strict timing control of experiments. BFD makes it easier for the GUI driven experiment control program to perform these tasks without the need for intimate knowledge of the hardware (i.e. the A/D converter chip used). The GUI experiment control program can operate without any changes despite possible hardware changes because BFD has provided a uniform interface to data control, acquisition and display.

Preliminary FIA Experiments. One of the more accurate gauges of polymer (P3MT) film reproducibility is the consistency of current and charge values consumed during polymerization of poly(3-methylthiophene). Initial studies gave results where the charge and current passed for the polymerization would vary as much as 30 to 40 percent. An improvement for the polymerization on an day to day basis was obtained by paying very close attention to the polymerization electrochemical cell design, and a consistent routine of 45 seconds for the polymerization time at a potential of +1.600 volts. When this protocol was followed, the discrepancy would amount to no more than about 8 percent. The particular details about this procedure has been elaborated elsewhere (*17*).

The next study was to test the stability of the P3MT films to repeated cycling. Figure 5 shows cyclic voltammograms for a film that was cycled ten times between the potential limits of -0.200 to +1.00 volts. The general features of the cyclic voltammogram are maintained which indicates electrochemical stability for the P3MT film. This procedure of multiple

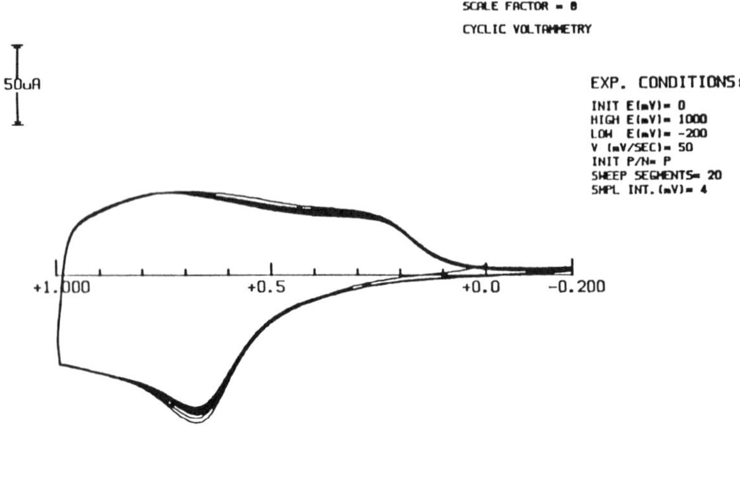

Figure 5: Repetitive Cyclic Voltammogram of P3MT in 0.10 M TBATFB Dissolved in Acetonitrile

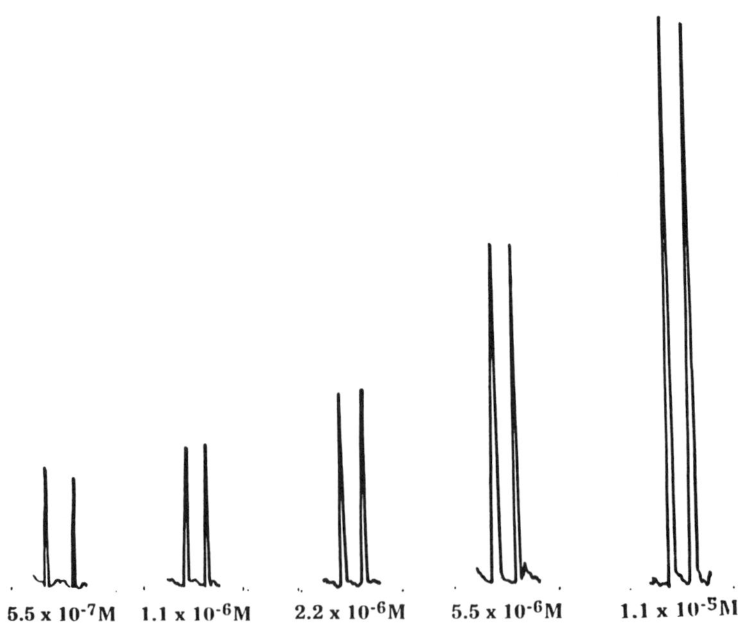

Figure 6: Typical Signal Response of P3MT Anion Sensor Towards Sodium Nitrite at Concentrations of 5.5×10^{-7}, 1.1×10^{-7}, 2.2×10^{-6}, 5.5×10^{-6} and 1.1×10^{-5} M in Flow Injection Analysis Mode

cyclic voltammograms is the basis for the "potential conditioning" of these polymer films and discussed separately (*17*).

Figure 6 provides the typical current response observed in FIA using the P3MT anion sensor. The anion under investigation was nitrite and the concentrations ranged from a low of 5.5 x 10^{-7} M to a high of 1.1 x 10^{-5} M. The film growth and conditioning was as outlined before, with the final potential of the film, before placement in the FIA mode, being set at -0.200 volts. In this study all detection was done at a potential of +1.00 volts versus Ag/AgCl. By plotting log concentration versus log current response, a linear calibration curve was obtained.

With analogous preparation and FIA experimental conditions, the response exhibited by nitrate was evaluated. The calibration curve for nitrate at a P3MT anion sensor is shown in Figure 7. The usual log-log relationship was observed between concentration and the detected current. The range of concentrations used in this study varied from 10^{-1} M to 10^{-8} M. The sensitivity of the P3MT anion sensor remains constant from 10^{-1} to 10^{-5} M. The response curve flattens between the concentrations of 10^{-6} to 10^{-8} M. Thus, the dynamic range is five orders of magnitude and the detection is 1 x 10^{-5} M.

Conclusions

Electrochemical conditions for the electropolymerization of 3-methylthiophene have been optimized. Responsive films for detection of anionic species in FIA were prepared under the following general conditions: 3-methylthiophene concentration (0.5 to 0.15 M), supporting electrolyte (tetrabutylammonium tetrafluoroborate) concentration (0.5 to 0.15 M), solvent (acetonitrile), applied potential (1.600 volts) and time of polymerization (30-60 seconds).

Electrochemical experiments involving cyclic voltammetry have indicated that the interaction between the P3MT film and anionic species in aqueous solutions is a non-faradaic process since no redox couples were observed. The exact mechanism for the observed amperometric response is under investigation. A mechanism such as a change in the

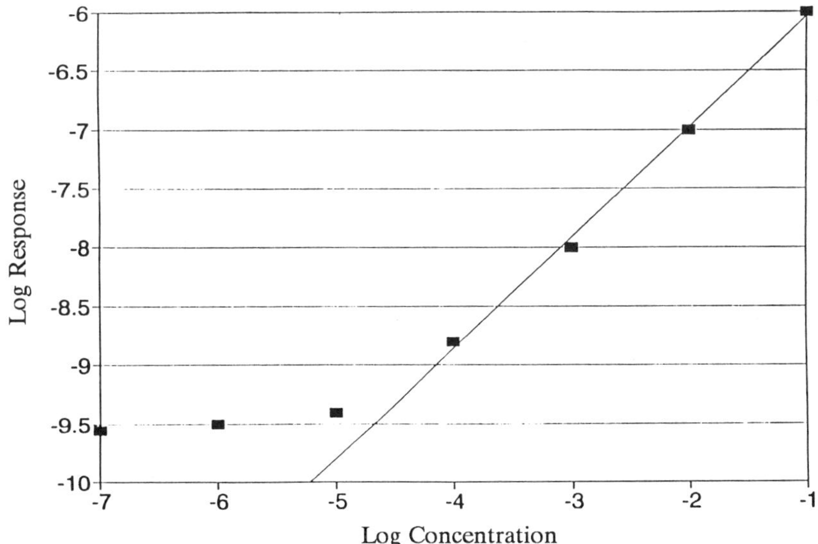

Figure 7: Calibration Curve of Sodium Nitrate Using Dual Working Electrode Potentiostat-Flow Injection Analysis System with the P3MT Anion Sensor

conductivity for the P3MT film is being evaluated. A trend was established for the halide current response to the P3MT anion sensor, which was the following: $F^- < Cl^- < Br^- < I^-$.

The P3MT anion sensor's current response to the analytes in this study was superior to the response exhibited by bare platinum. Calibration curves were constructed for several anionic species. Figures of merit from the calibration curves are a dynamic range of four to five orders of magnitude based on concentration. Detection limits of approximately 5×10^{-7} M, sensitivities are ion dependent, and R-squared values for the curves are larger than 0.99.

The development of the computer-controlled DWEP instrument package required efforts in both the hardware and software design. The hardware design utilized surface mount electronic component for their small size and reliability. The DWEP potentiostat was capable of measuring current levels at approximately 500 picoamperes to current magnitudes as large as 10 microamperes. This potentiostat also has the ability to compensate for background current over the same 500 picoamperes to 10 microamperes range. The software package allows for hardware control from a high level language, such as BASIC. The software also integrates the experimental conditions with the experimental data, all of which is accomplished within a "windows" like environment.

Literature Cited

1: Russell G.; Mark H.B., paper presented at Pittsburgh Conference, Atlanta, Georgia, 1989.
2: Russell G.; Mark H.B.; Ridgway T.H., paper presented at Pittsburgh Conference, New Orleans, Louisiana, 1992.
3: Baizer M.; Lund H., *Organic Electrochemistry, An Introduction and a Guide*, Marcel Dekker, New York, 1983.
4: MacDiarmid A.; Heeger A., *Molecular Metals*, Plenum, 1979.
5: Diaz A.; Kanazawa K., *J. Chem. Soc. Chem. Commun.*, **1979**, *635.*
6: Tourillon G.; Garnier F., *J. Phys. Chem.*, **1984**, *88*, p. 5281.
7: Kmetz A.R.; Von Willisen F.K., *Non-Emissive Electro-optic Displays*, Plenum, New York, 1976.
8: Horowitz G.; Tourillon G.; Garnier F., *J. Electrochem. Soc.*, **1984**, *131*, p. 151.
9: Yammanoto T., In Spring Electrochemical Society Meeting, Extended Abstracts, (82-1), Electrochemical Society, 1982; p. 987.
10: Gerischer H., *J. Electroanal. Chem.*, **1977**, *82*, p. 133.
11: Barbarin F.; Berthet G.; Blanc J.P., *Synth. Met.*, **1983**, *6*, p. 53.
12: Tourillon G.; Garnier F., *J. Electroanal. Chem.*, **1984**, *161*, p. 51.
13: Ikariyama Y.; Heineman W.R., *Anal. Chem.*, **1986**, *58*, p. 1803.
14: Karlberg B.; Pacey G., *Flow Injection Analysis: A Practical Guide*, Elsevier, New York, 1989.
15: Ruzicka J.; Hansen E., *Flow Injection Analysis*; 2nd Edition, Wiley and Sons, New York, 1989.
16: Blubaugh D.D., Ph.D. Thesis, University of Cincinnati, Cincinnati, Ohio, U.S.A., 1990.
17: Russell G., Ph.D. Thesis, University of Cincinnati, Cincinnati, Ohio, U.S.A., 1992.
18: Hoogvliet J.C.; Reijn J.M.; van Bennekom W.P., *Anal. Chem.*, **1991**, *63*, p. 2418.
19: Horowitz P.; Hill W., *The Art of Electronics*; 2nd Edition, Cambridge University Press, Cambridge, 1989.
20: (Ridgway, T.H., University of Cincinnati, personal communication, 1989.)

RECEIVED December 14, 1993

Chapter 12

Characterization of Stability of Modified Poly(vinyl chloride) Membranes for Microfabricated Ion-Selective Electrode Arrays in Biomedical Applications

Ernö Lindner[1], Vasile V. Cosofret, Tal M. Nahir, and Richard P. Buck

Department of Chemistry, University of North Carolina at Chapel Hill, Chapel Hill, NC 27599–3290

Microelectronically fabricated, semi-micro, planar, ion-selective electrodes on a flexible polyimide substrate (KAPTON) were tested with respect to sensor life time. The changes in the offset voltage of the potentiometric cells as well as in the membrane resistances were correlated with the dissolution of membrane ingredients into the contacting sample solution. The dissolved amounts were determined spectrophotometrically. The problems in the quantitative determination (slow dissolution rate, decomposition of the tested compounds during the experiment, very small concentrations in complex matrices) led to a new method for the determination of the free carrier in liquid membrane electrodes. The method is based on the linear relationship between the time constant of current-time transients of carrier based membranes (following a voltage step) and the free carrier concentration in the membrane.

Myocardial ischemia is characterized by the acute and total deprivation of blood flow within a specific, well defined heart region. The application of ion-selective electrodes to miniature electrode structures has permitted physiologists to observe the time course of ischemic events by monitoring the rise in extracellular potassium and the fall in intracellular pH. Typically, extracellular potassium concentration rises from a normal value of 3-4 mM to 12 mM or more, while extracellular pH falls by 0.5 to 1.0 units from a normal reading of 7.35-7.45. The changes in potassium concentration and pH, while regionally predictable in terms of the typical change observed, show local variability (*1*). The degree of ionic inhomogenity observed for a particular myocardial region at a specific time following ischemia is thought to be a contributing factor to the formation of arrhythmia-producing reentry circuits. These circuits disturb the synchronous movement of stimulation wave fronts in the heart and eventually lead to ventricular fibrillation and death.

[1]Current address: Institute for General and Analytical Chemistry, Technical University of Budapest, 1111 Budapest, Szent Gellért tér 4, Hungary

0097–6156/94/0556–0149$08.00/0

Obviously, the observation and quantification of more resolved regional understanding of ischemia that leads to reentry and ventricular fibrillation. For this purpose planar pH, pK and pCa sensors have been developed on a flexible polyimide KAPTON) substrate (*2,3*) to meet the requirements of the cardiologist. The scaling down of the sensor size led to customer designed multielectrode arrays with overall dimensions of 12 mm x 1 mm or 12 mm x 4.5 mm having eight or nine semi-micro ion-selective electrode spots integrated on their surfaces, respectively. The electrodes proved their advantageous properties (Nernstian slope before and after measurements in hostile environment, excellent stability and reproducibility in blood serum-like aqueous electrolyte solutions, etc.) in preliminary *in-vitro* and *in-vivo* tests.

However, by decreasing the overall dimensions of the sensing membrane (size and thickness), the loss of membrane ingredients becomes one of the most serious limitations in designing ionophore-based ion sensors with extended life time. Beside mechanical defects, surface contamination and membrane poisoning, the lifetime of liquid membrane electrodes is generally limited by the gradual extraction of membrane components (carriers, plasticizers, additives) into the test sample (*4*). The dissolution of membrane ingredients is an even more serious problem when the sensors are intended to be used under *in vivo* (acute or chronic) conditions. Enhanced dissolution is mainly due to the more lipophilic character of several body fluids (whole blood, plasma, serum etc.) compared to aqueous electrolyte solutions. Lipophilicity can be calculated theoretically (*5*) or measured quantitatively (*6*) and it is described by a "lipophilicity parameter" value (P). The required lipophilicity value (P) for direct measurements in blood or plasma is considerably higher than for aqueous solutions (*5,7*). Low lipophilicity values may result in substantial extraction of the membrane components as well as it may produce serious inflammatory reactions (*8*).

As part of our work in designing microelectronically fabricated semi-micro potentiometric sensors for acute and chronic applications, we investigated changes of membrane characteristics over time. Large number of membranes with different compositions were compared on the basis of their potentiometric behavior (*9*). A continuous drift in the offset voltage of the cell, a parallel increasing membrane resistance, and an increasing response time are all signs of loss of membrane ingredients (*10,11*). A newly introduced class of pH sensitive chromoionophores with extremely high molar absorbance values, namely the lipophilized Nile Blue derivatives, used first in ion sensitive optodes (*12*) and later in pH sensitive potentiometric sensors (*13*), seemed to offer an excellent possibility to follow bulk membrane transport processes (*14*) as well as to measure the loss of ionophore into different matrices through the decrease of membrane absorbance or through the appearance of some absorption bands in the contacting aqueous solutions.

The theoretical models to estimate the lifetime of potentiometric sensors (limited by the loss of membrane ingredients) are based on Fick's diffusion laws (*5,10*). The analysis of current-time transients during the application of a voltage step can be used for the determination of carrier diffusion coefficients in various membranes (*15-18*). However, due to characteristic changes in these transients as a function of carrier concentration, they can also be used to determine the carrier concentration in membranes following long term solution contact.

Experimental

Chemicals: For all experiments, deionized water (Barnstead, NANOpure II) and chemicals of puriss or pa grade were used. All ionophores and plasticizers as well as the lipophilic salt additives were products of Fluka AG (Buchs, Switzerland): Tridodecylamine (TDDA, Fluka 95292); ETH 5294 (Chromoionophore I, Fluka 27086); 2-Nitrophenyl octylether (o-NPOE, Fluka 73741); Bis(2-ethylhexyl) sebacate (DOS, Fluka 84818), Sodium tetraphenylborate (NaTPB, Fluka 72018), Potassium tetrakis(4-chlorophenyl borate (KTpClPB, Fluka 60591), Potassium tetrakis[3,5-bis(trifluoromethyl)phenyl] borate (KTm$(CF_3)_2$PB). As polymeric membrane materials, high molecular weight poly(vinyl chloride) (PVC-HMW, Fluka 81392), caboxylated PVC (PVC-COOH, with 1.8 % COOH groups, Aldrich 18,955), hydroxylated PVC (PVC-OH, a copolymer of 91 % vinyl chloride, 3% vinyl acetate and 6 % vinyl alcohol, Fluka 27827) as well as 1,4-diamino butane and piperazine modified aminated PVC (PVC-NH_2). The 1,4-diamino butane (DAB) and the piperazine (PIP) modified PVC samples were prepared in our laboratories (*19*) with the reaction scheme suggested by Ma and Meyerhoff (*20*). The hydroxylated PVC (PVC-OH) could not be used alone for membrane casting (as a consequence of the bad mechanical properties of the membrane), but only as a 1:1 mixture with PVC-HMW.

Membranes and Membrane Solutions: The solvent polymeric membrane cocktail composition was 1 wt % ionophore, 64-66 wt % plasticizer and 33 wt % PVC and 70 mole % (compared to the ionophore) lipophilic salt additive dissolved in tetrahydrofuran.

Fabrication of Planar Ion-selective Electrodes on Kapton Wafers (Site Preparations and Wire Connections): The fabrication sequence is based on integrated circuit technology and is described in detail in (*2,3*).

EMF measurements: Cell voltages were measured at room temperature in an air-conditioned laboratory at 22.5 ± 0.5 °C with an Orion pH Meter (Model 720A) connected to an Orion model 607 manual electrode switch. As a reference electrode an Orion model 90-02 Ag/AgCl double junction reference electrode was used throughout. The measured potential values were recorded with an ABB Goertz 420 strip charge recorder.

Buffer solutions: In the course of the work the following buffer solutions were used: (I) a Tris buffer with 140.0 mM Na^+ ion background and (II) a Britton Robinson buffer with or without 140 mM NaCl (see Table 10.47 in (*21*)). For details see (*13*).

Determination of the Internal Resistance of the Cells: AC bulk resistances were measured from impedance plots for each membrane using a Solatron 1186 Electrochemical Interface and Solatron 1250 Frequency Response Analyzer (FRA) controlled by a Hewlett-Packard 85B computer (*15*).

Current vs. time transients: Membranes with 1.77 cm^2 area were placed between two aqueous solution in an electrochemical cell designed for ionic transport measurements (*16,18*). Desired voltage steps were applied by a Solatron 1250 FRA through an EG&G 363 potentiostat controlled by a Zenith XT computer. Current vs. time data were collected by the same computer using a Data Translation DT 2801 board.

Determination of dissolved ionophore and lipophilic salt additives from different matrices: For the measurement of dissolved ionophore from different matrices, DOS plasticized membranes with ETH 5294 and Nile Blue A (free base) as ionophores were selected. For the quantitative determination of dissolved lipophilic salt additives (NaTPB, KTpClPB and KTm$(CF_3)_2$PB), colorless, TDDA-based membranes were used to avoid all possible interferences due to the absorbance of the dissolved ionophore. About 20-50 mg membrane pieces were placed into closed glass vials containing 5 ml aqueous buffer solution (pH values ranging from pH=2.5 to pH=12) and covered with aluminum foil (to avoid light initiated decomposition). The vials were shaken with an IKA-Vibramax-VXR shaker for as long as five days to ensure complete equilibration. The dissolved amounts of the chromoionophores and that of lipophilic salt additives were determined with the help of a calibration plot by measuring the absorbance values of the different aqueous samples at different time instances using a Hewlett Packard 8452A Diode Array Spectrophotometer.

Results and Discussions

The TDDA and ETH 5294 chromoionophore based macro and semi-micro planar pH sensors showed practically the same analytical features in a number of respects (*2*). However, by comparing the long term stability of the planar micro electrodes a considerable E^o change (intercept shift) was observed for membranes containing ETH 5294 as ionophore while no shift in the offset voltage was found within 37 days when TDDA was used as ionophore (Figure 1a). The alteration of the membrane resistances of the corresponding membranes varied similarly with time to the changes of in the E^0 values. Relatively small increases in the resistances were measured with the TDDA based electrodes, while the resistances of ETH 5294 based membranes considerably increased (Figure 1b). The slopes of the calibration curves in the physiological pH range were practically constant in both cases. It was assumed that the measured differences in the long term behavior of the TDDA and ETH 5294 based pH sensitive membranes correlated with the dissolution and/or decomposition of the chromoionophore.

The dissolution of membrane ingredients into an aqueous sample is controlled by the distribution coefficient between the water and the membrane phase. The distribution coefficient is a function of the lipophilicity (P) which can be calculated theoretically or determined experimentally by thin layer chromatography (*4*). On the basis of the experimentally determined $logP_{TLC}$ values an opposite trend is expected. ETH 5294 has a much higher $logP_{TLC}$ value ($logP_{TLC}$ = 18.0) (*22*) compared to TDDA ($logP_{TLC}$ =11.6) (*22*). Accordingly, the decomposition of the ionophore may be the determining step of the long term behavior and possible response deterioration of ETH 5294 based membranes.

The photochemical decomposition of a number of chromoionophores as well as of different tetraphenylborate derivatives were followed by measuring the absorbance decrease of optode membranes (*12*). It was shown that the decomposition of ETH 5294 could be extensive when the membranes were subject of intensive sunlight. However, decomposition might be negligible when the membranes were irradiated only periodically while recording of the spectrum (by exclusion of diffuse light). The decomposition or dissolution of tetraphenylborate derivatives from optode membranes into a dilute HCl solution (the decomposition is assumed to be proton catalyzed) were measured indirectly through the changes in the concentration ratio of the protonated and deprotonated form of the chromoionophore. The dissolved amounts were in decreasing sequence: NaTPB >> KTpClPB > $KTm(CF_3)_2PB$.

By measuring the dissolved amount of lipophilic salt membrane additives in the contacting solution at different pHs we found the same sequence (see Table 1). However, the dissolved amounts were generally higher at high pH values.

In contrast to the above, the dissolved amount of the chromoionophore was found always larger at low pH values (Figures 2 and 3). Unfortunately, the data shown in Figure 2 can not be interpreted unambiguously. The absorbance measured at 636 nm may be due to the dissolved (i) ionophore itself, (ii) Nile Blue A impurities, originally present in the ionophore (*23*) or (iii) decomposition products of the ionophore having the same or similar chromophore groups as in ETH 5294.

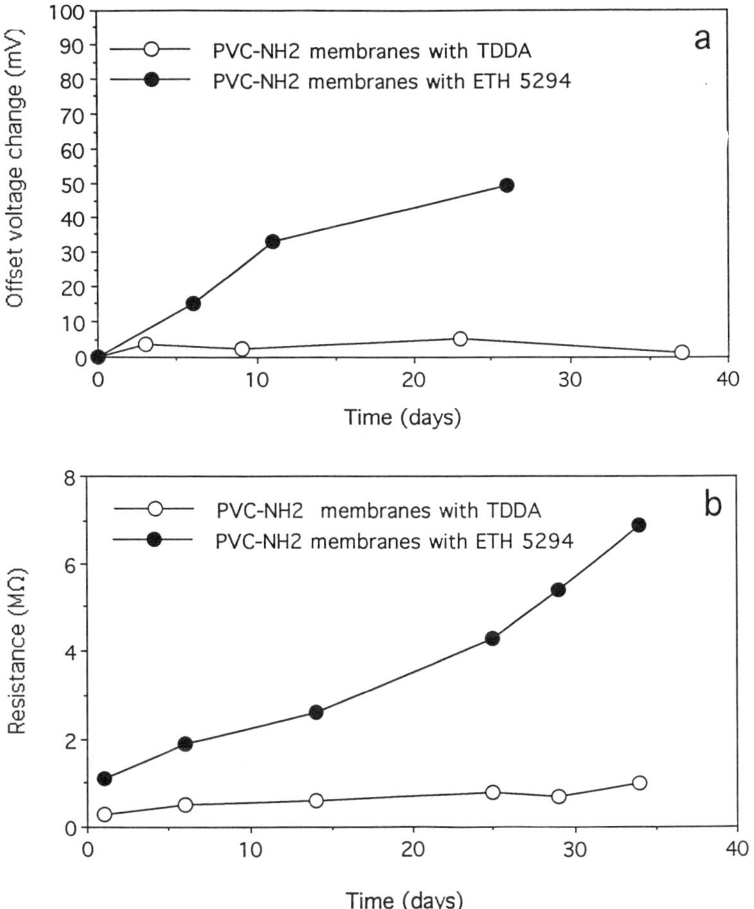

Figure 1. Changes in the membrane characteristics of a TDDA and an ETH 5294 based membranes as a function of time. a.) The shift in the intercept (E^o) of the pH vs. mV calibration curve as a function of solution contact time. b.) The increase of the membrane resistance as a function of conditioning time.

Table 1. Dissolved amounts of lipophilic salt membrane additives into buffer solutions of different pH values after 7 hours equilibration time. The data are in relative % compared to the total amount of additives incorporated originally into the membrane. Membrane composition: PVC-HMW, DOS, TDDA and the corresponding additive.

Additive	pH			
	4.0	7.0	10.0	12.0
NaTPB	6.9	8.0	11.0	15.7
KTpClPB	0.1	0.5	0.8	3.7
KTm(CF_3)$_2$PB	0.2	0.2	0.6	1.5

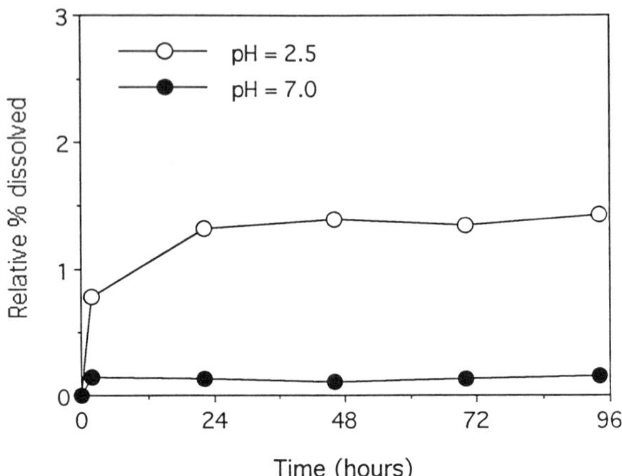

Figure 2. Dissolution of ETH 5294 ionophore from PVC-HMW/DOS membranes in Britton Robinson buffer solutions of different pHs. The data are given in relative % compared to the total amount of ionophore incorporated into the membrane.

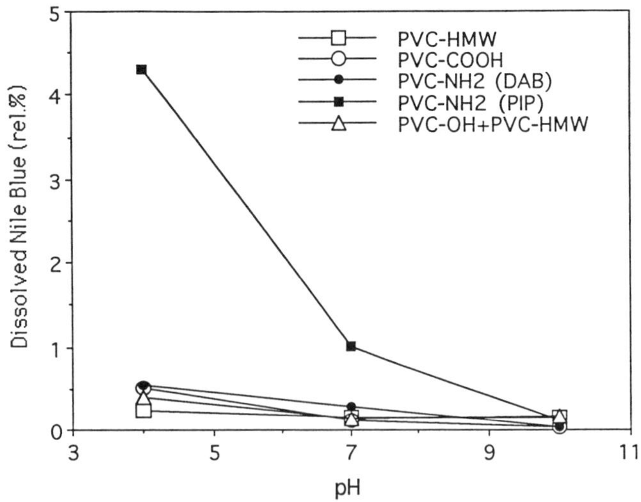

Figure 3. Dissolved amounts of Nile Blue A (free base) from different PVC matrices after one day equilibration time. The data are given in relative % compared to the total amount of ionophore incorporated into the membrane.

To minimize the above problems, ETH 5294 was replaced by Nile Blue A (free base) in the following experiments. ETH 5294 is assumed to differ from Nile Blue A only by its lipophilicity. Accordingly, larger amounts were expected to be dissolved. In contrast to the expectation the dissolved amounts were about the same order of magnitude. The polarity of the plasticizer (DOS or o-NPOE) and that of the membrane matrix had generally no important effect on the dissolved amount. The only serious exception was the piperazine-based aminated PVC matrix at low pH values where considerably larger amounts were leached from the membrane. In general, the dissolved amounts were found to be smaller and smaller by changing the membrane matrix from PVC-NH_2, through PVC-COOH, to PVC-HMW and PVC-OH. By comparing the dissolution rate or dissolved amounts of ionophores from membranes cast with KTpClPB or KTm$(CF_3)_2$PB, but with the same proton carrier, the loss of ionophore was almost always smaller in the latter case, suggesting a possible ion pair formation. Such association may stabilize the chromoionophore against decomposition, as well as it can increase the partition coefficient compared to the completely dissociated form of the ionophore. Some discrepancies were seen (an opposite trend) only at high pH values with PVC-HMW and PVC-COOH based membranes. The major problem in the quantitative evaluation of the dissolution results, calculated on the basis of absorbance values measured in the contacting bathing solutions, is caused by the decomposition of the chromoionophore (Nile Blue A) in aqueous solutions at all investigated pHs. The rate of decomposition and dissolution had the same order of magnitude. The rate of decomposition could only be slowed down by complete exclusion of light.

Since the direct experimental determination of the dissolution rate of the highly lipophilic membrane ingredients is extremely difficult we were interested in possible methods which might be applied despite exposure of the membranes to very complex samples like whole blood, blood serum or plasma. As it was shown earlier (*16*) the current-time transients of ionophore loaded, fixed-site membranes show a sharp break at characteristic time constant under a constant applied voltage:

$$\tau^{1/2} = \frac{FAc_{average}\left(D_{carrier}\pi\right)^{1/2}}{2\left(V_{applied} / R_{\infty}\right)} \tag{1}$$

where F is the Faraday constant; A is the membrane cross-section area; $c_{average}$ is the average concentration of the carrier within the membrane; $D_{carrier}$ is the diffusion coefficient of the carrier; $V_{applied}$ is the applied voltage to induce the current time transient and R_{∞} is the high frquency resistance.

The $\tau^{1/2}$ vs $c_{average}/I$ plot (where $I=V_{applied}/R_{\infty}$) gave an excellent correlation and was used to determine the diffusion coefficient of the carrier in DOS plasticized membranes (*16*). However, the same plot can be considered as a calibration curve for concentration and could be used to determine the actual carrier concentration after extended solution contact.

It is important to note that Eqn. (1) can be accurately used only if the current is equal to $V_{applied}/R_{\infty}$. In mobile site membranes, with some lipophilic salt additives, the resistance is relatively small (approximately 100 kΩ in our membranes), the interfacial potential drop may become substantial and the current will gradually decrease between t=0 and the characteristic current break (*18*). Under these circumstances, a new time constant (τ_n) is more appropriate. Following the derivation of the Sand equation (Eqn. 1), this value is defined as the time until the interfacial reactant (carrier) is completely depleted. Using the approximate solution in (*18)* this period is given by τ_n when

$$\frac{V_{applied}F}{2RT}\left[1 - e^{Q^2 D\tau_n} erfc\left(Q\sqrt{D\tau_n}\right)\right] = 1 \qquad (2)$$

is satisfied. Q is given by $2RT/F^2ADR_\infty c_{average}$. The error due to the use of Eqn. (1) for low resistance membranes (generally loaded with mobile sites) may be substantial. For example, when a 0.1 V potential step is applied to membranes containing 1 mM ionophore, τ underestimates τ_n by 17%, at a resistance of 100 kΩ.

To demonstrate the possibility of the determination of free carrier concentration in liquid membranes by using current-time transients, the chronoamperometric curve of a freshly prepared ETH 5294-based, mobile-site membrane was compared with the current vs. time curve of the same membrane kept in a dry closed vial for 15 days (Figure 4). As it is seen in the figure, the time needed for the complete depletion of ETH 5294 at the interface ($\tau^{1/2}$) is much shorter for the older membrane. This is an evidence for the lower concentration of free neutral carrier in the membrane.

It has been shown previously, that this membrane fails at low pH values (*13*). This was explained as an effect of the more extensive protonation of the ionophore due to the influx of solution anions. Furtheremore, as we showed above, the dissolution and/or decomposition of the ionophore is considerably larger at low pH values. The dramatic decrease in $\tau^{1/2}$ (decrease of the free carrier concentration), shown in Figure 4, was very similar to that noted when TDDA-based membrane was transfered from a pH=5.0 to a pH=2.0 solution (*24*). The depletion of the free neutral carrier in the membrane is the result of two processes: (i) protonation, (ii) dissolution and or decomposition. In this case the protonation is assumed to be predominant.

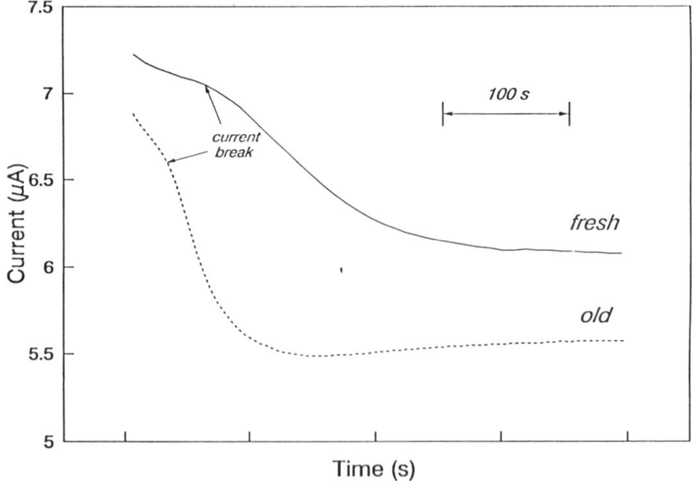

Figure 4. Current vs time transients of a freshly prepared and an old membrane during a 2 V constant voltage step. The experiment was carried out in a pH=5.0 buffer solution .

Acknowledgment

This work was supported by the North Carolina Biotechnology Center and the Duke-North Carolina NSF Engineering Research Center for Emerging Cardiovascular Technologies.

Literature Cited

(*1*) Gettes, L.S.; Cascio, W.E., in *The Heart and Cardiovascular System*, Fozzard, H.A., Ed.,Second edition, Raven Press, Ltd., New York, **1992**, p. 2021
(*2*) Lindner, E.; Cosofret, V.V.; Ufer, S.; Buck, R.P.; Kusy, R.P.; Ash, R.B.; Nagle, H.T., *J.Chem. Soc. Faraday Trans.,* **1993**, *89*, 361-367.
(*3*) Lindner, E.; Cosofret, V.V.; Ufer, S.; Johnson, T.A.; Ash, R.B.; Nagle, Neuman, M.R.; Buck, R.P., *Fresenius' Journal of Anal. Chem.*, sent for publication December, **1992**.
(*4*) Oesch, U.; Anker, P.; Ammann, D.; Simon, W., in *Ion-Selective Electrodes,4*, Pungor E., Ed.; Analytical Chemistry Symposia Series, Vol. 22, Akadémiai Kiadó Budapest, Hungary and Elsevier Sci. Publ. Amsterdam, The Netherlands, **1985**, 81-112.
(*5*) Hansch, C.; Leo, A., *Substituent Constants for Correlation Analysis in Chemistry and Biology,* Wiley & Sons, New York, **1979**
(*6*) Dinten, O.; Spichiger, U. E.; Chaniotakis, N.; Gehrig, P.; Rusterholtz, B.; Morf, W.E.; Simon, W., *Anal. Chem.*, **1991**, 63, 596-603
(*7*) Rosatzin, T.; Holy, P.; Seiler, K.; Rusterholtz, B.; Simon, W., *Anal. Chem.*, **1992**, 64, 2029-2035.
(*8*) Marchant, R. E.; Anderson, J.M.; Castillo E.; Hiltner, A., *J. Biomed. Mat. Res.* **1986**, 20, 153-168.
(*9*) Lindner, E.; Cosofret, V. V.; Kusy, R. P.; Buck, R.P.; Rosatzin, T.; Schaller, U.; Simon, W.; Jeney, J.; Tóth, K.; Pungor, E., *Talanta,* sent for publication November, **1992**
(*10*) Oesch, U.; Simon, W., *Anal. Chem.*, **1980**, 52, 692.
(*11*) Huser, M., Gehrig, P.M.; Morf, W.E.; Simon, W.; Lindner, E.; Jeney, J.; Tóth, K.; Pungor, E.,*Anal. Chem.*, **1991**, 63, 1380-1386.
(*12*) Seiler, K., *Ionenselektive Optodenmembranen*; Fluka AG; Buchs; Switzerland, **1991**.
(*13*) Cosofret, V. V.; Nahir, T. M.; Lindner, E.; Buck, R. P., *J. Electroanal. Chem.* **1992**, 327, 137-146.
(*14*) Nahir, T. M.; Buck, R. P., *Helv. Chim. Acta.*, **1993,** 76, 407-415.
(*15*) Iglehart, M.L.; Buck, R. P.; Pungor, E., *Anal. Chem.*, **1988**, 60, 290-295.
(*16*) Iglehart, M.L.; Buck, R. P.; Horvai, G.; Pungor, E., *Anal. Chem.*, **1988**, 60, 1018-1022.
(*17*) Iglehart, M.L.; Buck, R. P.,*Talanta*, **1989**, 36, 89-98.
(*18*) Nahir, T. M.; Buck, R. P., *J. Electroanal. Chem.* **1992**, 341, 1-14.
(*19*) Cosofret, V. V.; Lindner, E.; Buck, R. P., Kusy, R. P.; Whitley, J., *J. Electroanal. Chem.* **1993**, 345, 169-181.
(*20*) Ma, S. C., Meyerhoff, M. E., *Mikrochim Acta*, 1990 (I),197.
(*21*) Perrin, D. D.; Dempsey B., *Buffers for pH and Metal Ion Control,* Chapman and Hall, London, **1974.**
(*22*) Simon, W.; Spichiger, U.E., *International Laboratory*, **1991**, 21, 35-43
(*23*) Lindner, E.; Rosatzin, T., Jeney, J., Cosofret, V. V.; Simon, W., Buck, R. P., *J. Electroanal. Chem.* in press (JEC02651, February 24, 1993).
(*24*) Nahir, T. M.; Buck, R. P.; *Talanta*, submitted April, 1993

RECEIVED January 10, 1994

Chapter 13

Biosensors Based on Ultrathin Film Composite Membranes

Charles R. Martin, Barbara Ballarin, Charles Brumlik, and Del R. Lawson

Department of Chemistry, Colorado State University, Fort Collins, CO 80523

This paper introduces a new approach for designing chemical sensors. This approach is based on a concept borrowed from the membrane-based separations area - ultrathin film composite membranes. Ultrathin film composite membranes consist of an ultrathin (less than ca. 100 nm-thick) polymer skin coated onto the surface of a microporous support membrane. These composite membranes have made a tremendous impact on the field of membrane-based separations because they can offer high permeate flux without sacrificing chemical selectivity. These two qualities (high permeate flux and high chemical selectivity) are also required in polymeric barrier layers in chemical sensors. Therefore, the ultrathin film composite membrane concept should be applicable to sensor design. In this paper we present proof of this concept by showing the response characteristics of a prototype glucose sensor based on an ultrathin film composite membrane.

We introduce a new and general approach for designing chemical sensors. This approach is based on a concept borrowed from the membrane-based separations area (1) - ultrathin film composite membranes (2-7). Sensors based on such composite membranes should have a number of potential advantages including fast response time, simplicity of construction, and applicability to a number of molecular recognition chemistries and signal transduction schemes. In order to demonstrate the feasibility of this new approach for sensor design, we have prepared and evaluated a prototype electrochemical glucose sensor. This particular sensor was chosen to demonstrate the feasibility of this new sensor-design concept because the molecular recognition and signal transduction chemistries are well established (8-11). A number of other sensing schemes (12-16) could, however, have been chosen.

Concepts

The development of ultrathin film composite membranes was one of the most

0097–6156/94/0556–0158$08.00/0

important breakthroughs in the membrane-separations area (2-5). Such composites consist of an ultrathin (less than ca. 100 nm-thick) chemically-selective "skin" bonded to the surface of a microporous support membrane (3-6). The desired chemical separation occurs within the ultrathin skin and the thinness of this skin insures that the flux of permeate across the membrane is high. The microporous support provides the requisite mechanical strength. Ultrathin film composite membranes can provide high chemical selectivity, high permeate flux, and good mechanical strength. This combination of properties would be impossible to achieve in a homogeneous membrane (3-5).

In general, barrier layers in sensors must provide some degree of chemical selectivity, yet must also allow for high rates of analyte flux (so as to minimize sensor response time). These membrane requirements (high selectivity and high flux) are identical to the requirements in the membrane separations area (1). Hence, if ultrathin film composites are ideal in this area, these composites should also be ideally-suited for sensor applications.

Two general types of sensors based on ultrathin thin film composites can be conceptualized. These sensor-types are differentiated by the degree of selectivity required of the composite membrane. The first, and experimentally more difficult, type would be based on a membrane which has molecular specificity for the analyte species. That is, in this type of sensor, molecule-recognition chemistry would be built into the ultrathin film such that only the analyte molecule is extracted and transported by the composite membrane. The second, and experimentally easier, sensor type would be based on an ultrathin film which provides some rudimentary form of selectivity only (i.e. passes small molecules but not large molecules or neutral molecules but not charged molecules, etc.). This "prefilter" membrane would transport the analyte molecules into an internal sensing solution which would contain the molecule recognition chemistry and the transducer for translating this chemistry into a measurable electrical signal.
The prototype glucose sensor described here is an electrochemical example of a "prefilter" membrane device. The internal sensing solution contains glucose oxidase, an electron-transfer mediator (ferrocene-carboxylate, FcC^-), and a working, reference, and counter electrode (Figure 1). When glucose enters this inner solution (from the analyte solution), it is oxidized by the glucose oxidase; this oxidation process leaves the flavin adenine dinucleotide (FAD) cofactor associated with the enzyme in its reduced state ($FADH_2$) (17). $FADH_2$ is reoxidized by the oxidized form of the mediator. Note that the mediator is present within the device in its reduced form. The oxidized form is generated by scanning the working electrode potential through the mediator oxidation wave. When no $FADH_2$ is present (i.e. no glucose in the analyte solution), a normal diffusional voltammogram is obtained (see Figure 2, curve A). When $FADH_2$ is present, the oxidized mediator generated, oxidizes the $FADH_2$; the mediator, in turn, gets re-reduced. This causes the mediator wave to adopt a characteristic catalytic shape (8,11,17) (Figure 2, curve B). The difference between the maximium currents in the catalytic (with glucose) and diffusional (no glucose) waves is proportional to the concentration of glucose in the analyte solution. This recognition and transduction chemistry is well known and has been incorporated into other prototype glucose sensors (8,17).

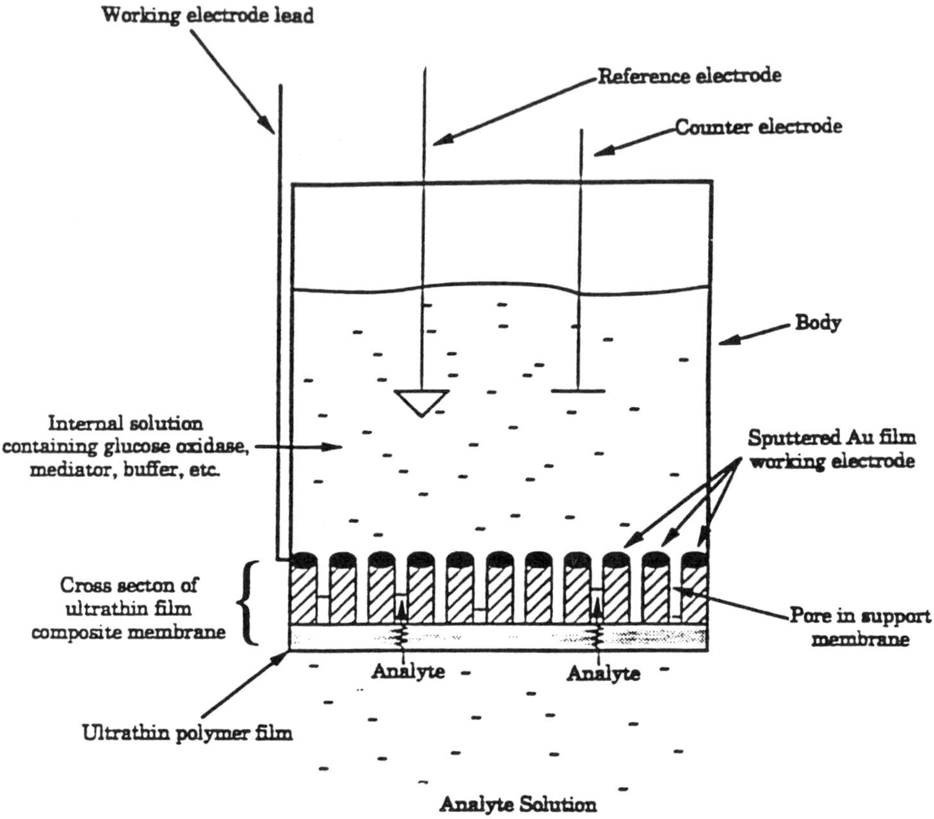

Working electrode lead
Reference electrode
Counter electrode
Body
Internal solution containing glucose oxidase, mediator, buffer, etc.
Sputtered Au film working electrode
Cross secton of ultrathin film composite membrane
Pore in support membrane
Analyte
Analyte
Ultrathin polymer film
Analyte Solution

Figure 1 Schematic diagram of the prototype ultrathin film composite membrane-based glucose sensor.

0.5
0.0
V (vs. Ag/AgCl)
A
200 µA
B

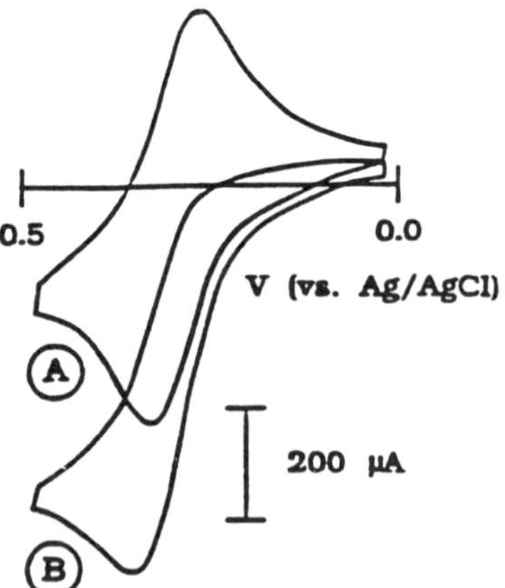

Figure 2 Cyclic voltammogram for the mediator (A) in the absence of glucose and (B) in the presence of glucose (7 mM).

Experimental

Device Fabrication. The support membrane for the ultrathin film composite was an Anopore (Alltech Associates) microporous alumina filter; these membranes are 55 µm thick, contain linear, cylindrical, 250 nm-diameter pores and are ca. 65 % porous (18). One face of the membrane was rendered electronically conductive by sputtering (19,20) an Au film (20-30 nm-thick) across the membrane surface. This Au film served as the working electrode; electrical contact was made by using Ag-epoxy to attach a Cu wire to the Au surface (19) (Figure 1). This Au film is too thin to block the pores at the membrane surface (5,21); this is important because analyte must pass through the membrane into the internal sensing solution (Figure 1).

The surface of the membrane opposite to the Au film was then coated with an ultrathin (ca. 50 nm) skin of a poly(dimethylsiloxane); this was accomplished by using a novel interfacial polymerization method developed in these laboratories (22). Briefly, the alumina support membrane is placed on a wet filter paper, which acts as a source of water vapor. The upper face of the membrane is then exposed to dimethlydichlorosilane vapor. This causes a thin skin of poly(dimethylsiloxane) to form across the upper surface of the membrane. The films used in the sensors described here were on the order of 50 nm in thickness.

We have found that in sensors based on this simple homopolymer, the mediator, FcC^-, leaches from the internal solution into the analyte solution. In order to mitigate this problem, the films were subsequently cross-linked and sulfonated. Cross-linking was accomplished by exposing the film to a 5 % (v/v) solution of trichloromethylsilane, in ethanol. Sulfonation was accomplished by exposing the film to a 2 % (v/v) solution of 2-(4-chlorosulfonylphenyl)ethyltrimethoxy silane, also in ethanol. These silanes attack the ends of the homopolymer chains and thus introduce the desired chemical functionality into the polymer film. The details of these materials science aspects of this sensor will be discussed in a future paper (22).

Finally, the Au/Anopore/polymer composite membrane was glued to the end of a glass tube (I.D. = 0.64 cm), which forms the body of the sensor (Figure 1). The internal solution - a small volume of pH 7.0 phosphate buffer (0.05 M) containing 200 units per mL of glucose oxidase (Sigma; Type VII) and saturated (ca. 3 mM) with FcC^- - was then added. A Ag/AgCl reference and gold wire counter electrode were immersed within this internal solution (Figure 1). As indicated in Figure 1, one of the beauties of this new approach for making sensors is that a totally self-contained sensing device is obtained; i.e. external electrodes are not required. In a sense, this design is like that of an ion-selective electrode (ISE) in that this glucose sensor contains an internal reference electrode. However, an ISE still usually needs an external reference electrode. The sensor design described here (Figure 1) requires no external electrodes.

Response Characteristics of the Prototype Sensor. The response characteristics of these devices were probed using a variety of experiments; the most straight forward was a simple calibration experiment. The sensor was placed in a known volume (30.0 mL) of pH = 7.0 phosphate buffer which initially contained no glucose. The potential of the Au film working electrode was then scanned through the mediator oxidation wave to obtain a cyclic voltammogram for the mediator confined within the internal solution. A typical voltammogram is shown as Curve A in Figure 2. Known volume increments of glucose solution were then added to the external "analyte" solution. The analyte solution was

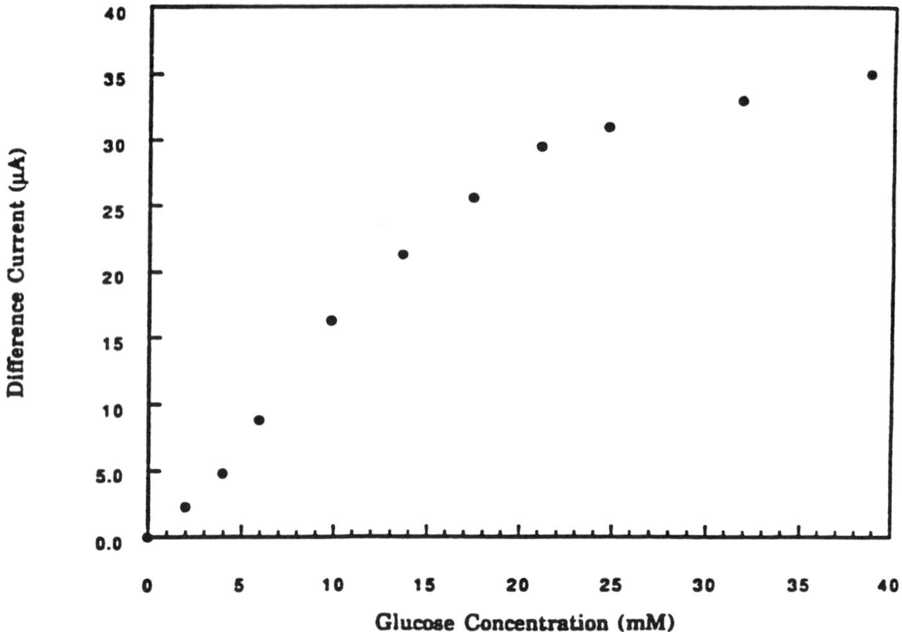

Figure 3 Calibration curve for the prototype ultrathin film composite glucose sensor. The regression coefficient for the linear region (up to 22 mM) is 0.995.

stirred continually. A voltammogram for the mediator was then obtained as before; a typical voltammogram is shown as Curve B in Figure 2. Note that the wave now has a catalytic appearance (9) due to the reaction between the oxidized mediator and $FADH_2$; this electrochemistry has been described in detail by others (11,17,24-26). Calibration curves were obtained by plotting the difference between the maximum currents in Curves A and B vs. the concentration of glucose in the analyte solution (Figure 3).

The response time of the device was also investigated. This was accomplished via a potential-step experiment. The sensor was placed in a vigorously-stirred buffer solution that was initially devoid of glucose. The potential of the Au film working electrode was stepped from 0 V (no mediator oxidation) to +0.5 V, where the oxidation of the mediator in the internal solution occurred at the diffusion-controlled rate. (All potentials are reported vs. Ag/AgCl). The resulting current-time transient associated with mediator oxidation was recorded on an X-Y recorder. The potential was then returned to 0 V to re-reduce the mediator. A second (identical) potential step was then conducted. However, in this case, the external (analyte) solution was spiked with glucose 4.5 sec. after initiating the potential step. The current-time transient was again recorded. The response time of the sensor was obtained from the difference between the two transients.

Oxygen present in the analyte solution presents a potential problem for enzyme/mediator-based sensors of this type. This is because O_2 can also oxidize $FADH_2$. Thus, if the O_2 concentration in the analyte solution changes during an analysis, the response of the device to glucose (via the mediator-oxidation route) will also change. Various schemes for circumventing this problem have been devised (10). We expected that our sensors would be less susceptible to changes in O_2 concentration in the analyte phase, because the electrochemistry occurs in the internal solution which is always exposed to air. To test this premise, we obtained voltammograms for the mediator with the sensor in contact with an air-saturated and a degassed analyte solution. As will be shown below, this sensor is nearly insensitive to changes in O_2 concentration in the analyte solution.

Results and Discussion

The primary function of the ultrathin film composite membrane in this simple prototype sensor is to confine the internal solution components. This necessitates that the ultrathin polymer film is completely defect-free. We have probed for defects in the films used here by looking for glucose oxidase in the analyte solution. If defects of the size of the glucose oxidase molecule (160,000 - 186,000 M.W. or ca. 4.3 nm in diameter) (27-29) were present in these films, the enzyme would leak freely into the analyte solution. However, no glucose oxidase could be detected (electrochemically) in the analyte solution, even after overnight exposure of the sensor to this solution. These results show that the ultrathin films used here have essentially no (i.e. an undetectable number of) defects that are larger than ca. 4.3 nm. While not evaluated in these preliminary investigations, this lack of transport of proteins should also be useful in keeping proteins that might be present in the analyte solution from entering the internal solution. Hence, the ultrathin film composite membranes used here have a rudimentary *sized-based selectivity*.

Chemical selectivity was also built into these films so as to minimize transport of the FcC^- through the film; this was accomplished by sulfonating and cross-linking the films. While this approach dramatically lowered the rate of FcC^- transport, trace concentrations of FcC^- could be detected (electrochemically) in the analyte solution after several hours of exposure of the sensor to this solution. This problem could be further mitigated, or perhaps eliminated, by increasing the molecular weight and the anionic charge of the mediator. Alternatively, the film chemistry could be changed. For example, recent work has shown that $FADH_2$ can give its electrons directly to the doped form of polypyrrole; i.e. a mediator is not necessary (30). We have recently developed an interfacial polymerization method to synthesize ultrathin film composite membranes based on polypyrrole and its derivatives (3). This creates the exciting possibility of making a mediator-free glucose sensor of the thin film composite type.

The mediator-based chemistry used here to detect glucose is well know and interference from other molecules, that might be present in the analyte solution, have been studied in detail (10). For this reason (and because the primary object of the work, to date, has been to provide proof of concept for a general sensor design) we have not yet investigated the effects of these interferences on the response of this prototype sensor. It is important to point out, however, that a number of these potential interferents are anionic (10). Because the rates of anion transport in the sulfonated film used here is low (vide supra), this film should provide some level of protection against these anionic interferents.
A calibration curve for glucose is shown in Figure 3. As is typical for enzyme-based sensors of this type, this curve shows a region of linear response at lower concentrations and a region of flattened response at higher concentrations (9,11,17,24-27,31). The reasons for these response characteristics have been discussed by others (9,11,17,24-27). The extent of the linear region can be

adjusted by varying the concentration of mediator and glucose oxidase in the internal solution. The calibration curve shown in Figure 4 is linear throughout the physiological concentration range for glucose (4 to 5 mM for healthy patients and up to 20 mM for diabetics) (32,33).

The effect of the concentration of O_2 in the analyte solution on the response of this prototype sensor is explored in Figure 4. Figure 4A shows voltammograms for the mediator when the electrode is in contact with a degassed analyte solution and Figure 4B shows analogous voltammograms when the electrode is in contact with an air-saturated analyte solution. In both cases mediator voltammograms for analyte solutions devoid of glucose (symmetrical wave) and in the presence of 2.0 mM glucose (catalytic wave) (9) are shown. As indicated previously, the analytical signal is just the difference between the maximum currents in the catalytic and conventional waves. In the air-saturated solution, this difference is 24.5 μA; in the degassed solution, this difference is 25.5 μA.

These data show that this new glucose sensor is nearly insensitive to changes in O_2 concentration in the analyte phase. Indeed, the change in O_2 concentration explored in Figure 4 represents the largest conceivable change (no O_2 to the air-saturated value) that could be encountered in, for example, an in vivo experiment. In spite of this large change in the concentration of O_2, the analytical signal changed by only 4 per cent (11,17). Analogous results were obtained at glucose concentrations as high as 30 mM. In essence, the O_2 present in the internal solution (where the electrochemistry occurs) acts as a buffer against changes in concentration of O_2 in the external solution. Furthermore, if the internal solution is exposed to air, the rapid equilibration of this solution with O_2 in the air insures a constant concentration of O_2 in the internal solution. If, for a particular application, the device needed to be hermetically sealed, one could envision introducing an O_2-generating electrode into the internal solution to insure a constant O_2 concentration in this solution.

Finally, one of the most important potential advantage of a sensor based on an ultrathin film composite membrane is fast response time. Figure 5 shows that extremely fast response is, indeed, observed with these simple prototype devices. The curve labeled "no glucose" is a current-time transient associated with a potential-step oxidation of the mediator within the internal solution, when the external (analyte) solution is devoid of glucose. A typical chronoamperometric decay of current with time is observed (34). The curve labeled "with glucose" is associated with an analogous potential step; however, in this case, the external solution was spiked with glucose 4.5 sec. after initiating the step. A steady-state signal to glucose is observed in less than two seconds. This is one of the fastest responses of any enzyme-based glucose sensor to be described in the literature to date (11,35-37).

Conclusions

We have demonstrated a new concept in chemical sensor design - sensors based on ultrathin film composite membranes. We believe that this design is generic in that it should be amenable to other molecular recognition schemes (38), other film chemistries (3-5,7), and other signal transduction processes (12-16). With regard to other film chemistries, we have developed four new interfacial polymerization methods for forming ultrathin film composite membranes (3-5,7). With these methods, films based on almost any conceivable chemistry could be

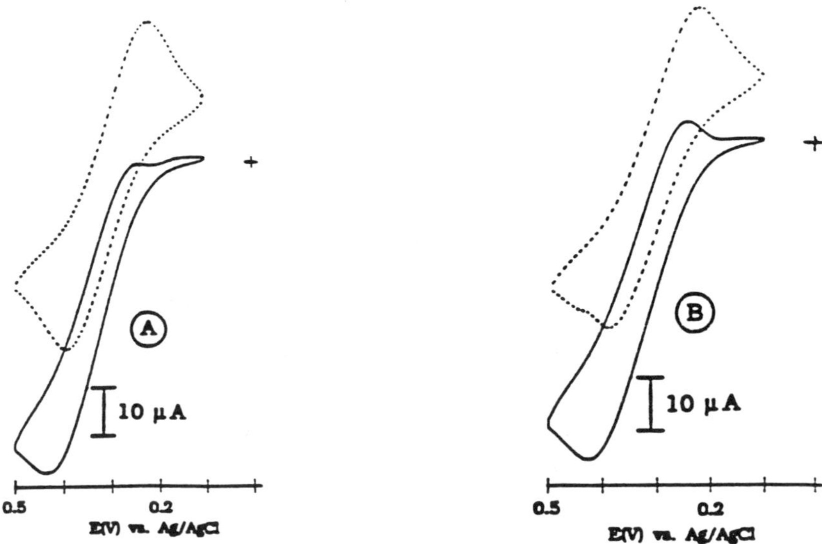

Figure 4 Cyclic voltammograms showing mediator voltammetric waves in deaerated (A) and air-saturated (B) analyte solutions.

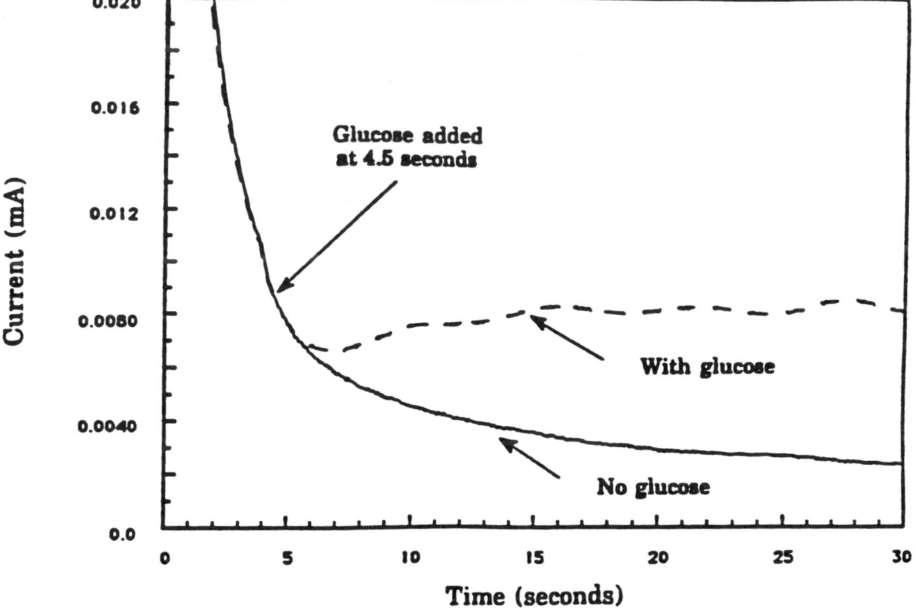

Figure 5 Chronoamperometric experiment showing the response time of the prototype glucose sensor.

fabricated for sensors of this type. Alternative transduction schemes might include fiber optic-based transducers (12- 16). For example, the molecular-recognition chemistry might produce a colored species which could be monitored by a fiber optic probe inserted into the internal sensing solution (16,39,40).

Furthermore, because the internal solution and transduction system can be easily removed from the body of the device (Figure 1), this sensor design creates the interesting prospect of using a single sensor body to make a variety of specific sensors. This design also allows for replenishing and sampling of the internal solution. This capability could be built into the device by adding solution inlet and outlet lines. The internal solution could also be subjected to additional chemical analyses to detect species which partitioned through the membrane. In this sense, the sensor would resemble a microdialysis device (41). Finally, sensors based on thin film composites should also be easy to miniaturize. One approach might be to coat the thin film onto a microporous hollow fiber. As part of our membrane separation work we are developing methods to coat such hollow fibers with ultrathin polymer films (43). We believe that the ultrathin film composite membrane is a promising and versatile approach for sensor design.

Acknowledgments

This work was supported by the Office of Naval Research and the Air Force Office of Scientific Research. Barbara Ballarin was supported by the Ministero Dell'Universita' e Della Ricerca Scientifica e Tecnologica (Rome).

Literature Cited

(1) Baker, R. W.; Cussler, E. L.; Eykamp, W.; Koros, W. J.; Riley, R. L.; Strathmann, H. *Membrane Separation Systems - A Research & Development Needs Assessment*; Department of Energy: Washington, D.C., 1990; *Vol. 1,2*; DE90-011770.

(2) Henis, J. M. S.; Tripodi, M. K. *J. Membr. Sci.* **1981**, *8*(3), 233-246.

(3) Liang, W.; Martin, C. R. *Chem. Mater.* **1991**, *3*(3), 390-391.

(4) Liu, C.; Martin, C. R. *Nature (London)* **1991**, *352*(6330), 50-52.

(5) Liu, C.; Espenscheid, M. W.; Chen, Wen-J.; Martin, C. R. *J. Am. Chem. Soc.* **1990**, *112*(6), 2458.

(6) Meares, P. In *Membranes in Gas Separation and Enrichment (4th BOC Priestly Conference)*; Williams, A., Ed.; The Royal Society of Chemistry: London, 1986; Vol. 62, pp. 1-25.

(7) Liu, C.; Chen, Wen-J.; Martin, C. R. *J. Membr. Sci.* **1992**, *65*, 113-128.

(8) Pickup, J. C.; Claremont, D. J. In *Hormone and Metabolic Research Supplement Series (Implantable Glucose Sensors - The State of the Art)*; Pfeiffer, E. F.; Kerner, W., Eds.; Georg Thieme Verlag Stuttgart: New York, 1988; Vol. 20, pp. 34-36.

(9) Foulds, N. C.; Lowe, C. R. *Anal. Chem.* **1988**, *60*(22), 2473-2478.

(10) Bindra, D. S.; Zhang, Y.; Wilson, G. S.; Sternberg, R.; Thevenot, D. R.; Moatti, D.; Reach, G. *Anal. Chem.* **1991**, *63*(17), 1692-1696.

(11) Hale, P. D.; Boguslavsky, L. I.; Inagaki, T.; Karan, H. I.; Lee, H. S.; Skotheim, T. A.; Okamoto, Y. *Anal. Chem.* **1991**, *63*(7), 677-682.

(12) Gunasingham, H.; Tan, Chin-H.; Seow, J. K. L. *Anal. Chem.* **1990**, *62*, 755-759.

(13) Schelp, C.; Scheper, T.; Buckmann, F.; Reardon, K. F. *Anal. Chim. Acta* **1991**, *255*, 223-229.
(14) Hughes, R. C.; Ricco, A. J.; Butler, M. A.; Martin, S. J. *Science* **1991**, *254*(October 4), 74-80.
(15) Krohn, D. A. *Fiber Optic Sensors: Fundamentals and Applications*, 1st ed.; Instrument Society of America: Research Triangle Park, NC, 1988.
(16) Rao, B. S.; Puschett, J. B.; Matyjaszewski, K. *J. Appl. Polym. Sci.* **1991**, *43*, 925-928.
(17) Cass, A. E. G.; Davis, G.; Francis, G. D.; Hill, H. A. O.; Aston, W. J.; Higgins, I. J.; Plotkin, E. V.; Scott, L. D. L.; Turner, A. P. F. *Anal. Chem.* **1984**, *56*, 667-671.
(18) Furneaux, R. C.; Rigby, W. R.; Davidson, A. P. *Nature* **1989**, *337*(6203), 147.
(19) Van Dyke, L. S.; Martin, C. R. *Langmuir* **1990**, *6*, 1118.
(20) Chapman, B. *Glow Discharge Processes*, 1st ed.; John Wiley & Sons, Inc.: New York, 1980.
(21) Colthup, N. B.; Daly, L. H.; Wiberley. *Introduction to Infrared and Raman Spectroscopy*, 3rd ed.; Academic Press: New York, 1990.
(22) Liang, W.; Stowe, S.; Martin, C. R. Unpublished results, March 1, 1992.
(23) Goldstein, J. I.; Newbury, D. E.; Echlin, P.; Joy, D. C.; Fiori, C.; Lifshin, E. *Scanning Electron Microscopy and X-Ray Microanalysis*, 1st ed.; Plenum Press: New York, 1981.
(24) Lange, M. A.; Chambers, J. Q. *Anal. Chim. Acta* **1985**, *175*, 89-97.
(25) Iwakura, C.; Kajiya, Y.; Yoneyama, H. *J. Chem. Soc., Chem. Commun.* **1988**, 1019-1020.
(26) Jonsson, G.; Gorton, L.; Petterson, L. *Electroanalysis* **1989**, *1*, 49-55.
(27) Degani, Y.; Heller, A. *J. Phys. Chem.* **1987**, *91*(6), 1285-1289.
(28) Swoboda, B. E. P.; Massey, V. *The Journal of Biological Chemistry* **1965**, *240*(5), 2209-2215.
(29) Nakamura, S.; Hayashi, S.; Koga, K. *Biochim. Biophys. Acta* **1976**, *445*, 294.
(30) Koopal, C. G. J.; de Ruiter, B; Nolte, R. J. M. *J. Chem. Soc., Chem. Commun*, 1991, 1691.
(31) Malitesta, C.; Palmisano, F.; Torsi, L.; Zambonin, P. G. *Anal. Chem.* **1990**, *62*(24), 2735-2740.
(32) Lehninger, A. L. *Principles of Biochemistry*, 1st ed.; Worth Publishers, Inc.: New York, 1982.
(33) Ouellette, R. J. *Introduction to General Organic and Biological Chemistry*, 1st ed.; Macmillan Press, Inc.: New York, 1984; p. 470.
(34) Bard, A. J.; Faulkner, L. R. *Electrochemical Methods: Fundamentals and Applications*, 1st ed.; John Wiley & Sons: New York, 1980.
(35) Shimizu, Y.; Morita, K. *Anal. Chem.* **1990**, *62*(14), 1498-1501.
(36) Bartlett, P. N.; Tebbutt, P.; Tyrrell, C. H. *Anal. Chem.* **1992**, *64*(2), 138-142.
(37) Pishko, M. V.; Michael, A. C.; Heller, A. *Anal. Chem.* **1991**, *63*(20), 2268-2272.
(38) Cammann, K. In *Hormone and Metabolic Research Supplemental Series (Implantable Glucose Sensors - The State of the Art)*; Pfeiffer, E. F.; Kerner, W., Eds.; George Thieme Verlag Stuttgard: New York, 1988; Vol. 20, pp. 4-8.
(39) Moreno-Bondi, M. C.; Wolfbeis, O. S.; Leiner, M. J. P.; Schaffar, B. P. H. *Anal. Chem.* **1990**, *62*(21), 2377-2380.

(40) Mullen, K. I.; Carron, K. T. *Anal. Chem.* **1991**, *63*(19), 2196-2199.
(41) Lunte, C. E.; Scott, D. O.; Kissinger, P. T. *Anal. Chem.* **1991**, *63*(15), 773A-780A.
(42) Martin, C. R.; Parthasarathy, A.; Brumlik, C. J.; Collins, G.; Li, L. Presented at the CU Center for Separations Using Thin Films, Boulder, CO, Jan. 1992.
(43) Ballarin, B.; Brumlik, C. J.; Lawson, D. R.; Liang, W.; Van Dyke, L. S. and Martin, C. R. *Anal. Chem.* **1992**, *64*, 2647-2651.

RECEIVED February 2, 1994

Chapter 14

Viologen Derivative Containing Polysiloxane as an Electron-Transfer Mediator in Amperometric Glucose Sensors

Hiroko I. Karan[1], Hsing Lin Lan[2,3], and Yoshiyuki Okamoto[2,4]

[1]Department of Physical Sciences and Computer Science, Medgar Evers College, City University of New York, Brooklyn, NY 11225
[2]Department of Chemistry and Polymer Research Institute, Polytechnic University, 6 Metrotech Center, Brooklyn, NY 11201

Viologen derivatives which have higher oxidation potentials than that of FAD can mediate electrons efficiently from the FAD centers of glucose oxidase to the electrode surface. The oxidation potential of viologen derivatives are lower than those of ferrocene derivatives or quinones and hence sensors can operate at potentials where the oxidation of common interferents in biological fluid are minimized. Since viologen derivative are water soluble, insoluble viologen-containing siloxane polymer was prepared for this study to prevent the mediator diffusing away from the electrode surface into the bulk media. Sensors constructed from this polymer and glucose oxidase efficiently mediated electron transfer and showed linear response in clinically relevant ranges of glucose concentrations.

Amperometric glucose sensors based on glucose oxidase and non-physiological redox mediators use the following mechanism to shuttle electrons between the reduced flavin adenine dinucleotide center of the enzyme ($FADH_2$) and the electrode:

$$\text{glucose} + \text{GO(FAD)} \longrightarrow \text{gluconolactone} + \text{GO}(\text{FADH}_2)$$

$$\text{GO}(\text{FADH}_2) + 2\text{M}_{ox} \longrightarrow \text{GO(FAD)} + 2\text{M}_{red} + 2\text{H}^+$$

$$2\text{M}_{red} \longrightarrow 2\text{M}_{ox} + 2e^- \quad \text{(at the electrode)}$$

GO (FAD) and GO ($FADH_2$) represent the oxidized and reduced forms of glucose oxidase, respectively. The mediator M_{ox}/M_{red} is assumed to be a one-electron

[3]This chapter is based on the Ph.D. thesis of Hsing Lin Lan, written in 1992. Current address: Institute of Biomedical Engineering, National Yang Ming Medical College, Taipei 11221, Taiwan
[4]Corresponding author

0097–6156/94/0556–0169$08.00/0

couple. Sensors based on ferrocene/ferricinium redox couple (*1-3*), various quinones (*4, 5*), tetrathiafulvalene-tetracyanoquinodimethane (*6 - 8*), tungsten complexes (*9*) and rhutenium complexes (*10*) have recently been reported. These mediators have fairly high oxidation potentials (e.g., +100 to +400 mV vs. the saturated calomel electrode (SCE)), however, and must be operated at potentials where several common interferents in biological fluids, such as ascorbic acid, uric acid and acetaminophen, are directly oxidized at the electrode (*11, 12*). This additional response current can make it difficult to accurately determine the glucose concentration in a sample. To minimize this interferent effect several research groups have modified the surface of the glucose sensors with anionic polymer coating (*13, 14*), which can partially exclude negatively charged ascorbate and urate ions from the electrode surface. More recently, some research groups preoxidized the samples by applying peroxidase films to the electrode surfaces to eliminate interferents (*15*). We have previously reported an effective method to avoid these elecrtrooxidizable consituents by using viologen (4,4'-bipyridyl) derivatives as mediators (*16, 17*). These mediators can be synthetically tailored to lower their oxidation potentials sufficiently to eleminate the electrooxidation of interferents. However, viologen derivatives are water soluble and sensors based on these mediators suffer the inherent drawback that soluble mediating species diffuse away from the electrode surface into the bulk solution. This precludes the use of these sensors for on line systems or *in vivo* systems. Recently, studies have included systems in which mediating redox moieties are covalently attached to polymers such as polypyrrole (*18-20*), poly(vinylpyridine) (*21-23*), and polysiloxane (*24-27*). Here, we have synthesized a viologen- containing siloxane polymer, by covalently attaching a viologen derivative to siloxane copolymer. The electrochemical properties of the polymer were investigated and comparison of electron mediation efficiency among previously reported viologen derivatives and the viologen-containing polysiloxane was made.

Experimental

Materials

Methylhydro(55%)-dimethyl(45%)siloxane copolymer was obtained from Petrarch Systems (Bristol, PA). 4-Nitrobenzyl chloride, 4,4′-bipyridyl and chloromethyl-styrene were obtained from Aldrich (Milwaukee, WI). Graphite powder (product No. 50870) and paraffin oil (product No. 76235) were obtained from Fluka (Ronkonkoma, NY). Glucose oxidase (E.C. 1.1.3.4, type VII, from Aspergillus Niger) and glucose were obtained from Sigma (St. Louis, MO). Glucose solutions were prepared by dissolving appropriate amounts in 0.1M phosphate- 0.1M KCl buffer (PH 7.0); the glucose was allowed to reach mutarotational equilibrium before use (ca. 24 hr). All other chemicals were of reagent grade and were used as received.

Synthesis of a viologen-containing polymer

N-4-Nitrobenzylviologen was prepared by reacting 1:1 molar ratio of 4,4′-bipyridyl and p-nitrobenzyl chloride in acetonitrile; the solution was refluxed for 40 hr in a nitrogen atmosphere. The resulting precipitate was removed by filtration, purified by recrystallization from methanol-acetonitrile and dried under vacuum at 60 °C. The viologen-containing polysiloxane was prepared by reacting p-chloromethylstyrene (7.6g, 0.05 mole) with methylhydro(55%)-dimethyl(45%)-siloxane copolymer (6.1g) in the presence of H_2PtCl_6 and trace amount of hydroquinone in THF under nitrogen atmosphere at reflux temperature. After the hydrosilation reaction, the resulting viscous polymer (2.2g) was reacted with 1.2g of N-4-nitrobenzylviologen in 150 mL DMSO at 90 °C for 24 hr (Scheme I). The polymer obtained was precipitated by pouring the reaction mixture into chloroform and then purified by washing with water. The monomer and polymer synthesized were characterized by IR and NMR measurements.

Electrode construction
A modified carbon paste for the sensors was made by thoroughly mixing 50 mg of graphite powder with a measured amount of the viologen-containing polymer, the molar amount of viologen moiety was the same for all electrodes (36 μmol of viologen per gram of graphite powder). 5 mg of glucose oxidase and 10 μL of paraffin oil were then added, and the resulting mixture was blended into paste. The carbon paste electrode was constructed using a 1.0 mL plastic syringe which had previously been partially filled with unmodified carbon paste, leaving approximately a 2 mm deep well at the base of the syringe to be filled by the modified carbon paste. The electrode were polished by rubbing gently on piece of weighing paper, which produced a flat shiny surface with an area of ca. 0.025 cm^2. Electrical contact was achieved by inserting a silver wire into the top of the carbon paste.

Electrochemical methods
Cyclic voltammetry and stationary potential measurements were performed using a Princeton Applied Research Potentiostat (Model 173), a Universal Programmer (Model 175) and a Western Graphitec X-Y-T recorder (Model WX 2400). All experiments were carried out in a conventional electrochemical cell containing 0.1M phosphate buffer (PH 7.0) with 0.1M KCl at 23 ± 2 °C. All solutions were thoroughly deoxygenated by bubbling nitrogen through the solution for at least 10 min; a gentle flow of nitrogen was also used to facilitate stirring. In addition to the modified carbon paste working electrode, a Ag/AgCl reference electrode and a platinum wire auxiliary electrode were employed. After application of the potential, the background current was allowed to decay to a constant value (@ 10 min.) before samples of a stock glucose solution were added to the buffer solution.

Results and Discussion

Cyclic voltammetry
The electrochemical behavior of viologen derivatives has been studied in detail over several decades(*28*). The dibenzylviologens (DBV) are reported to have PH-independent redox potentials, E°´, of ca. - 0.5V ($DBV^{2+/+}$) and - 0.9V ($DBV^{+/0}$) vs. SCE (28, 29). Since the redox potential of FAD is substantially more positive (- 460mV vs. SCE using carbon paste electrodes) (*17*) than that of DBV(-510mV vs. SCE) (*17*) , it should be incapable of reoxidizing the reduced flavin. We have previously reported that placing electron-withdrawing substituents on the benzyl group of DBV or carrying out polymerization of viologen derivatives cause a substantial anodic shift in the E°´ value (*17*) (Table I). Figure 1 shows the

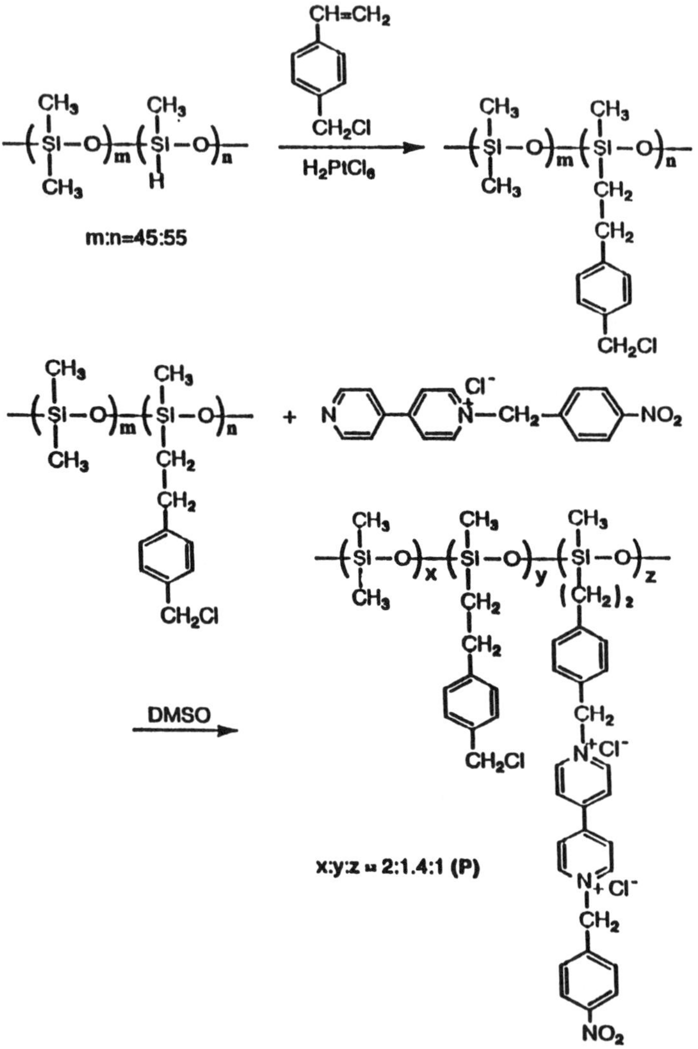
CH=CH2
CH2Cl
H2PtCl6
m:n=45:55
DMSO
x:y:z = 2:1.4:1 (P)

Scheme I Synthesis of viologen-containing polysiloxane.

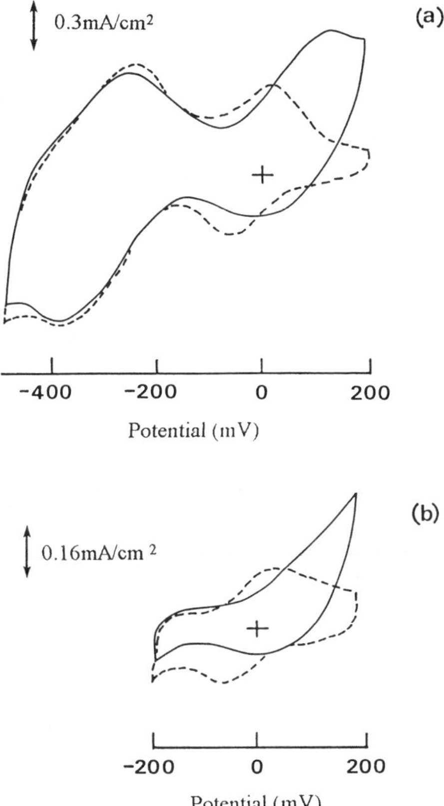

Figure 1 Cyclic voltammograms for viologen-containing polysiloxane (P)/ glucose oxidase/ carbon paste electrode (scan rate: 2mV/s) in PH 7.0 phosphate buffer (with 0.1M KCl) solution with no glucose present (dashed line) and in the presence of 0.1M glucose (solid line); scan range: (a) -500mV to +200mV; (b) -200mV to +200mV.

cyclic voltammogram for carbon paste electrodes (vs.Ag/AgCl) containing glucose oxidase and the viologen-containing polysiloxane (P). The voltammogram (Figure 1a) indicates two redox potentials at + 25mV and -240mV. Upon addition of glucose, the response of increasing oxidation current was only shown at the redox potential + 25mV. This was apparent when the scanning range was limited to the potentials from - 200mV to + 200mV (Figure 1b). This indicated that covalently attaching the viologen derivative to a flexible polymeric back bone such as polysiloxane increase anodic E°′ (Table I) and hence was capable of reoxidizing the reduced flavin centers of glucose oxidase.

Stationary potential measurements
Our previous study showed that sensors constructed with viologen derivatives (B, C, and D) and glucose oxidase efficiently responded to 31.5 mM glucose over a wide range of steady state applied potentials (from - 200mV to +300 mV) (Figure 2). These sensors also showed no response to ascorbate or urate at an applied potential of - 200mV vs. SCE, while at - 100mV there is a small response to ascorbate (Figure 3). Typical glucose calibration curves for B, C and D, and for polymer P are shown in Figures 4 and 5, respectively. It is apparent that at the potential - 100mV the monomeric mediator B is considerably more efficient than the polymeric mediators C and D. Perhaps attributable the greater diffusional mobility of the smaller molecule B compared with the bulky polymeric molecules. Sensors constructed with polymer P and glucose oxidase clearly showed linear response to clinically relevant glucose concentrations at the potential of -100mV vs. A/AgCl and electron mediation efficiency was similar to the polymeric viologen D. The time required to reach 95% of the steady-state current was slightly longer than 1 min., which makes this mediator less efficient than mediators such as ferrocene. However electron mediation efficiency may be improved by systematically tailoring the polymeric back bone and the pendant chain as was reported for ferrocene modified polymeric electron transfer mediators (*26*). This is currently under investigation in our laboratory.

Apparent Michaelis-Menten constants
The linear response range of the sensors can be estimated from a Michaelis-Menten analysis of the glucose calibration graphs in Figures 4 and 5. The apparent Michaelis-Menten constant K_M^{app} can be determined from the electrochemical Eadie-Hofstee form of the Michaelis-Menten equation (*30*):

$$j = j_{max} - K_M^{app} (j/C)$$

where j is the steady- state current density, j_{max} is the maximum current density measured under conditions of enzyme saturation and C is the glucose concentration. Figure 6 show the typical electrochemical Eadie Hofstee plot of the polymer P and Table II shows the apparent Michaelis-Menten constant determined from electrochemical Edie-Hofstee plots. Sensor response clearly begins to deviate from linearity even at very low glucose concentrations. The linear range of a similar sensor can be increased substantially (K_M^{app} > 200mM) by using an anion-exchange polymer coating on the electrode surface (*14*) and these coatings can also exclude anionic interferents from the surface of the electrode and can prevent sensor fouling in biological fluids (*14*).

Table I. Redox potential obtained from cyclic voltammetry at a carbon paste working electrode in 0.1M phosphate buffer with 0.1M KCl, PH 7.0; scan rate, 5 mV/s; concentration of viologen derivatives expcept polymer P, 10mM; concentration of FAD, 1 mM.

Species		$E^{\circ\prime}$ (mV vs. SCE)[a]	ΔE_p (mV)[a]
N, N′-dibenzyl viologen	(A)	- 510	140
N, N′-di(4-nitrobenzyl)viologen	(B)	-370	150
poly(o-xylylviologen dibromide)	(C)	-358	115
poly(p-xylylviologen dibromide)	(D)	-398	165
polymer P		~ -68 (-18*)	85*
FAD		-460	100

a $E^{\circ\prime}$ is taken as the average of the oxidation and reduction peak potentials; ΔE_p is the difference between the oxidation and reduction peak potentials.
* scan rate, 2 mV/s vs. Ag/AgCl.

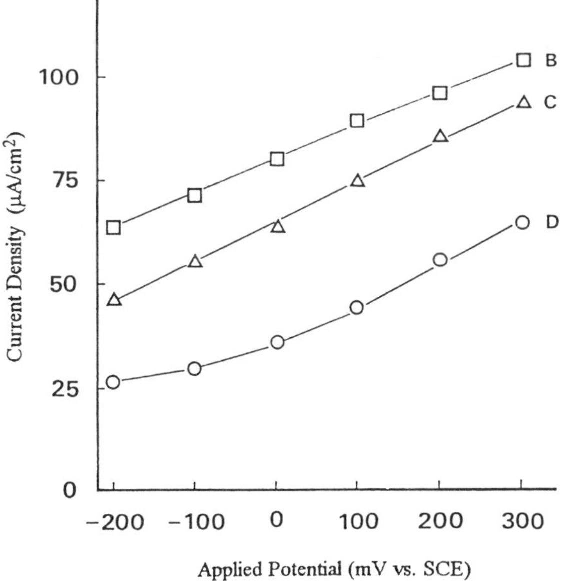

Figure 2 Steady state current response to 31.5mM glucose for the viologen-glucose oxidase-carbon paste electrodes for several applied potentials. Each point is the mean result for five electrodes. The mediators are indicated next to each curve. Reproduced with permission from reference 17.

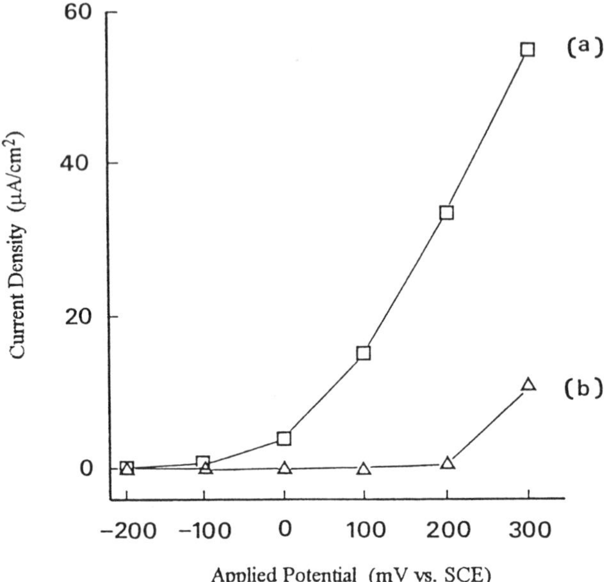

Figure 3 Steady state response of the N.N´-(4-nitrobenzyl)viologen-glucose oxidase-carbon paste electrode to (a) 10mM ascorbic acid and (b) 0.5mM uric acid at several applied potentials. Each point is the mean result fro five electrodes. Reproduced with permission from reference 17.

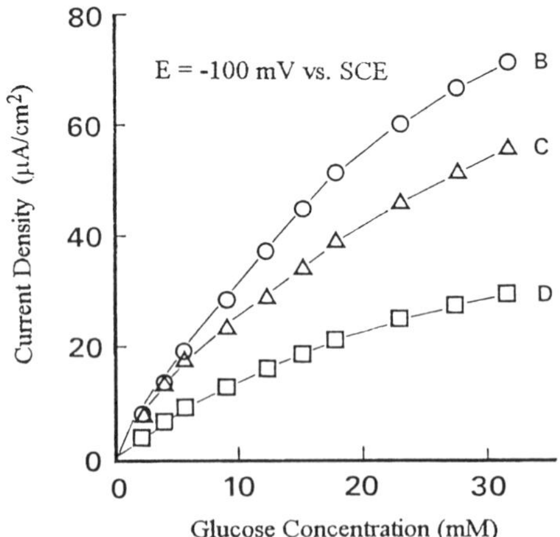

Figure 4 Glucose calibration graphs for the viologen-glucose oxidase-carbon paste electrodes at E = -100mV vs. SCE. Each point is mean result for five electrodes. The mediators are indicated next to each curve. Reproduced with permission from reference 17.

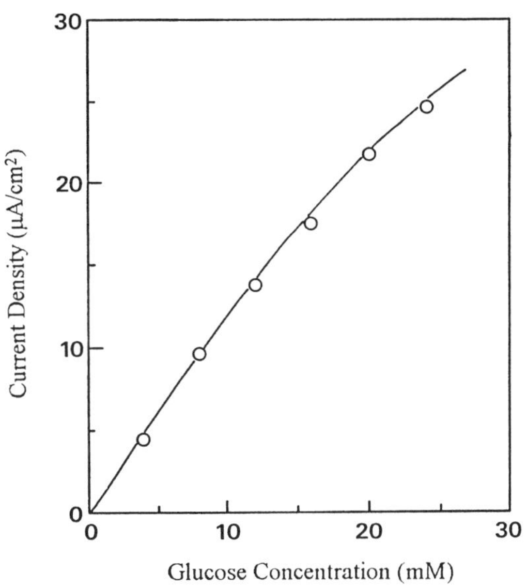

Figure 5 Glucose calibration curves for the viologen-siloxane polymer (P)/ glucose oxidase/ carbon paste electrodes at E = -100mV (vs. Ag/AgCl).

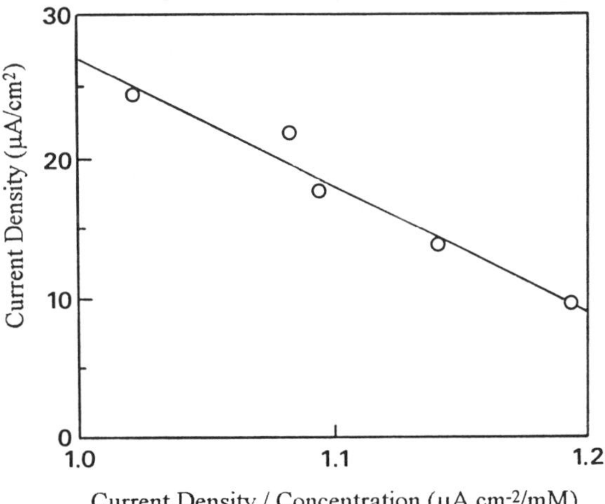

Figure 6 Typical electrochemical Eadie-Hofstee plot for the viologen-derivative containing polysiloxane/ glucose oxidase/carbon paste electrodes at E = -100mV (vs. Ag/AgCl).

Table II. Apparent Michaelis-Menten constants determined from electrochemical Eadie-Hofstee plots

Mediator	E = - 100mV vs. SCE	
	K_M^{app} (mM)	j_{max} (μA/cm^2)
B	45.0	176
C	18.1	79.4
D	24.2	50.5
polymer P	95*	121*

* E = - 100mV vs. Ag/AgCl

Conclusion
A viologen- derivative containing polymer was synthesized by covalently attaching N-4-nitrobenzylviologen to siloxane copolymer and the resulting polymer was water insoluble. The cyclic voltammogram of the sensor made from this polymer and glucose oxidase showed the redox potential, E°´, -18mV vs. Ag/AgCl. This is sufficiently positive to reoxidize FAD centers of glucose oxidase that have been reduced by reaction with glucose. The sensor efficiently mediated electron transfer from FAD centers in glucose oxidase to the surface of a conventional carbon paste electrode at an applied potential of -100mV where common biological interferents such as ascorbate, urate and acetaminophen are minimal. A linear response in clinically relevant ranges of glucose concentration was demonstrated.

Literature Cited

1. Cass, A.E.G., Davis, G., Francis, G.D., Hill, H.A.O., Aston, W. J., Higgins, I. J., Plotkin, E.V., Scott, L.D.L. and Turner, A. P.F., *Anal. Chem.*, **1984,** 56, 667.
2. Lange, M.A. and Chambers, J.Q., *Anal. Chim. Acta,* **1985,** 175, 89.
3. Wang, J., Wu, L. H., Lu, Z., Li, R., and Sanchez, J., *Anal. Chim. Acta,* **1990,** 228, 251.
4. Ikeda, T., Hamada, H., and Senda, M., *Agric. Biol. Chem.*, **1986**, 50, 883.
5. Kulys, J. J., and Cénas, N. K., *Biochim. Biophys. Acta,* **1983**, 744, 57.
6. Albery, W.J., Bartlett, P.N. and Craston, D.H., *J. Electroanal. Chem.*, **1985,** 194, 223.
7. Hale, P.D. and Wightman, R.M., *Mol. Cryst. Liq. Cryst.*, **1988**, 160, 269.
8. Gunasingham, H. and Tan, C.H., *Analyst,* **1990**, 115, 35.
9. Taniguchi, I., Miyamoto, S., Tomimura, S. and Hawkridge, F.M., *J. Electroanal. Chem.*, **1988,** 240, 33.
10. Crumbliss, A.L., Hill. H.A.O. and Page, D.J., *J. Electroanal. Chem.*, **1986,** 206, 327.
11. Wang, J. and Shan Lin, M., *Anal Chem.*, **1988,** 60, 499.
12. Pettersson, M., *Anal. Chim. Acta,* **1983**, 147, 359.

13. Harrison, D.J., Turner, R.F.B. and Bates, H.P., *Anal. Chem.*, **1988**, 60, 2002.
14. Gorton, L., Karan, H.I., Hale, P.D. Inagaki, T., Okamoto, Y. and Skotheim, T.A., *Anal. Chim. Acta*, **1990**, 228, 23.
15. Heller, A., *J. Phys. Chem.*, **1992**, 96, 3579.
16. Hale, P.D., Boguslavsky, L.I., Skotheim, T.A., Karan, H.I. , Lan, H.L. and Okamoto, Y., *Mol. Cryst. Liq. Cryst.*, **1990,** 190, 259.
17. Hale,P.D., Boguslavsky, L.I., Karan, H.I., Lan, H.L.. Lee, H.S., Okamoto, Y. and Skotheim, T.A., *Anal. Chim. Acta,* **1991**, 248, 155.
18. Belanger, D., Nadrean, D. and Fortier, J., *J. Electroanal. Chem.*, **1989,** 274, 143.
19. Yabuki,S., Shinohara, H. and Aizawa, M., *J. Chem Soc., Chem. Comm.*, **1989,** 945.
20. Trojanavicz, M., Matuszewski, W. and Podsiadla, *Biosensors Bioelectron.*, **1990**, 5, 149.
21. Degani, Y. and Heller, A., *J. Am. Chem. Soc.*, **1989**, 11, 2357.
22. Gregg, B.A. and Heller, A., *Anal. Chem.*, **1990**, 62, 258.
23. Pishko, M.V., Katakis, I., Lindquist, S.E., Ye, L., Gregg, B.A. and Heller, A., *Angew. Chem., Ed. Engl.*, **1990,** 29, 82.
24. Hale, P.D., Inagaki, T., Karan, H.I., Okamoto, Y. and Skotheim,. T.A., *J. Am. Chem. Soc.*, **1989**, 111, 3482.
25. Hale, P.D., Inagaki, T., Lee, H.S., Karan, H.I., Okamoto, Y. and Skotheim, T.A., *Anal. Chim. Acta,* **1990**, 228, 31.
26. Hale, P.D., Boguslavsky, L.I., Inagaki, T., Karan, H.I., Lee, H.S., Skotheim, T.A. and Okamoto, Y., *Anal. Chem.*, **1991**, 63, 677.
27. Karan, H.I., Hale, P.D., Lan, H.L., Lee, H.S., Liu, L.F., Skotheim, T.A. and Okamoto, Y., *Polym. For Adv. Technol.*, **1992**, 2, 229.
28. Bird, C.L. and Kuhn, A.T., *Chem. Soc. Rev.*, **1981**, 10, 49.
29. Sato, H. and Tamamura, T., *J. Appl. Polym. Sci.*, **1979,** 24, 2075.
30. Kamin, R.A. and Wilson, G.S., *Anal. Chem.*, **1980**, 52, 1198.

RECEIVED March 7, 1994

Chapter 15

Hydrogen Peroxide Electrodes Based on Electrical Connection of Redox Centers

of Various Peroxidases to Electrodes through a Three-Dimensional Electron-Relaying Polymer Network

Mark S. Vreeke and Adam Heller

Department of Chemical Engineering, The University of Texas at Austin, Austin, TX 78712

Hydrogen Peroxide has been shown to be efficiently electroreduced at an electrode modified with a hydrophilic, permeable film of horseradish peroxidase covalently bound to a 3-dimensional epoxy network having polyvinyl pyridine (PVP)-complexed $[Os(bpy)_2Cl]^{+2/+3}$ redox centers.[1] Four peroxide sensing cathodes based on peroxidases from Arthromyces ramosus, horseradish and bovine milk are compared. Their sensitivity at 0.0V (SCE) ranges from 0.1 - 1.0 A cm^{-2} M^{-1}, and their limiting currents relate to the enzyme's ability to complex with the redox epoxy network.

Electrochemical and optical hydrogen peroxide detection forms the basis for several medical diagnostic tests. Electrochemical detection offers the advantages of smaller required sample size and ease of integration into a flow system. A common electrochemical scheme uses an oxidase to catalyze the selective conversion of substrate equivalents to H_2O_2 equivalents. This conversion is followed by amperometric assay of the H_2O_2, e.g. by its oxidation on platinum at 700mV (SCE). At 700mV (SCE) electrooxidation of various reducing species in the biological samples can interfere with the assay.

Peroxidase enzymes (POD) catalyze the reduction of H_2O_2 by electron donors (HA) in the following reaction

$$2HA + H_2O_2 \xrightarrow{(POD)} 2H_2O + 2A \tag{1}$$

Amperometric peroxidase based H_2O_2 sensors have been made by using fast reversible redox couples (see Tables I and II). In these, the reducing member of the redox couple (essentially species HA in reaction 1) donates electrons to H_2O_2 and is oxidized (reaction 2)

0097–6156/94/0556–0180$08.00/0

Table I. Amperometric H_2O_2 Sensors Based on HRP Modified Electrodes

Electrode Surface	Mediator or Redox Matrix	Electrode Potential[a]	Sensitivity $Acm^{-2}M^{-1}$	Linear range mM	Comments	Reference
Glassy carbon	None	0.0	10^{-2}	-----	HRP covalently bound to a hydrophilic epoxy network. Polyvinyl pyridine-derived polyamine crosslinked with PEGDE.	1
Glassy carbon	Osmium bipyridine	0.0	1	0.1-100	HRP covalently bound to a hydrophilic redox epoxy network crossinked with PEGDGE	1
Spectrographic graphite	None[c]	0.05	0.175	0.1-500	BSA with glutaraldehyde cross-linking	2
Carbon paste	O-phenylene-diamine[d]	-0.15	N.A.[b]	3.1-200	Butanone peroxide was used as the substrate	3
Pt	Hexacyanoferrate (.01M)[d]	-0.05	N.A.[e]	5-1700	HRP was immobilized onto a nylon net	4
NMP^+TCNQ^-	None[f]	0.05	.168	-----	HRP entrapped with dialysis membrane	5
SnO_2	ferrocene carboxylic acid[d]	0.2	.04	0.01-1	HRP immobilized with glu-taraldehyde	6
Carbon paste	ferrocene[d]	0.05	N.A.[b]	0.1-10	Nafion coating was applied to the electrode to prevent loss of mediator	7
Graphite foil	potassium hexa-cyanoferrate(II)[d]	-0.02	0.03	< 600	Electrolyte was dioxane with 15% aqueous buffer	8

Continued on next page

Table I. Continued.

Carbon fiber	None	Note g	-----	40-5000	The biotin/avidin complex was used to obtain a surface layer of HRP	9
Pt, organic metal, or glassy carbon	potassium ferrocyanide	-----	N.A.[h]	-----	Membrane with albumin and glutaraldehyde	10
Spectrographic graphite or Carbon film	hexacyano-ferrate (II)[d]	0.0	N.A.[e]	0.1-1000	HRP immobilized on arylamino-derivatized controlled-pore glass, packed into a flow through reactor	11
Amino silylated glassy carbon	hexacyano-ferrate (II)[d]	0.0	N.A.[h]	-----	Glycerophosphate oxidase, HRP and BSA were covalently cross-linked on the glassy carbon surface.	12
Glassy carbon	hexacyano-ferrate (II)[d]	0.0	N.A.[h]	-----	Albumin, glutaraldehyde, HRP and oxidase (xanthine, uricase, glucose) matrix held close to the electrode with a dialysis membrane.	13
Gold or graphite	Several[i]	Note i	2.0[j]	0.05-6[j]	HRP was free in solution	14

a) potential vs SCE
b) macroporous electrode
c) uncertainty as to whether surface species created during electrode pretreatment are mediating
d) freely diffusing mediator
e) flow system
f) probably mediated by soluble component of organic metal or reaction product of organic metal
g) cyclic voltametry used to provide selective detection of oxygen generated by autocatalytic decomposition of hydrogen peroxide
h) HRP incorporated into a bienzyme system
i) mediators used and redox potential are: $[Ru(NH_3)_5py](ClO_4)_3$ = +28 $CpFeC_2B_9H_{11}$ = -80 aminomethylferrocene = +309 (2-aminoethyl)ferrocene = +185 ferrocene monocarboxylic acid = +275 1,1' dimethyl-3-(2-aminoethyl) ferrocene = +75
j) best reported result for ferrocene monocarboxylic acid

Table II. Amperometric H_2O_2 sensors utilizing biocatalysts other than HRP

Electrode Surface	Mediator or Redox Matrix	Electrode Potential[a]	Sensitivity $Acm^{-2}M^{-1}$	Linear range mM	Comments	Reference
Spectrographic graphite	None[b]	0.1	1	.27-2.46	Fungal peroxidase from Arthromyces ramosus immodilized with a difunctional carbodiimide	15
Pyrolytic graphite edge	None[d]	<.55	0.12	<70	Cytochrome C peroxidase was used free in solution	16
Gold	1,3'-dimethyl ferrocene ethylamine[c]	-0.1	1.76	-----	Cytochrome C peroxidase was immobilized on a nylon membrane	17
SnO_2	None	0.2	.0009	1-500	Heme nonapeptide (MW 1600)	18
Edge-oriented pyrolytic graphite	None	0.0	3.0	-----	Cytochrome C peroxidase (MW 34.1kDa) was adsorbed onto the surface. Electrode polishing protocol was found to impact on the stability and reproducibility of the electrocatalytic response.	19
Gold	None	-----	-----	-----	The surface was modified with methyl viologen. HRP reduction below -.7v and oxidation above -.4 to -.6v	20

a) potential vs SCE
b) uncertainty as to whether surface species created during electrode pretretment are mediating
c) freely diffusing mediator
d) aminoglycosides gentamycin or neomycin were used it bind cytochrome C peroxidase and facilitate electron transfer

$$2H^{+} + 2\ \text{redox couple}_{(red)} + H_2O_2 \xrightarrow{(POD)} 2H_2O + 2\ \text{redox couple}^{+}_{(ox)} \quad (2)$$

The oxidized redox couple is then cathodicly reduced at the electrode surface (reaction 3).

$$\text{redox couple}^{+}_{(ox)} \xrightarrow{\text{electrode at reducing potential}} \text{redox couple}_{(red)} \quad (3)$$

Horseradish peroxidase (HRP) is the most commonly used peroxidase for diagnostic testing (Table I). Other peroxidases (Table II) are used less frequently because they are less easily available or cost more. Tables I and II show the detection schemes vary in their method of immobilization of mediator and enzyme. At one extreme one finds systems based on the direct transfer of electrons from the electrode surface through surface bound mediators to HRP redox centers contacting the surface. At the other extreme one finds systems with freely diffusing mediators and enzyme.

Experimental

Electrode Construction. Horseradish peroxidase (HRP), lactoperoxidase (LOP), and peroxidase from the microorganism Arthromyces ramosus (ARP) were purchased from Sigma. Poly(ethylene glycol 400 diglycidyl ether), technical grade (PEGDGE) was purchased from Polysciences (Catalog No. 8211). The osmium redox polyamine (Figure 1) was synthesized as described previously.[21] Rotating disk electrodes were made using 3mm diameter vitreous carbon rods. These were press fitted into a Teflon rotating assembly leaving exposed a 0.07068 cm^2 surface of vitreous carbon. Electrode coatings were prepared by mixing the desired ratio of enzyme, polymer, and crosslinker solutions. An aliquot of the resulting cocktail was applied to the electrode and allowed to cure for a minimum of 24 hours at room temperature. Typical surface coverage ranged from 10 to 100 $\mu g\ cm^{-2}$.

Electrode Operation. The electrodes were operated at room temperature in modified Dulbecco's buffer at pH 7.4. A standard 3 electrode cell was used with a Pt counter electrode and a saturated calomel as reference. The chronoamperometric experiments were performed on an EG & G potentiostat/galvanostat Model 173 and recorded on a Kipp & Zonen XY recorder Model BD91. The cyclic voltammograms were run on an EG & G potentiostat/galvanostat Model 273A and recorded on a PC with software developed in this lab. The rotator used was a Pine Instruments AFMSRX with the ACMDI 1906C shaft.

Isoelectric Focusing. Isoelectric focusing experiments were performed using a Pharmacia multidrive XL power supply and a BioRad horizontal electrophoresis cell.

The focusing medium was an agrose gel supported on Pharmacia Gel Bond PAG film. Gels were stained using a 0.2% Coomassie brilliant blue solution.

Results and Discussion

H_2O_2 Detection based on 3-Dimensional Redox Epoxy Networks. Here we describe electrodes based on peroxidases from horseradish (HRP), Arthromyces ramosus (ARP), and bovine milk (LOP) immobilized in a three dimensional redox epoxy hydrogel on a current collector. In the case of HRP we used either the purified native enzyme or its sodium periodate oxidized derivative (HRPox). The redox hydrogel was made of a poly(4-vinyl pyridine) backbone partially complexed by osmium bipyridine redox centers and ethylamine groups. The osmium centers are electron donors (species HA in reaction 1), and serve to relay electrons from the electrode to the reactive centers of the peroxidase (Figure 2). The ethylamine groups allow the cross linking of the enzyme containing polymer film with a polyethylene glycol diglycidyl ether (reaction 4).

$$R'-NH_2 \; + \; \underset{\text{(epoxide, O bridging C–C)}}{C-C-R''} \longrightarrow R'-\underset{H}{N}-\overset{H}{\underset{|}{C}}-\overset{OH}{\underset{|}{C}}-R'' \qquad (4)$$

The three dimensional redox epoxy offers some of the advantages of both the freely diffusing systems and the immobilized systems. As in freely diffusing mediator based systems, not only the electrode adsorbed enzyme molecules, but also those remote from the electrode, yet connected by the redox polymer are electroactive. At the same time, there is no need to add mediator to the sample, and the mediator can not leach out or contaminate the sample.

POD enzymes are rather small (40 ≈ kd), and their active groups are positioned relatively close to the enzyme surface. This allows direct electron transfer between the POD enzymes and the electrodes. Figure 3 shows the dependence of the current on potential for HRP immobilized in a hydrogel similar to the electron relaying one but without the electron relaying osmium redox sites. Here only those enzyme molecules actually adsorbed on the electrode surface in electrical contact with it contribute to the signal. In contrast, when the enzyme is immobilized in the hydrogel with electron relaying osmium centers, there is a hundredfold increase in current, because enzyme molecules contacting relays are also "wired". (Figure 4)

Peroxidase Sensor Response. The dependence of the current response on H_2O_2 concentration for optimized HRP, HRPox, ARP, and LOP cathodes is shown in figure 5. The ARP and HRPox cathodes show a linear range from 0.1 to 100μM H_2O_2. The limit of detection for HRPox is 10nM. Though the ARP cathode is slightly less sensitive then the HRPox cathode, it has the advantage of increased dynamic range. The LOP cathode does not exhibit the good sensor response of the other enzymes. Its linear range is the narrowest, and the sensor response deteriorates more rapidly than the other sensors. The HRP based sensor shows only a 10% loss

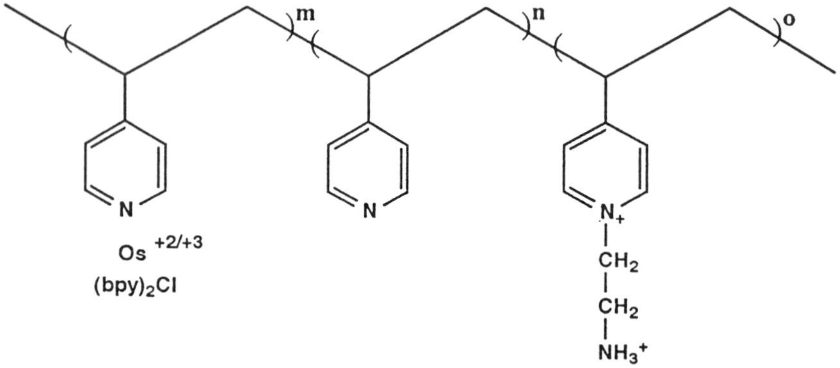

Figure 1. Composition of the electron-relaying redox polymer (m=1; n=3.35; o=0.6). After cross-linking with PEGDGE, it forms a 3-dimensional network that is able to relay electrons to covalently bound peroxidase.

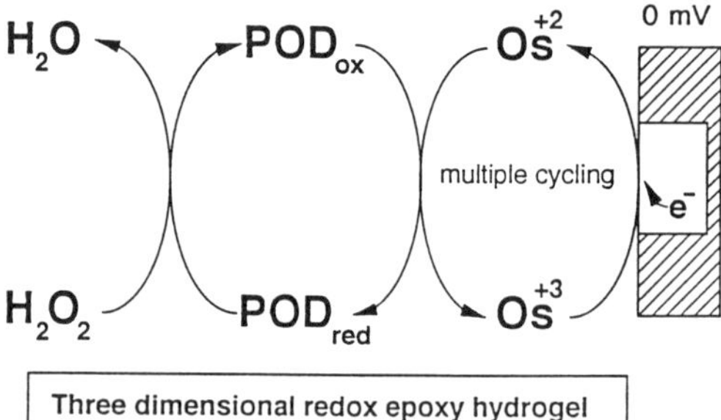

Figure 2. Redox cycles occurring in the 3-dimensional redox epoxy hydrogel. POD represents any of the following enzymes: native horseradish peroxidase, $NaIO_4$ treated horseradish peroxidase, lactoperoxidase, or Arthromyces ramosus peroxidase.

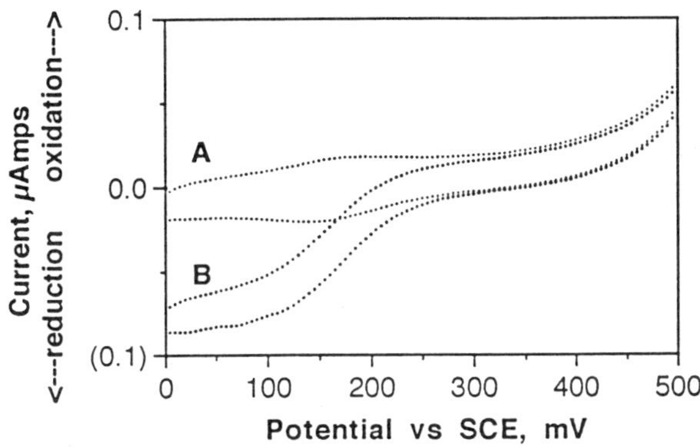

Figure 3. Dependence of current on potential for a $NaIO_4$ oxidized horseradish peroxidase immobilized in a 3-dimensional epoxy hydrogel free of electron relaying osmium redox centers. (A) no H_2O_2; (B) 0.1mM H_2O_2 Conditions: aerated pH 7 physiological phosphate buffer solution; scan rate 2.5 mV s^{-1}; 500 RPM.

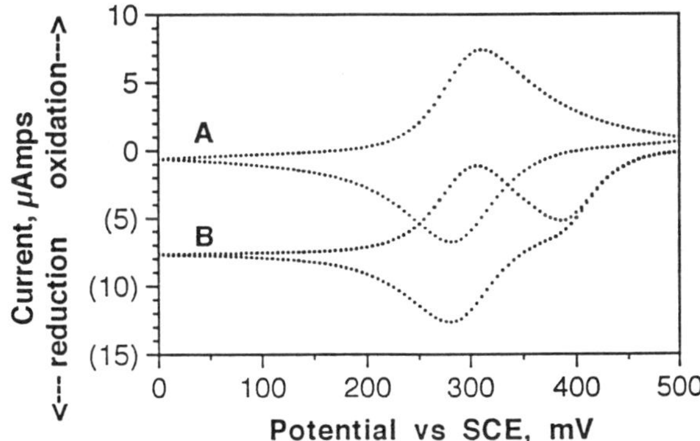

Figure 4. Electrode as in figure 3, but with osmium electron relaying redox centers. (A) no H_2O_2; (B) 0.1 mM H_2O_2; Conditions are as in figure 3.

over the course of 3 days continuous operation, while the LOP sensor looses 10% of its output in hours.

The weight fraction of POD in the redox epoxy film effects the sensor response. As the amount of enzyme is initially increased in the film the current increases, reaching a maximum at 8 to 50% weight POD. As more enzyme is added current then decreases. The shape of the curve implies that at low enzyme loading the sensor is limited by the number of catalytic sites. As enzyme catalytic sites are added sensor response increases. The decreasing current at high POD concentration can be related to work on the Glucose oxidase system. We know that as the fraction of an electrically insulating protein increases in the film, the film's electron diffusion coefficient decreases. This results in a decrease of the current.[22]

Figures 6 to 9 show this effect. The ARP (figure 6), LOP (figure 7) and HRP (figure 8) reach their respective maximum currents at 20 to 45%, 35 to 50% and 20 to 40% enzyme loading. These percentages are similar to those found in anodes made with the same redox polymer but with glucose oxidase.[23] It is interesting to note the difference in enzyme loading at the maximum current for the two HRP enzymes. For $NaIO_4$ oxidized HRP the current maximum is found at 8-20% enzyme loading (figure 9) while for native HRP a 20-40% enzyme loading is optimal (figure 8). $NaIO_4$ treatment of the glycoprotein is a standard procedure for generation of aldehydes by the oxidation of sugar residues. The aldehydes produced can be covalently bound to the redox polymer, which is a polyamine, in a reaction where multiple Shiff bases are formed. Formation of a dense system of covalent bonds implies tight binding of the enzyme and its "wiring" redox polymer. It results in effective electrical connection of a large fraction, perhaps most, of the enzyme molecules present. Thus the current rises rapidly and to a high level as enzyme is added, then becomes limited by the network's current carrying capacity when the fraction of insulating enzyme becomes excessive.[22]

Electrostatic Interaction of Polymer and Enzyme. We account for the differences between the sensors by the different electrostatic interactions between their polymer and enzyme molecules. Strong electrostatic interaction between the two leads to tight coupling of the enzyme and the "wiring" polymer, and thus to a shorter average distance for electron transfer. Because the redox polymer is a polycation, the greater the negative charge of the enzyme at neutral pH, the higher the current. This explanation is consistent with the interpretation of the behavior of the response of different FAD enzyme sensors made with the present redox polymer.[24] The charge on an enzyme at neutral pH is readily observed in isoelectric focusing (IEF) experiments. Figure 10 shows IEF runs for the 4 enzymes. ARP, HRP and LOP are respectively negatively, slightly positively and positively charged at pH 7. Native HRP focuses as two separate isoenzymes with very close pI's. The HRPox forms a complex mixture, and does not focus to a single spot, but is more negative than HRP.

Conclusion

Comparison of the results of the isoelectric focusing experiments with the limiting currents supports the proposition that an attractive electrostatic interaction contributes to sensor performance: the order of increasing negative charge for

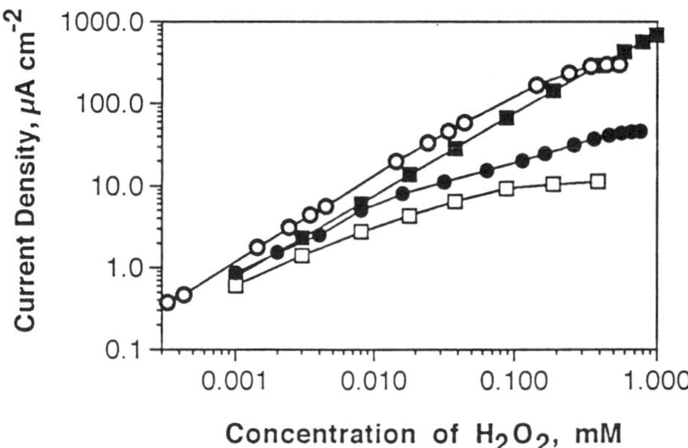

Figure 5. Dependence of current density on hydrogen peroxide concentration for cathodes based on different peroxidases. (open circles) $NaIO_4$ treated horseradish peroxidase; (closed circles) native horseradish peroxidase; (open squares) lactoperoxidase; (closed squares) Arthromyces ramosus. Each electrode contains approximately 10μg osmium redox polymer, 1μg polyethylene glycol diglycidyl ether crosslinker and 1 to 4μg peroxidase. Conditions: aerated pH 7 physiological phosphate buffer solution; 1000 RPM.

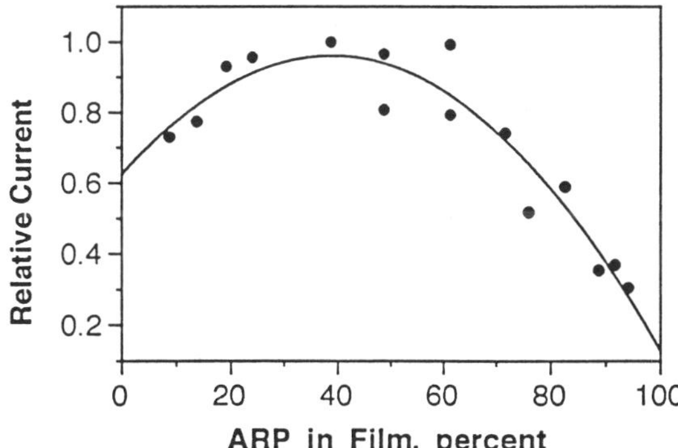

Figure 6. Dependence of normalized current on the weight fraction of Arthromyces ramosus peroxidase (ARP) in the film. The osmium redox polymer and crosslinker amounts were held constant at approximately 10 and 1μg. Conditions: aerated pH 7 physiological phosphate buffer solution; 1000 RPM.

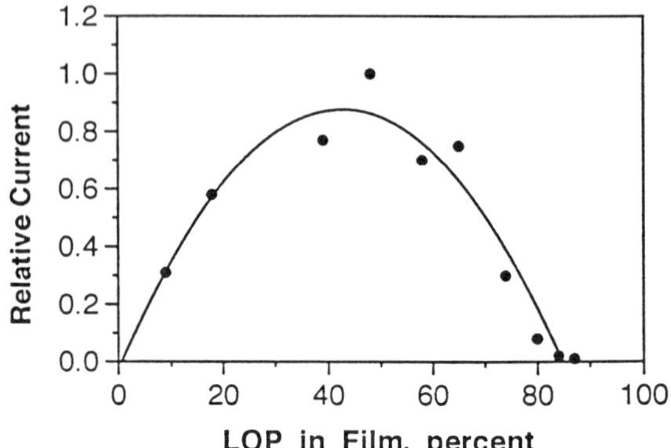

Figure 7. Dependence of normalized current on the weight fraction of lactoperoxidase (LOP) in the film. The osmium redox polymer and crosslinker amounts were held constant at approximately 10 and 1μg. Conditions: aerated pH 7 physiological phosphate buffer solution; 1000 RPM.

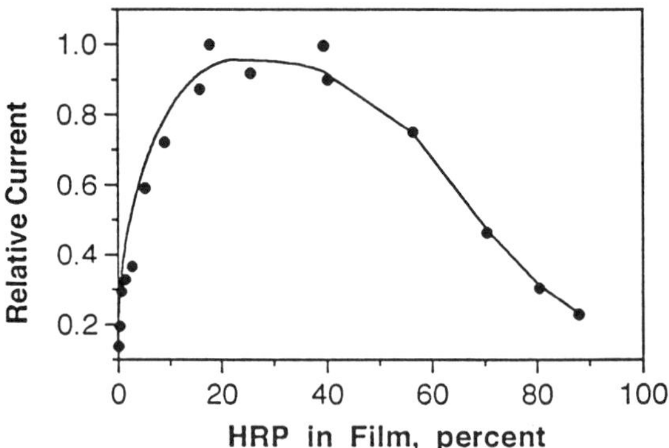

Figure 8. Dependence of normalized current on the weight fraction of horseradish peroxidase (HRP) in the film. The osmium redox polymer and crosslinker amounts were held constant at approximately 10 and 1μg. Conditions: aerated pH 7 physiological phosphate buffer solution; 1000 RPM.

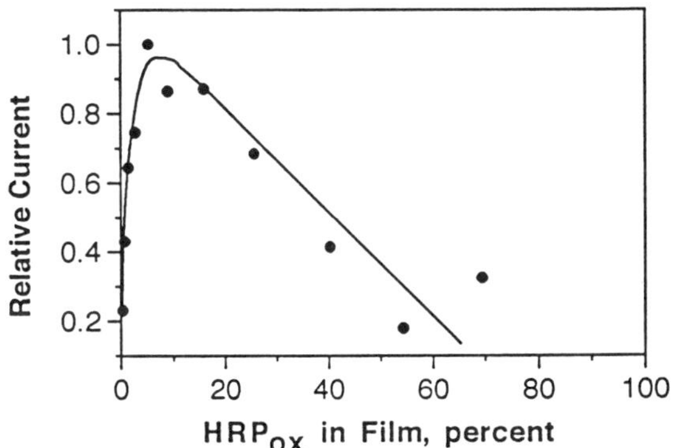

Figure 9. Dependence of normalized current on the weight fraction of NaIO4 treated horseradish peroxidase (HRPox) in the film. The osmium redox polymer and crosslinker amounts were held constant at approximately 10 and 1μg. Conditions: aerated pH 7 physiological phosphate buffer solution; 1000 RPM.

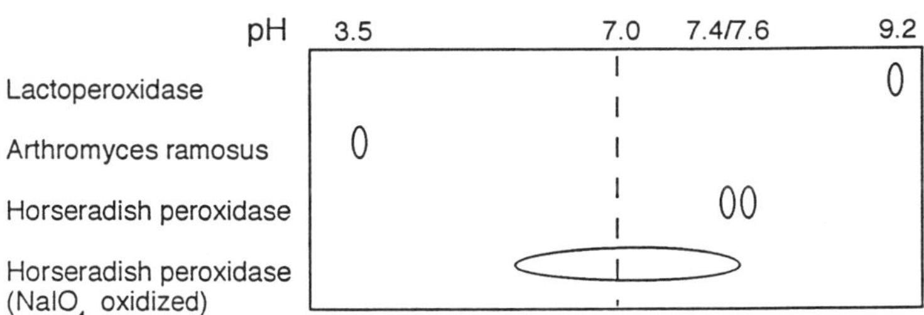

Figure 10. Isoelectric focusing of the 4 enzymes. The agrose gel was loaded with 3.5 to 9.5 pH ampholite to set up the gradient.

LOP<HRP<HRPox<ARP parallels the increase in limiting currents when the peroxidases are "wired" with a polycationic redox polymer.

Acknowledgments

The work described was supported by the Office of Naval Research, National Science Foundation and Welch Foundation.

Literature Cited

(1) Vreeke, M., Maidan, R., Heller, A., *Anal. Chem.* **1992**, *64*, 3084-90.
(2) Jonsson, G.; Gorton, L. *Electroanalysis* **1989**, *1*, 465-8.
(3) Wang, J.; Fricha, B.; Naser, N.; Romero, E. G.; Wollenberger, U.; Ozsoz, M.; Evans, O. *Anal. Chim. Acta* **1991**, *254*, 81-88.
(4) Cosgrove, M.; Moody, G. J.; Thomas, J. D. R. *Analyst* **1988**, *113*, 1811-15.
(5) Kulys, J. J.; Samalius, A. S.; Svirmickas, G. -J. S. *FEBS Letters* **1980**, *114*, 7-10.
(6) Tatsuma, T.; Okawa, Y.; Watanabe, T. *Anal. Chem.* **1989**, *61*, 2352-5.
(7) Sanchez, P. D.; Ordieres, A. J. M.; Garcia, A. C.; Blanco, P. T. *Electroanalysis* **1991**, *3*, 281-5.
(8) Schubert, F.; Saini, S.; Turner, A. P. F. *Anal. Chim. Acta* **1991**, *245*, 133-8.
(9) Pantano, P.; Morton T. H.; Kuhr, W. G. *J. Am. Chem. Soc.* **1991**, *113*, 1832-3.
(10) Kulys, J. J.; Pesliakiene, M. V.; Samalius, A. S. *Bioelectrochem. Bioenerg.* **1981**, *8*, 81-8.
(11) Lundback, H.; Olsson, B. *Analytical Letters* **1985**, *18(B7)*, 871-89.
(12) Kojima, J.; Morita, N.; Takagi, M. *Analytical Sciences* **1988**, *4*, 497-500.
(13) Kulys, J. J.; Laurinavicius, V. -S. A.; Pesliakiene, M. V.; Gureviciene, V. V. *Anal. Chim. Acta* **1983**, *148*, 13-8.
(14) Frew, J. E.; Harmer, M. A.; Allen, H.; Hill, O.; Libor, S. I. *J. Electroanal. Chem. Interfacial Electrochem.* **1986**, *201*, 1-10.
(15) Kulys, J. J.; Schmid, R. D. *Bioelectrodchem. Bioenerg.* **1990**, *24*, 305-11.
(16) Armstrong, F. A.; Lannon A. M. *J. Am. Chem. Soc.* **1987**, *109*, 7211-2.
(17) Cooper, J. M.; Alvarez-Icaza, M.; McNeil, C. J.; Bartlett, P. N. *J. Electroanal. Chem. Interfacial Electrochem.* **1989**, *272*, 57-70.
(18) Tatsuma, T.; Watanabe, T. *Anal. Chem.* **1991**, *61*, 1580-5.
(19) Paddock, R. M.; Bowden, E. F. *J. Electroanal. Chem. Interfacial Electrochem.* **1989**, *260*, 487-94.
(20) Razumas, V. J.; Gudavicius, A. V.; Kulys, J. J. *J. Electroanal. Chem. Interfacial Electrochem.* **1983**, *151*, 311-5.
(21) Gregg, B. A.; Heller, A. *J. Phys. Chem.* **1991**, *64*, 5970-75.
(22) Surridge, N., this volume.
(23) Gregg, B. A.; Heller, A. *J. Phys. Chem.* **1991**, *64*, 5976-80.
(24) Katakis, I., Vreeke, M., Ye, L., Aoki, A., Heller, A. "Biochemical Technology, Advances in Molecular and Cell Biology Series", Proceedings of Mosbach Symposium, JAI Press, Inc., **1993**.

RECEIVED July 26, 1993

BIOCOMPATIBILITY AND BIOMIMETICS

Chapter 16

New Biocompatible Polymer

Application for Implantable Glucose Sensor

Kazuhiko Ishihara[1], Nobuo Nakabayashi[1], Kenro Nishida[2], Michiharu Sakakida[2], and Motoaki Shichiri[2]

[1]Institute for Medical and Dental Engineering, Tokyo Medical and Dental University, Kanda-surugadai, Chiyoda-ku, Tokyo 101, Japan
[2]Department of Metabolic Medicine, Kumamoto University School of Medicine, Honjo, Kumamoto 860, Japan

To develop a new biocompatible polymer with attention to the formation of biomembrane-like structure on the surface, 2-methacryloyloxyethyl phosphorylcholine(MPC) copolymers with n-butyl methacrylate(BMA) were synthesized. The poly(MPC-*co*-BMA) showed excellent blood compatibility, that is, inhibition of cell adhesion and activation and reduction of protein adsorption from human blood. On the other hand, phospholipid adsorption from human plasma on the poly(MPC-*co*-BMA) surface was enhanced with an increase in MPC composition in the copolymer. This finding suggested that the phospholipid adsorption on the poly(MPC-*co*-BMA) plays an important role in the reduction of thrombogenicity of the copolymer. Moreover, the poly(MPC-*co*-BMA) membrane has good permeability for solute. Therefore, it can be applied to a medical membrane required both biocompatibility and permeability. The surface of a needle-type glucose sensor with electron mediator was covered with the poly(MPC-*co*-BMA) membrane and its activity *in vivo* was investigated. The output signal from the sensor was stabilized by the covering and the glucose concentration in subcutaneous tissue could be detected for 14 days when the sensor was applied to human patient.

When artificial materials contact living organisms, serious responses such as thrombus formation, unfavorable immunoresponse, capsulation, etc., are observed. This is a very important response for living but induces many problems in the treatment of patients using the artificial medical devices. Therefore, biocompatibility, particularly blood compatibility, is the most important property required for biomedical materials.

The molecular design of biocompatible polymers is classified into four categories

0097–6156/94/0556–0194$08.00/0

as shown in Table 1 based on the approach to regulation of blood materials interactions (*1*). Since the normal endothelial surface does not induce thrombus formation, we considered that the tailor-made polymers which can regulate phospholipid adsorption from plasma and produce a biomembrane-like surface have excellent blood compatibility (*2*). This concept is newer and is based on the properties of natural phospholipids themselves. It is assumed that polymers having a strong affinity for phospholipids could be used to construct a biomembrane-like structure by adsorbing phospholipids from blood and organizing them on the polymer surface. Phospholipid molecules have a self-organizing property with each other and form a bilayer membrane structure (*3*). Therefore, polymers having a phospholipid polar group are expected to have a strong affinity for phospholipid molecules. A methacrylate with a phospholipid polar group in the side chain, 2-methacryloyloxyethyl phosphorylcholine (MPC, Fig. 1), and its copolymers with various methacrylates and styrene were prepared (*4-6*).

MPC can be dissolved in alcohol and easily polymerized with other vinyl monomers by conventional radical copolymerization using a radical initiator. Moreover, the MPC copolymers obtained are soluble in alcohol but insoluble in water, depending on the MPC composition. This is one of the good characteristics required for biomedical polymers for surface modification of medical devices. Among these copolymers, poly(MPC-*co-n*-butyl methacrylate (BMA))s exhibit excellent blood compatibility as shown by reduction of platelet adhesion and aggregation and suppression of protein adsorption (*7-10*).

In this review, the blood compatibility of poly(MPC-*co*-BMA) against human whole blood, the mechanism of nonthrombogenicity observed on the polymer surface, and application of the poly(MPC-*co*-BMA) for an implantable glucose sensor are described.

BLOOD COMPATIBILITY OF MPC COPOLYMERS

In Table 2, the whole blood coagulation time on the polymer measured by the Lee-White method is summarized (*11*). The results clearly indicated reduced thrombogenicity on the MPC copolymers. The coagulation time on glass was 8.4 ± 0.46 min and that on poly(BMA)(no MPC) was 9.6 ± 1.3 min. However, there was no significant difference ($P>0.05$). With a coating of poly[2-hydroxyethyl methacrylate)(HEMA)] and MPC copolymers, the coagulation time was significantly increased compared with glass and a poly(BMA) coating case($P<0.01$). On the surface of poly(MPC-*co*-BMA), the longest times for coagulation, 28 ± 2.6 min, were observed.

The blood-material interactions were also investigated by a microsphere-column method (*7,9,11*). After human whole blood without an anticoagulant was passed through the columns packed with polymer-coated beads for 15 min and the column was rinsed with phosphate bùffered solution(PBS), the color of the column with poly(BMA) turned red, that is, serious thrombus formation occurred in the poly(BMA) column. However, the poly(MPC-*co*-BMA) columns were clear even when the MPC mole fraction in the copolymer was 0.13. Fig. 2 shows scanning

Table 1. Molecular design of biocompatible polymer

Type of materials	Basic concept	Typical polymer	Approach
Synthetic polymer	Hydrophobic surface Zero critical surface tension theory	Polytetrafluoroethylene (PTFE) Polydimethylsiloxane Polyethylene	Physicochemical approach
	Hydrophilic surface Zero interfacial free energy concept	Poly(2-hydroxyethyl methacrylate) Poly(acrylamide)	
	Heterogenic surface Microdomain concept	Block-type copolymer Graft-type copolymer Segmented polyurethane	Interfacial chemical approach Biochemical approach
	Molecular cilia mechanism	Poly(ethylene glycol)	
	Negatively charged surface	Sulfonated polymer	
Synthetic polymer + Biopolymer	Pseudomembrane formation	Expanded PTFE, Dacron	Biological approach
	Heparinization	Heparin released polymer Heparin immobilized polymer	
	Immobilization of fibrinolysis enzyme	Urokinase immobilized polymer	
Synthetic polymer + Biological molecules	Albumin adsorbed surface	Polyurethane with alkyl group	
	Phospholipid adsorbed surface Biomembrane-like surface formation	Phospholipid polymer	

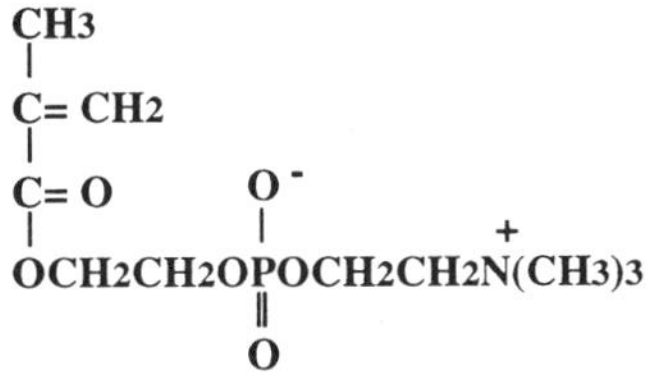

```
CH3
|
C= CH2
|
C= O      O-
|         |            +
OCH2CH2OPOCH2CH2N(CH3)3
          ||
          O
```

Figure 1. Chemical structure of 2-methacryloyloxyethyl phosphorylcholine (MPC).

Table 2. Whole blood coagulation time determined by the Lee-White method

Sample	MPC mole fraction	Coagulation time [a] (min)	t-Test [b]
Glass	—	8.4 ± 0.46	*P>0.05 (Glass vs. Poly(BMA))
Poly(BMA)	—	9.6 ± 1.3	**P<0.01 (Poly(BMA) vs. Poly(HEMA))
Poly(HEMA)	—	21 ± 1.2	* (Poly(HEMA) vs. Poly(MPC-co-MMA))
Poly(MPC-co-MMA) [c]	0.18	21 ± 0.58	** (Poly(MPC-co-MMA) vs. Poly(MPC-co-BMA))
Poly(MPC-co-BMA)	0.26	28 ± 2.6	

a) ± value indicates standard deviation and number of samples is 3 for each polymer.

b) * means not significant difference between these values(P>0.05), whereas ** means significant difference(P<0.01).

c) MMA is methyl methacrylate

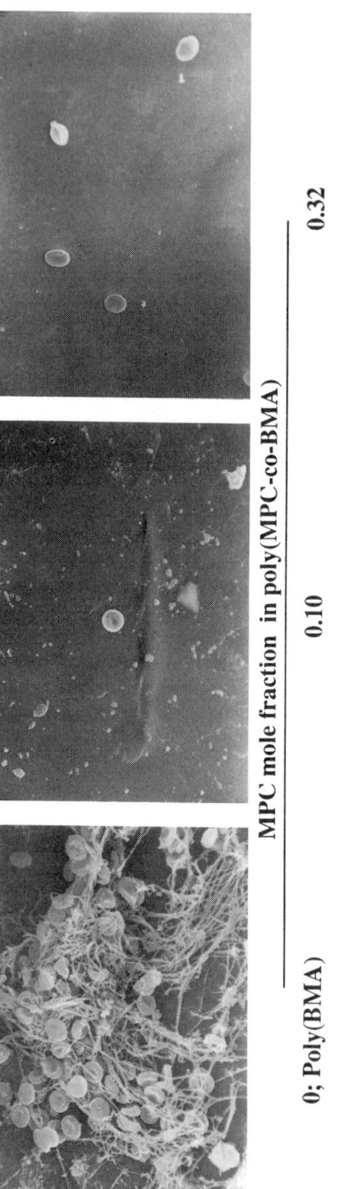

Figure 2. SEM picture of poly(MPC-co-BMA) surface after contact with human whole blood in the absence of anticoagulant for 15 min. (from ref. 11 with permission).

electron microscopic(SEM) pictures of the polymer-coated beads after contact with human whole blood (*11*). On the surface of poly(BMA), a fibrin net completely covered the bead surfaces and many blood cells were adhered. On the other hand, no fibrin deposition could be found on poly(MPC-*co*-BMA)s. On the MPC copolymer with a 0.32 MPC mole fraction, a few blood red cells without deformation were found. These results clearly show that MPC copolymers have excellent nonthrombogenic properties and MPC moieties in the copolymer are an important element in the nonthrombogenicity of the copolymers.

The interesting property of the MPC copolymer is its affinity for phospholipids (*5,12-14*). The amount of a phospholipid, dipalmitoylphosphatidylcholine (DPPC), adsorbed on MPC copolymers was larger than that on polystyrene, poly(BMA) and poly(HEMA) and increased with increasing MPC moiety when the MPC copolymers were contacted with a liposomal solution of DPPC (*5*). This tendency was the same as that of the adsorption of phospholipid from human plasma which is indicated in Fig. 3. Thus, the affinity of poly(MPC-*co*-BMA) for the phospholipids could be observed even in the plasma. The DPPC molecules adsorbed on the poly(MPC-*co*-BMA) surface assumed an organized structure like that for a bilayer membrane, which was confirmed by differential scanning calorimetry(DSC) and X-ray photoelectron spectroscopy(XPS) when the poly(MPC-*co*-BMA) membrane was immersed in the solution containing DPPC (*12,14*) It is therefore concluded that the MPC copolymers stabilized the adsorption layer of phospholipids on the surface. Stabilization of the liposomal structure in water by a water-soluble MPC copolymer was also found (*13*).

Little albumin and γ-globulins were adsorbed on the MPC copolymer surface treated with DPPC, which is in sharp contrast to the fact that on a poly(BMA) surface, pretreatment with DPPC was not effective for suppression of protein adsorption (*14*). The difference in protein adsorption on these polymer surfaces reflects the difference in the orientation of the DPPC which covered the polymer surface. In Fig. 3, the adsorbed amounts of phospholipids, phosphatidylcholine (PC) on the polymers from human plasma is indicated (*11*). The amount of phospholipids adsorbed on hydrophobic poly(BMA) was almost the same as that on hydrophilic poly(HEMA). On poly(MPC-*co*-BMA), a larger amount of phospholipids was adsorbed compared with poly(BMA), and the amount adsorbed increased with an increase in the MPC mole fraction of the copolymer.

The amount of total proteins adsorbed on poly(HEMA) from human plasma was significantly smaller than that on poly(BMA) ($P<0.01$), and it was at the same level as on poly(MPC-*co*-BMA) with a 0.13 MPC mole fraction as shown in Fig. 4 (*11*). However, on poly(MPC-*co*-BMA) with a 0.26 MPC mole fraction was quite small compared with that on other polymers tested. It is clearly demonstrated that there are weak interactions between the surface and proteins.

The protein composition and distribution on the blood contacting surface of MPC copolymers from human plasma has been determined by radioimmunoassay and an immunogold labeling technique (*8*). In Fig. 5, proteins existed on the plasma contacting surface after 60 min contact with poly(MPC-*co*-BMA) and glass were determined by radioimmunoassay. On all materials, not only major components of plasma pro-

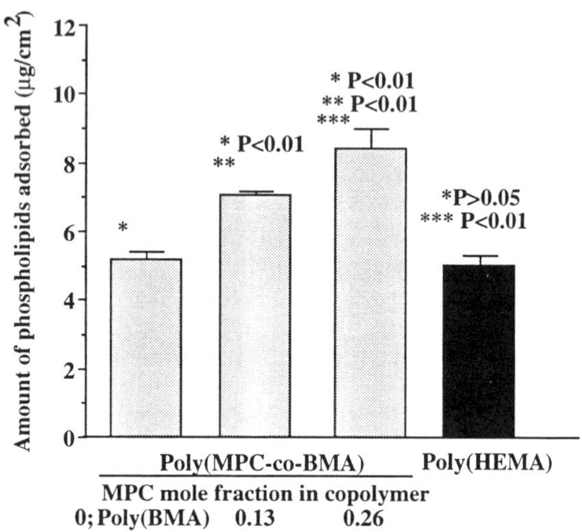

Figure 3. Amount of phosphatidylcholines adsorbed on poly(MPC-co-BMA) and poly(HEMA) from human plasma after 60 min. (from ref. 11 with permission).

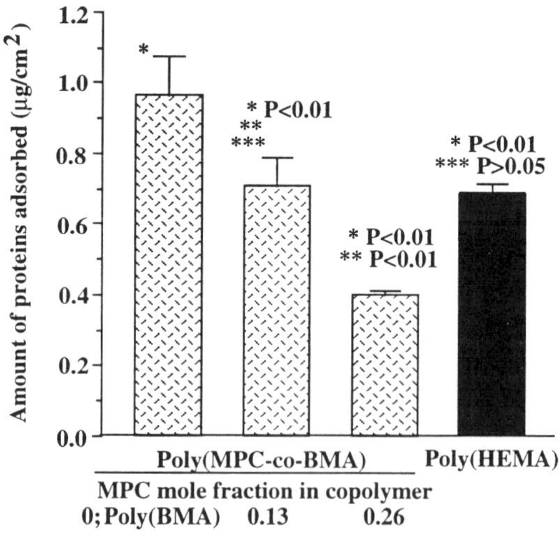

Figure 4. Amount of proteins adsorbed on poly(MPC-co-BMA) and poly(HEMA) from human plasma after 60 min. (from ref. 11 with permission).

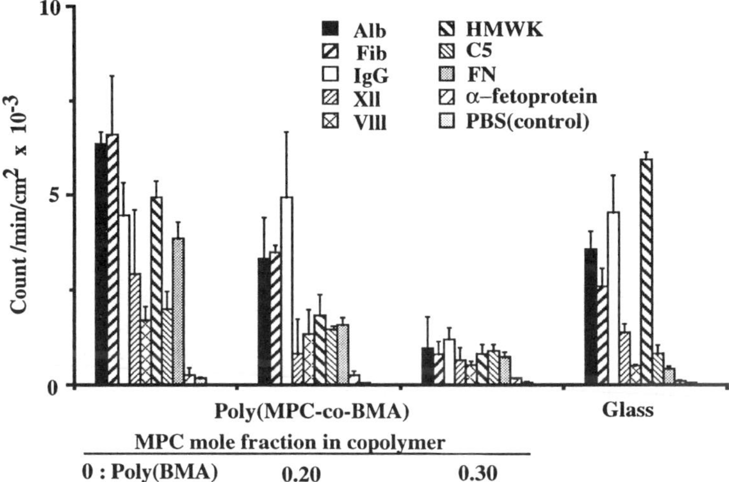

Figure 5. Amount of protein adsorbed at the plasma contacting surface of various materials from human plasma after 60 min. Alb; albumin, Fib; fibrinogen, IgG; γ-globulin, XII(Hageman factor); coagulation factor, VIII; coagulation factor VIII, HMWK; high-molecular-weight kininogen, C5; complement C5, FN; fibronectin, PBS; phosphate buffered solution(pH 7.4).

teins, albumin(Alb), fibrinogen(Fib), and γ-globulin(IgG) but also minor components were observed. Protein adsorption was reduced with an increase in the MPC composition in the poly(MPC-*co*-BMA). In the case of the copolymer with 0.30 MPC composition, all adsorbed protein was drastically reduced compared with that on glass. In particular, the adsorbed amount of high-molecular-weight kininogen(HMWK) decreased significantly.

The amount of protein adsorbed on a poly(MPC-*co*-BMA) surface from a solution containing a single protein was determined spectrophotometrically (*10,15,16*) The amounts of albumin (initial concentration, 4.5 g/dl) or IgG (initial concentration, 1.6 g/dl) adsorbed on poly(MPC-*co*-BMA) with a 0.25 MPC mole fraction were very little. Moreover, it is also found that the conformation of albumin adsorbed on poly(MPC-*co*-BMA) does not change even after 60 min contact (*16*). This is due to extremely weak interactions between proteins and poly(MPC-*co*-BMA) and reversible adsorption of proteins on the MPC polymer surface.

The distribution of proteins adsorbed on the surfaces was determined by the immunogold labeling technique. Fig. 6 shows the typical result, that is, fibrinogen adsorption patterns on poly(BMA) and poly(MPC-*co*-BMA) from 100% plasma. Little fibrinogen adsorbed in a dispersed pattern on poly(MPC-*co*-BMA). However, a large amount of fibrinogen was adsorbed and some aggregates were observed on poly(BMA), suggesting strong interactions between the fibrinogen molecules adsorbed on poly(BMA). These observations are in agreement with studies of the detection of fibrinogen on expanded polytetrafluoroethylene surfaces (*17*). In addition, others have shown "networks" of fibronectin on polyethylene surfaces and have speculated that these networks result from incomplete or altered adsorption due to intermolecular association of fibronectin.

If nonthrombogenic materials are to be designed, it may be necessary to decrease protein adsorption, particularly proteins such as coagulation factors and complement components on the blood contacting surface. The surface concentration of all proteins including fibrinogen, coagulation factors and complement component decreased with an increase in the MPC moiety and appeared to absorb on the surfaces in a uniform and evenly distributed manner. Phospholipid polymers for biocompatibility and reduced thrombogenicity may be potentially useful for these applications.

MECHANISM OF BLOOD COMPATIBILITY OBSERVED ON MPC COPOLYMER SURFACE

By comparison of the adsorption behavior of phospholipids on MPC copolymers with that of proteins, a mechanism of nonthrombogenicity observed on MPC copolymers is considered as shown in Fig. 7 (*11*). Since the molecular size of phospholipids is smaller than that of proteins and the molar concentration of phospholipids is larger than that of proteins, the diffusion of phospholipid molecules from plasma to the polymer surface occurs more easily than that of proteins. Moreover, the MPC copolymers have both a strong affinity for phospholipids (*5,12-14*) and protein adsorption-resistant properties in buffered aqueous protein solution (*10,15,16*). From these findings, it is concluded that the phospholipids in plasma are adsorbed immedi-

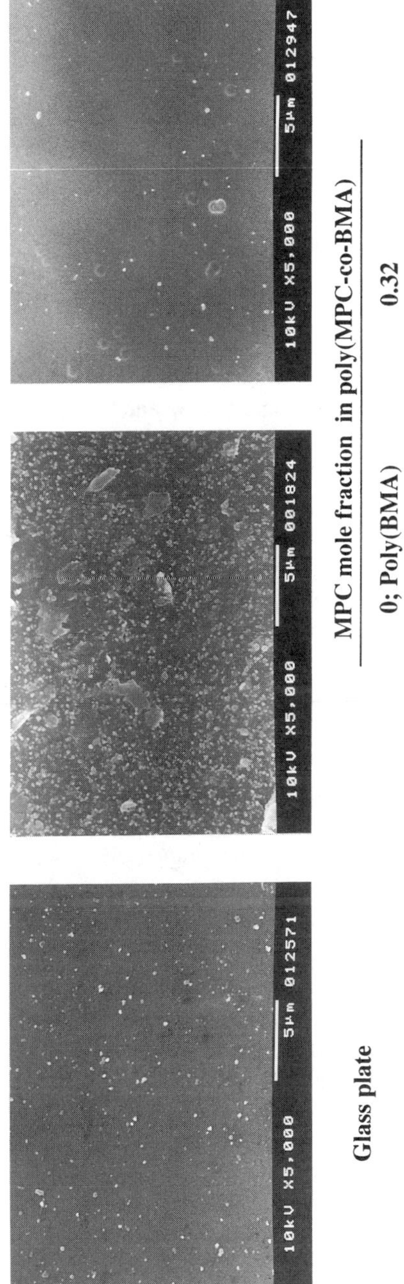

Figure 6. SEM picture of adsorption pattern of fibrinogen on various materials after 60 min adsorption from human plasma.

ately on the surface and form a stable adsorbed layer with a biomembrane-like surface through their self-organizing properties. As suggested by Harris (*18*) and Hayward (*19*), the phosphorylcholine moiety may be the most important in reducing thrombogenicity. The biomembrane-like surface interacts minimally with proteins and cells and therefore inhibits thrombus formation.

APPLICATION OF MPC COPOLYMERS FOR IMPLANTABLE GLUCOSE SENSOR OF ARTIFICIAL ENDOCRINE PANCREAS

Using the MPC copolymers, improvements in biocompatibility of medical membranes such as a cellulose hemodialysis membrane (*20,21*), a polyolefin membrane for an oxygenator (*22*), and a covering membrane for an implantable biosensor are underway (*23*). The MPC copolymers have nonthrombogenicity, protein adsorption resistance, and good solute permeability.

In 1982, one of the authors(M.S.) and coworkers succeeded in miniaturizing a glucose monitoring system into a needle-type device and also in developing a wearable artificial endocrine pancreas (*24*). The wearable artificial endocrine pancreas was then applied to ambulatory diabetic patients and they first demonstrated that perfect glycemic control for longer period could be obtained with the system. However, for a long-term clinical use of this wearable artificial endocrine pancreas, there are several problems to be solved, especially regarding the glucose sensor. Among them, a stable, long-life glucose sensor is most crucial. We considered that the major reason for sensor decay with time when *in vivo* continuous monitoring is undertaken might be due to protein adsorption on the membrane surface covering on the sensor, resulting in insufficient availability of oxygen in the glucose sensor inserted in subcutaneous tissue where oxygen tension is very low and easily changed. Therefore, to solve the oxygen limitation problems, a ferrocene-mediated glucose sensor that does not require oxygen has been developed. To prepare a stable, long-life glucose sensor (*25*), the sensor surface is covered with a newly designed membrane having excellent biocompatibility, a poly(MPC-*co*-BMA) membrane (*26*).

The principal arrangement of the ferrocene-mediated needle-type glucose sensor is shown in Fig. 8, and the performance of the sensor is summarized in Table 3 compared with a poly(vinyl alcohol)(PVA) covered ferrocene-mediated sensor and a standard glucose sensor. The oxygen dependency was improved by the use of ferrocene as a mediator. There was not significant difference between the sensor covered with a poly(MPC-*co*-BMA) membrane and that with a PVA membrane in the absence of plasma proteins. In serum albumin solution(7 wt % in phosphate buffered solution, pH 7.4 at 37°C), though the output current of the ferrocene-mediated sensor with a PVA membrane decreased to 62% of the initial value after 14 days immersion, that observed on the sensor covered with a poly(MPC-*co*-BMA) membrane remained above 92% of the initial value.

Fig. 9 shows the time dependence of the relative sensor output and the relative response time of the ferrocene-mediated needle-type glucose sensor covered with a poly(MPC-*co*-BMA) membrane or a PVA membrane when the sensor was inserted continuously into the subcutaneous tissue of rats. By using the mediator, sensor

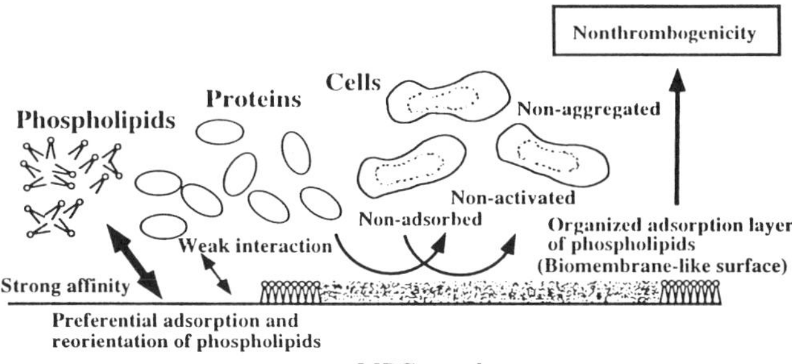

Figure 7. Schematic representation of mechanism of nonthrombogenicity observed on poly(MPC-co-BMA). (from ref. 11 with permission).

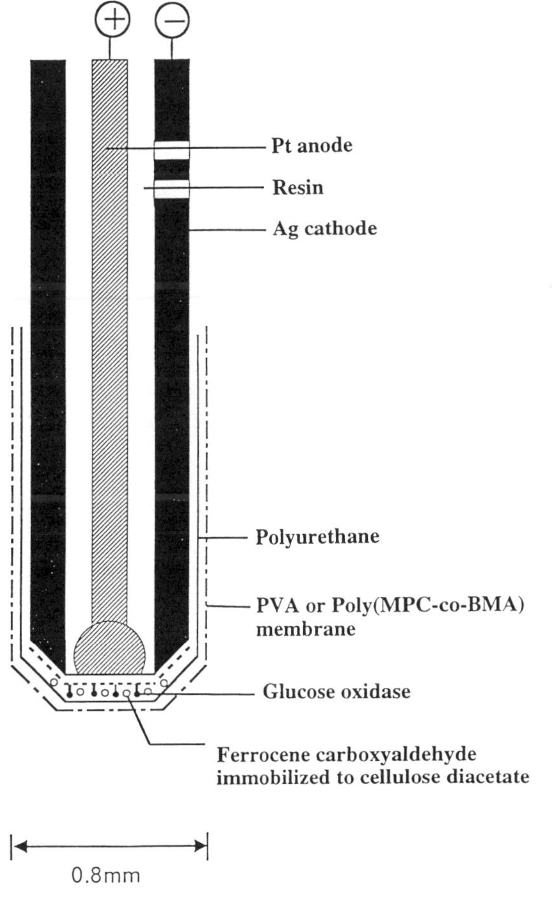

Figure 8. Structure of a ferrocene-mediated needle-type glucose sensor.

Table 3. Performance of conventional and ferrocene-mediated needle-type glucose sensors covered with various polymers *in vitro*

	Conventional type	Ferrocene-mediated type	
Covering membrane	PVA	PVA	Poly(MPC-co-BMA)
Baseline drift (%/day)	1.1 ± 0.1	0.8 ± 0.1	1.0 ± 0.1
Response time(T90%) (min)	2.0 ± 0.2	2.1 ± 0.3	2.5 ± 0.4
Linearity in glucose concentration (mg/dl)	0 - 500	0 - 500	0 - 500
Oxygen dependency (%/ mmHg)	0 - 0.1 (25 - 150 mmHg) 1.8 - 2.2 (2 - 25 mmHg)	No change (0 - 150 mmHg)	No change (0 - 150 mmHg)

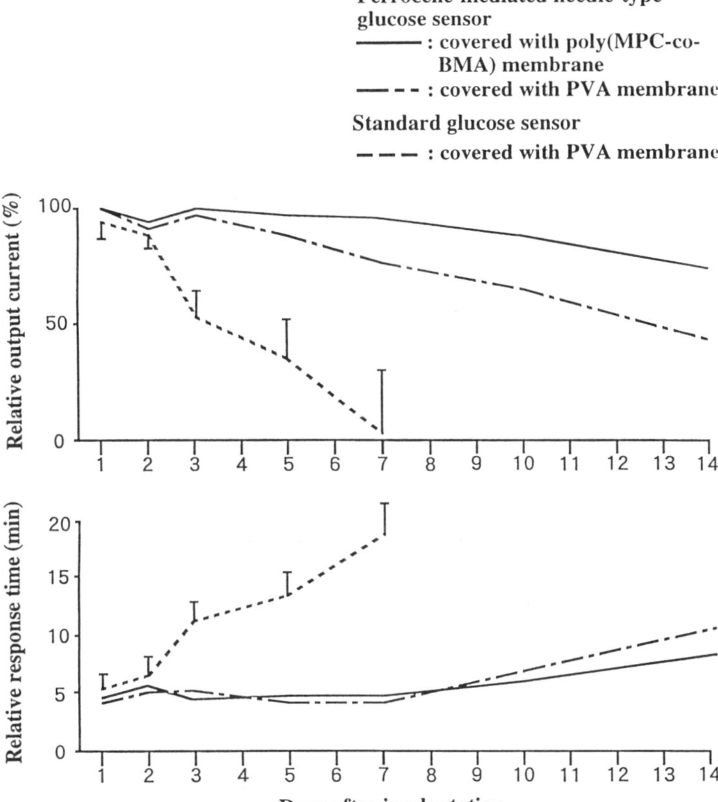

Figure 9. Time course of relative output currents and response times of needle-type standard glucose sensor covered with PVA membrane and these of needle-type ferrocene-mediated glucose sensors covered with PVA membrane or poly(MPC-co-BMA) membrane when these sensors were inserted into subcutaneous tissue of normal rat.

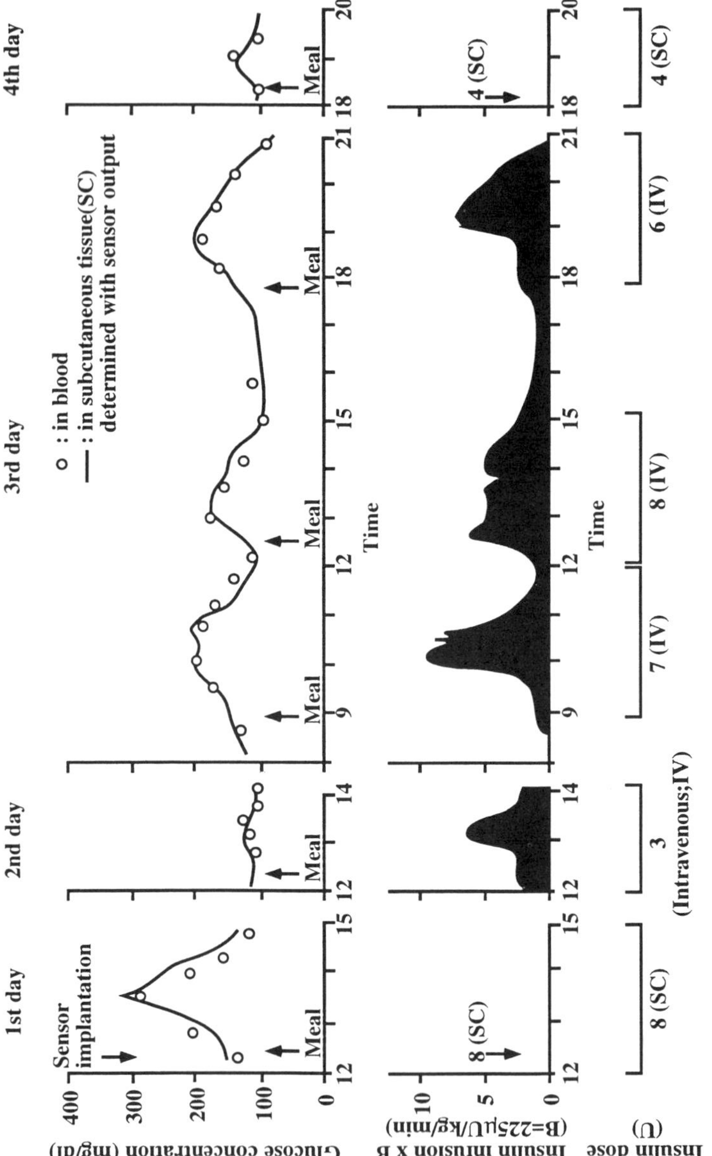

Figure 10. Continuous monitoring of subcutaneous tissue glucose concentrations and blood glucose regulation in an insulin requiring NIDDM patient with a ferrocene-mediated needle-type glucose sensor covered with poly(MPC-co-BMA) membrane.

performance was improved considerably. In particular, the ferrocene-mediated glucose sensor covered with a poly(MPC-*co*-BMA) membrane showed excellent stability compared that with the PVA membrane case, that is, the output current after 7 days was 94% of initial value and 74% even after a 14-day continuous immersion. The scanning electron microscopic observation revealed that protein adsorption and cell adhesion were inhibited on the sensor surface covered with a poly(MPC-*co*-BMA) membrane. The ferrocene-mediated needle-type glucose sensor covered with poly(MPC-*co*-BMA) membrane could be applied to continuous glycemic monitoring in a healthy human subject for 14 days. The subcutaneous tissue glucose concentrations could be monitored and followed nicely to the changes in blood glucose excursions for up to 7 days without any calibrations and to 14 days with in situ calibrations. In the final stage, we made the wearable artificial endocrine pancreas using the ferrocene-mediated needle-type glucose sensor covered with poly(MPC-*co*-BMA).

Fig. 10 shows a typical examination of the operation of a wearable artificial endocrine pancreas with the new implantable glucose sensor covered with a poly(MPC-*co*-BMA) membrane in a diabetic patient. Physiological glycemic control was obtained after meal intake, and physiological daily glycemic control was also obtained. These results strongly suggested that perfect glycemic control in diabetic patients can be obtained for a longer period with the artificial endocrine pancreas by repeating the replacing of thc fcrroccnc-mediated needle-type glucose sensor covered with the poly(MPC-*co*-BMA) membrane every 4th day.

Acknowledgments

Since a part of this study was supported by a Grant-in-Aid for Scientific Research from the Ministry of Education, Japan (02205030, 03205030, 04205030), Nissan Science Foundation(1990) and Terumo Scientific Foundation(1992), one of the authors(K.I.) would like to express his appreciation for these sources of support.

Literature Cited

1. Ishihara, K. In Polymer Materials for Biomedical Applications; Tsuruta, T.; Hayashi, T.; Kataoka, K; Kimura, Y., Ishihara, K., Eds.; CRC Press: Boca Raton, FL, 1992; Chapter 3.1.

2. Nakabayashi, N.; Ishihara, K.; Watanabe, A.; Kojima, M. In Polymers and Biomaterials; Feng, H.; Han, Y.; Huang, L., Eds.; Elsevier Science Publishers: Amsterdam, 1991; pp343-351.

3. Gennis, R.B. In Biomembranes: Molecular structure and function, ; Springer-Verlag, New York, NY, 1989.

4. Ishihara, K.; Ueda, T.; Nakabayashi, N.; *Polym.J.*, **1990**, *22*, 355.

5. Kojima, M.; Ishihara, K.; Watanabe, A.; Nakabayashi, N.; *Biomaterials*, **1991**, *12* , 121.

6. Ueda, T.; Oshida, H.; Kurita, K.; Ishihara, K.; Nakabayashi, N.; *Polym.J.*, **1992**, *24*, 1259.

7. Ishihara, K.; Aragaki, R.; Ueda, T.; Watanabe, A.; Nakabayashi, N.; *J.Biomed.Mater.Res.*, **1990**, *24*, 1069.
8. Ishihara, K.; Ziats, N.P.; Tierney, B.P.; Nakabayashi, N.; Anderson, J.M.; *J.Biomed.Mater.Res.*, **1991**, *25*, 1397.
9. Ueda, T.; Ishihara, K.; Nakabayashi, N.; *Kobunshi Ronbunshu,* **1991**, *48*, 289.
10. Ueda, T.; Ishihara, K.; Nakabayashi, N.; *Seitai Zairyo(J.Jpn.Soc.Biomat.)*, **1991**, *9*, 288.
11. Ishihara, K.; Oshida, H.; Endo, Y.; Ueda, T.; Watanabe, A.; Nakabayashi, N.; *J.Biomed.Mater.Res.*, **1992**, *26*, 1543.
12. Ueda, T.; Watanabe, A.; Ishihara, K.; Nakabayashi, N.; *J.Biomat.Sci.Polym.*, **1991**, *3*, 185.
13. Ishihara, K.; Nakabayashi, N.; *J.Polym.Sci. Part A, Polym.Chem.,* 1991, 29, 831.
14. Ishihara, K.; Aragaki, R.; Yamazaki, J.; Ueda, T.; Nakabayashi, N.;*Seitai Zairyo(J.Jpn.Soc.Biomat.)*, **1990**, *8*, 231.
15. Ishihara, K.; Ueda, T.; Nakabayashi, N.; *Kobunshi Ronbunshu*, **1989**, *46*, 591.
16. Ishihara, K.; Ueda, T.; Saito, N.; Kurita, K.; Nakabayashi, N.;*Seitai Zairyo(J.Jpn.Soc.Biomat.),* **1991**, *9*, 243.
17. Ziats, N.P.; Topham, N.S.; Pankowsky, D.A.; Anderson, J.M.; *Cells & Mater.*, **1991**, *1*, 73.
18. Harris, P.I.; Hall, B.; Bird, R.R.; Chapman, D., In Horizons in Membrane Biotechnology; Nicolau, C.; Chapman, D., Eds., Wiley-Liss, New York, NY, 1990, pp.1-12.
19. Hayward, J.A.; Chapman, D.; *Biomaterials*, **1984**, *5*, 135.
20. Ishihara, K.; Nakabayashi, N.; Fukumoto, K.; Aoki, J., *Biomaterials*, **1992**, *13*, 145.
21. Fukumoto, K.; Ishihara, K.; Takayama, R.; Nakabayashi, N.; Aoki, J.; *Biomaterials,* **1992**, *13*, 235.
22. Ishihara, K.; Taguchi, N.; Ueda, T.; Nakabayashi, N.; *Polym.Preprints Jpn*, **1992**, *41*, 484.
23. Ishihara, K.; Ohta, S.; Yoshikawa, T.; Nakabayashi, N.; *J.Polym.Sci. Part A, Polym.Chem.,* **1992**, *30*, 929
24. Shichiri, M.; Kawamori, R.; Yamasaki, Y.; Hakui, N.; Abe, H.; *Lancet* , **1982**, *II*, 1129.
25. Sakakida, M.; Ichinose, K.; Fukushima, H.; Nishida, K.; Kajiwara, K.; Hashiguchi, Y.; Taniguchi, I.; Shichiri, M.; *Artif. Organs Today,* **1992**, *2*, 145.
26. Nishida, K.,; Sakakida, M.; Hashiguchi, Y.; Uehara, M.; Uemura, T.; Kajiwara, K.; Shichiri, M.; Ishihara, K.; Nakabayashi, N.; *Jinko Zoki(J.Jap.Artif.Organs)*, **1993**, *22*, 1090.

RECEIVED February 10, 1994

Chapter 17

Biocompatibility of Perfluorosulfonic Acid Polymer Membranes for Biosensor Applications

R. F. B. Turner[1,2] and C. S. Sherwood[1]

[1]Biotechnology Laboratory, University of British Columbia, 237–6174 University Boulevard, Vancouver, British Columbia V6T 1Z3, Canada
[2]Department of Electrical Engineering, University of British Columbia, 434–2356 Main Hall, Vancouver, British Columbia V6T 1Z4, Canada

Perfluorosulfonic acid (Nafion) polymer membranes have been shown to yield exceptional reproducibility and performance of glucose sensors in whole blood *in vitro*, demonstrating that sensor criteria are well satisfied by this material. This paper presents a summary of recent studies which indicate that Nafion polymer also possesses a high degree of biocompatibility, making it a promising candidate material for implantable sensors. In particular, we have employed *in vitro* cell culture methods in order to obtain a quantitative evaluation of Nafion polymer biocompatibility in terms of cell growth rates, nutrient metabolism, recombinant protein production and gross morphology. The results obtained *in vitro* augment and support preliminary *in vivo* animal studies describing the qualitative evaluation of the chronic biological response to Nafion polymer implants.

The development of implantable biosensors suitable for long-term *in vivo* clinical monitoring remains an elusive but compelling goal (*1*,*2*). Perhaps the greatest obstacle is the need to satisfy, simultaneously, both the biocompatibility requirements as well as the performance requirements of the sensor. This need imposes an extremely challenging set of criteria on the selection or development of encapsulating membrane materials. In general, conventional polymer membrane materials that ensure satisfactory sensor performance do not offer sufficient biocompatibility, and *vice versa* (*3*).

It is important to realize that biocompatibility issues are not only relevant in respect of the well being of the host, but also in respect of the requirements of the sensor itself. More specifically, the required chemical interactions between the sensor and the body must not be interfered with by interactions—either chemical or physical—between the membrane material and contacting/adhering cells. For example, an encapsulating membrane of an electrochemically based sensor must maintain appropriate mass transport conditions for the analyte and electrolyte species, and must exclude species that could interfere with the electrochemistry, or denature or inhibit the activity of immobilized enzymes. Stability of mass transport conditions is especially critical, since any change in the permeability of the membrane or the surrounding tissues can affect the sensor calibration. Satisfactory stability can not be achieved without a biocompatible encapsulation material.

0097–6156/94/0556–0211$08.00/0

One of the authors and co-workers have reported exceptional reproducibility and performance of glucose biosensors in whole blood *in vitro* with the use of perfluorosulfonic acid (Nafion) polymer membranes (*4,5*). Nafion polymer has since been used by several other investigators as a membrane/encapsulation material for biosensor applications (*6-8*). Nafion membranes have also been used to suppress ascorbate interference during acute *in vivo* determination of various neurotransmitters (*9,10*). This work demonstrates that the sensor membrane criteria are readily satisfied by Nafion, at least in the short term, making it a promising candidate material for implantable sensors. Very little, however, is presently known about its biological compatibility which, clearly, would determine the utility of this material in applications requiring prolonged exposure.

Nafion polymer and other closely related perfluorinated ionomers were originally developed by E.I. du Pont de Nemours & Co., Inc. in the early 1960's for use as superacid catalysts in organic syntheses. Since then, numerous patents have been issued describing other applications including chlor-alkali membranes, electrode coatings, fuel cells and other electrochemical processes. It is primarily the ion-exchange and permselectivity properties of Nafion polymer that have motivated the increasing level of interest in this polymer for use in chemical sensor and biosensor applications. Nafion is commercially available both as a thermally cured solid sheet and, more importantly for sensor and other biotechnology applications, in a dissolved form that can easily be solvent cast in the laboratory to give thin films. Nafion is a linear copolymer derived from tetrafluoroethylene and perfluorosulfonic acid monomers, and has the following general structure:

$$[(CF_2CF_2)_n\ CFCF_2]_x$$
$$|$$
$$O$$
$$|$$
$$(CF_2CF)_m\text{-}O\text{-}CF_2CF_2SO_3H$$
$$|$$
$$CF_3$$

$n/m \cong 7$ for a 1200 e.w. resin

Structurally, this polymer is a close relative to Teflon except that the saturated fluorocarbon backbone is randomly substituted with perfluorinated side chains, each terminating in a superacidic sulfonic acid group. Despite the structural predominance of hydrophobic fluorocarbon backbone units, bulk Nafion films actually exhibit a hydrophilic character due to sulfonate ion clustering which gives rise to hydrophilic domains distributed throughout the fluorocarbon phase. Detailed studies of the morphology, as well as the chemical, thermal and mechanical properties of Nafion membranes have been reported elsewhere (*11*). The results presented here summarize recent efforts to assess the biocompatibility properties of Nafion polymer.

In Vivo Biocompatibility Investigations

The first *in vivo* evidence of the biological inertness of Nafion polymer came from the original product safety testing by du Pont, although these tests were limited to acute toxicity (from oral ingestion) and skin irritation tests (*12*). Recently, however, one of the authors and previous co-workers carried out the first chronic animal implant studies involving Nafion polymer. These results have been published elsewhere (*13*) and hence will be described here in summary only.

Briefly, sterile samples of Nafion (experimental group) and medical implant grade silicone rubber (control) were surgically implanted at several sub-cutaneous, intravenous and/or intraperitoneal sites in male Sprague-Dawley rats. The samples and surrounding tissues were explanted after 1-200 days; these specimens were then fixed and processed for either histological examination by light microscopy, or scanning electron microscopy. For example, Figure 1 shows light photomicrographs of typical

(A)

(B)

Figure 1. Partial transverse sections of intraperitoneal implants and surrounding tissues recovered after 105 days: (**A**) Nafion film solution-cast on thermally cured Nafion sheet (the arrows identify the solution-cast film which has separated from the underlying Nafion sheet substrate); (**B**) Medical Implant-Grade Silastic silicone rubber (tissue capsule only). The double-headed arrows correspond to 0.21 mm on original specimens. (Reproduced with permission from reference 13. Copyright 1991 Butterworth–Heinemann Ltd.)

samples recovered from intraperitoneal (omentum) sites following 105 days *in situ*. Here, commercially obtained (thermally cured) Nafion sheet material was used as a supporting "substrate" onto which solution-cast, non-thermally cured Nafion films were coated prior to implantation. The capsule thickness (or extent of fibrosis), the absence of inflammatory cells, and the degree of vascularization and morphology of the surrounding tissues are all indicative of a very mild chronic biological response to a foreign material. Similar findings were reported in regard to implants recovered from other abdominal cavity and sub-cutaneous sites (*13*). Note that Figure 1 also shows some typical artifacts that arise due to the specimen preparation procedures: (A) the dehydrated tissue capsules often contracted and pulled the attached solution-cast film away from the solid Nafion substrate; (B) the silicone rubber (control) specimens were highly susceptible to microtome damage during sectioning, often detaching and separating from the section. However, the surrounding tissue capsule, which is the most informative part of the specimen, remained intact in both specimen types.

Within the scope of this study, it was concluded that the chronic *in vivo* biological response to Nafion polymer was qualitatively similar to that of implant grade silicone rubber. Both materials typically resulted in the formation of a thin fibrous tissue encapsulation, but surrounding tissues were well vascularized and free of inflammation in all cases. Although this study provided encouraging evidence of the potential biocompatibility of Nafion polymer, it yielded only qualitative results and it was not possible to isolate the material response *per se* from the response evoked by mechanical irritation, surgical procedures, geometric effects, *etc*. Furthermore, as noted above, it was difficult to isolate (with total confidence) the response due to the solution-cast film alone—and hence to directly evaluate the morphological form of Nafion polymer used in biosensor applications—and to make appropriate comparisons with control materials. It was therefore decided to pursue further studies employing *in vitro* cell culture methods in order to obtain additional quantitative data under conditions that could be more carefully controlled.

In Vitro Biocompatibility Investigations

Cell culture techniques have been used extensively in biocompatibility studies involving polymeric materials (*14-17*). Cell culture systems have been shown to be highly sensitive to toxic moieties and well correlated with animal studies (*15-17*). We have adapted these techniques as described below in order to investigate the *in vitro* biocompatibility of Nafion polymer. The results presented here represent the progress to date on continuing studies directed toward characterizing the properties of Nafion polymer for biosensor applications.

Materials and Methods. The cell line employed in this study was a recombinant derivative of BHK-21/C-13 (Syrian Hamster) fibroblast, which expresses the amino-terminal lobe of human serum transferrin (hTF) (*18*), kindly provided by Professor R.T.A MacGillivray. Nafion polymer was purchased from Aldrich (Milwaukee, WO) as a 5% solution. Dulbecco's Modified Eagle's Medium (DMEM), trypsin-EDTA, Fetal Bovine Serum (FBS) and Nunclon sterile six-well TC-polystyrene culture plates were purchased from Gibco/BRL (Burlington, ON). Glucose, Trypan Blue stain and goat-antihuman transferrin antibody were purchased from Sigma Chemical (St. Louis, MO). FITC-conjugated sheep-antihuman transferrin was obtained from Biodesign International (Kennebunkport, ME). Fluoricon polystyrene capture particles and Fluoricon multiwell assay plates were obtained from Idexx Corp. (Portland, ME). Calcium free, phosphate buffered saline (PBS) solution (pH 7.4, $\mu \approx 0.17$ M) was prepared using distilled, de-ionized H_2O with NaCl (0.8 g/dL) and KCl (2.7 mM) as electrolytes; all inorganic chemicals were of reagent grade and all solutions were sterile filtered (0.2 µm pore size). Glucose assays were carried out using a Beckman

Instruments (Fullerton, CA) Glucose Analyzer 2; Lactate concentrations were measured using a Yellow Springs Instruments (Yellow Springs, OH) Model 27 Lactate Analyzer; transferrin immunoassays were performed using a Pandex Laboratories (Mundelein, IL) Fluorescence Concentration Analyzer. Microscopy and image analysis were performed using a Wild (Heerbrugg, Switzerland) inverted light microscope and the OPTIMAS personal computer based image analysis system from BioScan (Edmonds, WA).

Experimental cultures were divided into three groups based on the cell-contacting material provided as a substrate for cell attachment: i) *Control*, where cells were grown on the untreated Nunclon Tissue Culture Polystyrene (TCP); ii) *Sham*, where cells were grown on TCP treated with a solvent mixture of a similar composition to the Nafion polymer casting solution; and iii) *Nafion*, where cells were grown on Nafion polymer membranes cast from the 5% solution on TCP by solvent evaporation such that the resulting hydrated Nafion film thickness is approximately 20±5 μm. Prior to inoculation, the Nafion Group culture wells were rinsed 3 times with 2 ml of serum-free DMEM in order to convert the Nafion film from the acid form to the sodium form of the polymer. Several six-well culture plates of each surface type were inoculated with the same number of cells obtained from the same stock culture. In each case, cultures were raised in DMEM supplemented with 5% fetal bovine serum (FBS) with 1000 mg/L glucose; cultures were incubated at 37°C under an atmosphere of 5% CO_2 with a balance of humidified air. After each 24 h period, at least three cultures from each group were terminated; the supernatant medium was frozen for later analyses for glucose, lactate and transferrin; the cells were detached using 0.5% trypsin-EDTA solution and counted using either a Speirs-Levy eosinophil-type haemocytometer or an Elzone (Elmhurst, IL) Particle Data Analyzer, which also provides the statistical distributions of cell size/volumes; cell viabilities were determined by haemocytometer using the Trypan Blue dye exclusion method.

Results and Discussion. The first, and perhaps most obvious, comparison that can be made between Nafion polymer and the other surfaces is in terms of growth rates of the cultured cells, which must adhere to the surface in order to divide and grow. Figure 2 shows the growth profiles obtained over 96 h in terms of the average number of cells recovered from each culture per unit volume of culture medium. Each of the experimental groups exhibits exponential growth, following a brief lag phase, however the rate of growth of the Nafion group appears to be slightly lower relative to that of the Control and Sham groups. Note that, within experimental error, the Control and Sham groups exhibit similar profiles, indicating that the observed result for Nafion is not due to solvent residues or other artifacts arising from the Nafion casting process. In all experiments, the fraction of non-viable cells recovered were less than 2% of the total fraction recovered and there was no discernable difference in viable cell numbers between the Nafion and Control or Sham groups.

A quantitative measure of the exponential growth rates was obtained from logarithmic plots of the growth profiles as shown in Figure 3. Using the data presented in Figure 2, the growth rate on Nafion is approximately 10% less than that on the TCP and Sham surfaces as determined from the slopes of least-squares lines fitted to the experimental data. This result is typical of that obtained in more than 10 similar experiments and is considered to be a relatively small difference of the order of that which can be observed when comparing different cell lines grown on the same material. We have performed similar experiments (data not shown) using an alternative cell line (L-929 mouse fibroblasts) in which the growth profiles were quantitatively different from those involving the BHK cells used here, but exhibited qualitatively similar variation between Nafion and Control/Sham groups. We also note that, at the end of some 96 h cultures, the cells from all three experimental groups were replated onto fresh TCP surfaces and these subcultures exhibited growth profiles

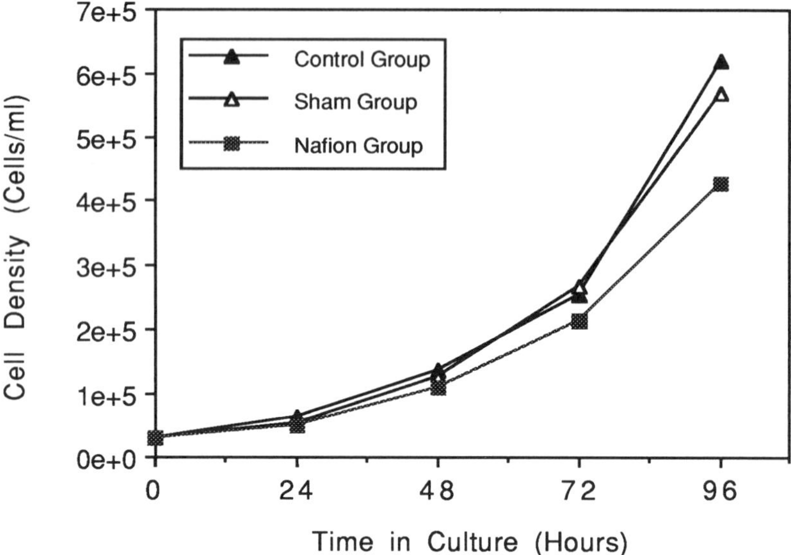

Figure 2. Comparison of growth profiles over 96 h for recombinant BHK cells grown on the three experimental surfaces. Each data point represents an average of six counts on each of three separate cultures from each group.

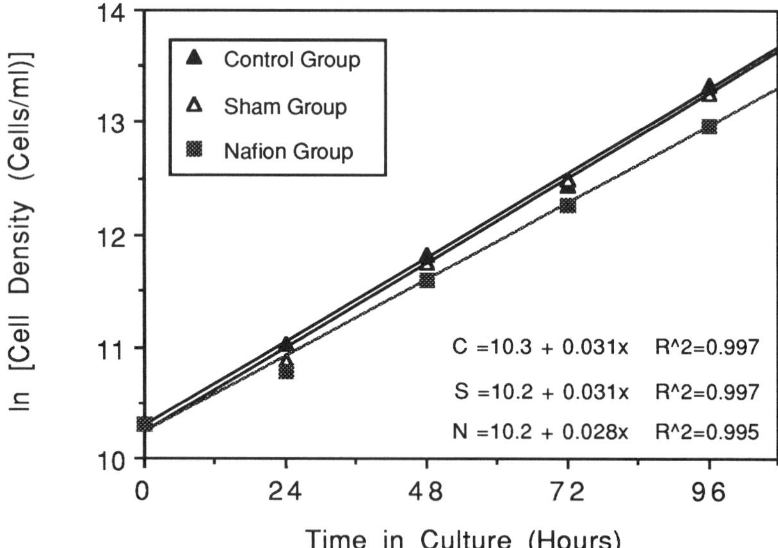

Figure 3. Logarithmic plot of growth profiles over 96 h for recombinant BHK cells grown on the three experimental surfaces. The equations given describe the linear regression lines for the Control (C), Sham (S) and Nafion (N) groups respectively; the R^2 values refer to coefficients of determination.

that were indistinguishable from each other (*i.e.* the Nafion group "returned" to the behavior of the Control group). Hence, the conditions imposed by growth on a Nafion surface did not apply any discernable selection pressure on the experimental inoculum. In contrast to these results for Nafion, the growth profile of BHK cells grown on a poly(vinyl chloride) surface is shown in Figure 4 along with a control culture similar to that used in the Nafion comparison; note that the BHK cells do not exhibit a positive growth rate on this eminently non-biocompatible surface.

The growth rate of anchorage-dependent cells on any material depends not only on direct interactions between the cells and the surface, but also on indirect interactions between the cells and adsorbed proteins such as fibronectin and vitronectin from the serum component of the culture medium. The data presented in Figures 2 and 3 were obtained from cells cultured in 5% FBS. We also performed similar experiments where the serum supplement was varied between 2% and 10% in order to determine whether such indirect interactions could wholly or partially account for the lower growth rates observed for the Nafion group. The essential results are summarized in Figure 5, which shows the cell counts after 24 h and indicates that the variation between the Nafion and Control groups is clearly serum-dependent, diminishing as the serum level is increased; in other similar experiments, where the serum level was taken as high as 20%, the differences between Control and Nafion groups effectively disappears. This result is also consistent with the results reported by others (*19,20*) who have shown that Nafion surfaces are less effective in adsorbing fibronectin and vitronectin from the culture medium.

A number of cultures were examined visually, using an inverted light microscope, and by computer image analysis in order to compare the morphologies of cultured cells in each of the experimental groups (data not shown). The number of attached and growing cells (characterized by their flattened, multifaceted, fibroblast-like morphology), exhibited profiles that closely resembled the overall growth profiles displayed in Figure 2, with the Nafion group slightly lagging. However, no statistically significant difference between the Nafion and Control groups was found in the number of dividing cells (characterized by their smooth spherical morphology) throughout the culture period of 96 h. Although the dividing cells represent by far the smallest fraction (<5%) in both groups, these observations are suggestive of the possibly dominant role of physical chemical, as opposed to biochemical, interactions being responsible for the observed differences in overall growth rates.

The glucose utilization in each of the three experimental groups was examined by analyzing the culture supernatants to determine the total glucose and lactate concentrations at each of the 24 h intervals represented in Figures 2 and 3. Figure 6 shows the lactate production data plotted against the corresponding glucose consumption data. Note that the glucose utilization characteristic of cultured cells is commonly seen to vary throughout a culture in a manner similar to that shown. The significant point here is that the data from all three experimental groups lie on essentially the same characteristic, indicating that no discernable difference was observed between the Nafion and Control/Sham groups with respect to glucose utilization via glycolysis. Hence, no cytotoxic moieties, either surface-associated or medium-associated, capable of affecting glucose metabolism (at least via the major glycolytic pathway ending in lactate) were detected by the cells growing on Nafion surfaces. Apropos of this, it should be pointed out that no traces of solubilized Nafion were found (at levels detectable by fluorine NMR), either in the culture supernatants or whole cell lysates, after 96 h in culture.

A substantial portion of the energy derived from glycolysis is utilized by the cells for protein synthesis. It is difficult to resolve subtle changes in the levels of total protein synthesis, and was not evaluated (thus far) in this study. However, the recombinant hTF secreted into the culture supernatant by the BHK cells employed here does provide a readily quantifiable measure of one relatively sensitive component of

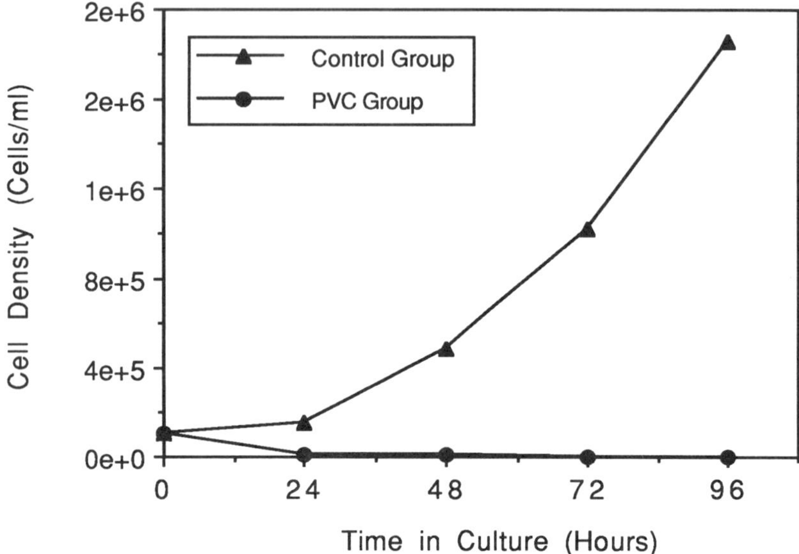

Figure 4. Comparison of growth profiles over 96 h for recombinant BHK cells grown on poly(vinyl chloride) (PVC) and Control (TCP) surfaces. Cell counts for the PVC group were obtained by haemocytometer.

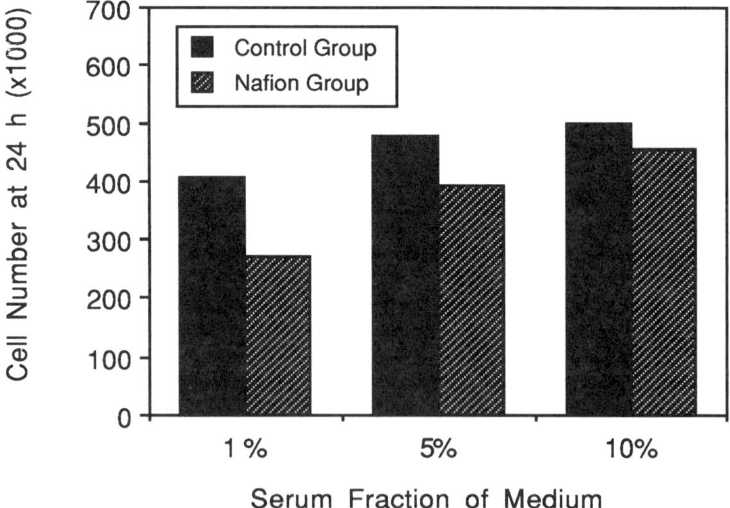

Figure 5. Dependence of cell counts after 24 h on serum (FBS) supplement level for recombinant BHK cells grown on Control (TCP) and Nafion surfaces. All cell counts were obtained by haemocytometer.

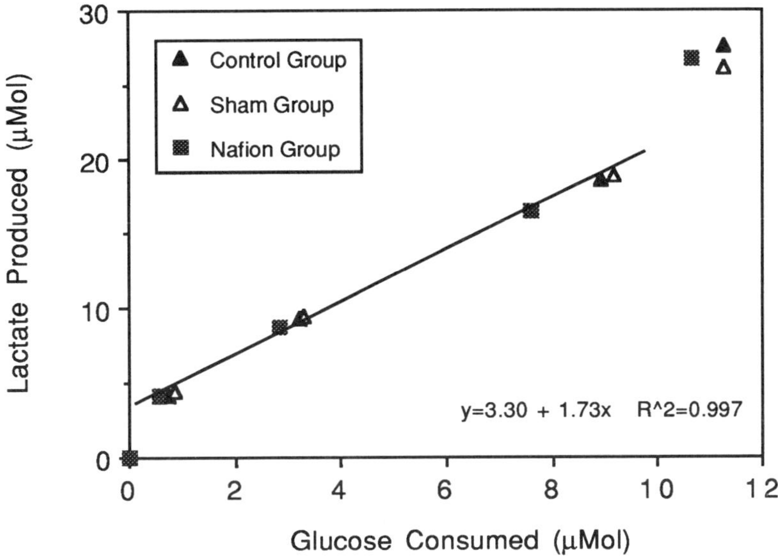

Figure 6. Combined lactate production *versus* glucose consumption data for each of the three experimental groups. Data from all three groups (at 24, 48 and 72 h) were pooled and equally weighted in the calculation of the regression line.

protein synthesis, and hence a potential indicator of the possible effects of chemical or surface environmental factors (due to the presence of the Nafion film) affecting intracellular processes. Figure 7 shows the yield on glucose of hTF in each of the experimental groups over the 96 h culture period. The profiles are again qualitatively similar. Quantitatively, however, the yield is significantly higher from the Nafion group, particularly through the early stages of the culture, with the yield of the Nafion group appearing to converge toward that of the Control/Sham groups as the culture approaches a confluent monolayer (slightly beyond 96 h). The cell counts and glucose consumption profiles indicate that the cultures remain in an exponential growth phase at 96 h, and hence the apparent convergence was not an artifact of glucose limitation.

It is impossible at this point to interpret the initially higher hTF yields seen in the Nafion group in terms of any specific mechanism of interaction between the cells and Nafion polymer. Indeed, heterologous protein expression and secretion by cultured cells has been shown to vary considerably in response to many environmental and/or medium conditions (*21,22*) which has yet to be explained or characterized in depth. Nevertheless, the effect seen here was consistently observed in more than 10 separate experiments and thus represents at least a transient biological effect of Nafion polymer. It should be noted that, here again, when replated at the end of 96 h onto fresh TCP surfaces, the subculture derived from the Nafion group exhibited hTF secretion that was indistinguishable from that of the subcultures derived from the Control/Sham groups.

The differences in hTF expression are perhaps minor, but in the context of a *biocompatibility* study, are difficult to explain or dismiss and hence warrant some further investigation. The apparently transient nature of this behavior, together with other data suggesting that cells attach and spread less efficiently on Nafion surfaces, have prompted us to investigate the possibility of a polymer-related stress response. Several studies have in fact shown that a variety of environmental stressors can induce such a response, manifested by elevated expression of the HSP families of so-called

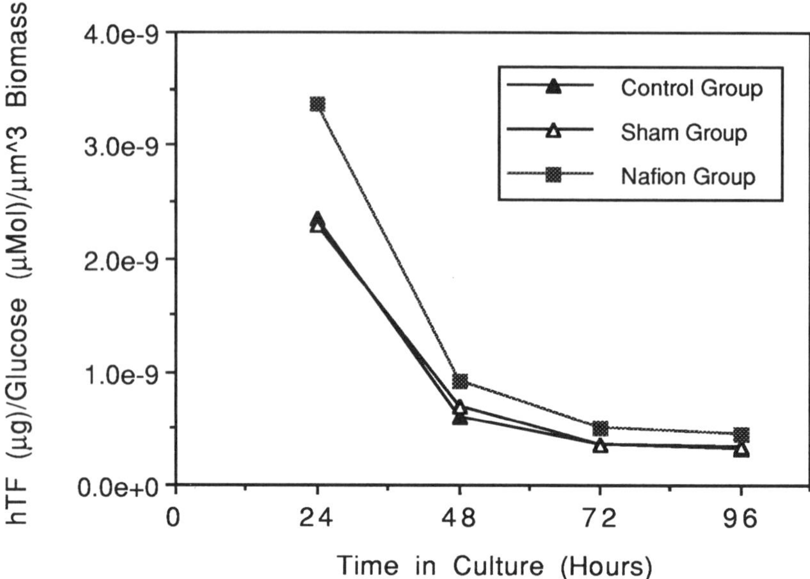

Figure 7. Recombinant hTF yield on glucose as a function of time for each of the three experimental groups. All hTF assays have been corrected for cell number and cell volume to give yields per μm^3 of biomass.

stress proteins (*23,24*), and higher recombinant protein expression has occasionally been observed to accompany this HSP induction. These studies are now in progress and can not be fully reported here, although preliminary results have thus far revealed no detectable stress response.

Conclusions

We have presented a summary of experiments to date directed toward assessing the biocompatibility of solution-cast Nafion polymer membranes. *In Vitro* cell culture methods were employed to obtain a quantitative comparison between Nafion polymer and biocompatible TC-polystyrene (control) surfaces in terms of cell growth rates, nutrient metabolism, recombinant protein production and gross morphology. It was found that BHK fibroblasts exhibit similar glucose utilization characteristics, but slightly slower growth rates, on Nafion surfaces (*cf.* TCP). The morphologies of cells growing on Nafion were qualitatively similar to those growing on TCP and no significant differences were observed in the numbers of dividing cells on either surface, however the numbers of spread cells were slightly lower for the Nafion surfaces. The recombinant hTF yields from the Nafion group were consistently higher immediately following inoculation, but rapidly converged to those of the Control/Sham groups as the cultures approached confluency; further investigation of this behavior is part of ongoing work in this laboratory on the characterization of Nafion polymer for biosensor and biotechnology applications. In general, the variations between Nafion and Control/Sham groups seen thus far are minor and, collectively, these results suggest that Nafion polymer is indeed sufficiently biocompatible to warrant further development of implantable biosensors based on Nafion polymer encapsulating/ dialysis membranes. Finally, we point out that these results augment and support the conclusions of earlier preliminary *in vivo* biocompatibility studies involving chronic Nafion polymer implants.

Acknowledgments

The authors wish to acknowledge the financial support of the Natural Sciences and Engineering Research Council of Canada. We are also extremely grateful to Dr. R.T.A. MacGillivray of the Department of Biochemistry at UBC for providing the recombinant cell line used in these studies, and to Dr. J.M. Piret of the Biotechnology Laboratory at UBC for valuable discussions.

Literature Cited

1. Guilbault, G.G.; Luong, J.H. *Chimia* **1988**, *42*, 267-271.
2. Czaban, J.D. *Anal. Chem.* **1985**, *57*, 345A-352A.
3. Regnault, W.F.; Picciolo, G.L. *J. Biomed. Mater. Res.* **1987**, *21*, 163-180.
4. Harrison, D.J.; Turner, R.F.B.; Baltes, H.P. *Anal. Chem.* **1988**, *60*, 2002-2007.
5. Turner, R.F.B.; Harrison, D.J.; Rajotte, R.V.; Baltes, H.P. *Sensors and Actuators* **1990**, *22(B1)*, 561-564.
6. Fortier, G.; Vaillancort, M.; Bélanger, D. *Electroanalysis* **1992**, *4*, 275-283.
7. Crespi, F.; Möbius, C. *J. Neuroscience Methods* **1992**, *42*, 149-161.
8. Jin, L.; Ning, B.; Ye, J.; Fang, Y. *Mikrochimica Acta* **1991**,*1*, 115-120.
9. Gerhardt, G.A.; Oke, A.F.; Nagy, F.; Moghaddam, B.; Adams, R.N. *Brain Res.* **1984**, *290*, 390-395.
10. Kristensen, E.W.; Kuhr, W.G.; Wightman, R.M. *Anal. Chem.* **1986**, *59*, 1752-1757.
11. *Perfluorinated ionomer membranes*; Eisenberg, A.; Yeager, H.L., Eds.; ACS Symposium Series 180, American Chemical Society: Washington, DC, 1982.
12. E.I. du Pont de Nemours and Co., *Nafion perfluorinated membranes—safety in handling and use*, Bulletin E-38524, Polymer Products Department: Wilmington, DE, 1983.
13. Turner, R.F.B.; Harrison, D.J.; Rajotte, R.V. *Biomaterials* **1991**, *12*, 361-368.
14. Guess, W.L.; Rosenbluth, S.A.; Schmidt, B.; Autian, J. *J. Pharm. Sci.* **1965**, *54*, 1545.
15. Johnson, H.J.; Northrup, S.J.; Seagraves, P.A.; Atallah, M.; Garvin, P.J.; Lin, L.; Darby, T.D. *J. Biomed. Matr. Res.* **1985**, *9*, 489-508.
16. Rae, T. Tissue; In *Techniques of Biocompatibility Testing*; Williams, D.F., Ed.; CRC Series in Biocompatibility, CRC Press, Boca Raton, FL, 1986, Vol. II, Ch. 3.
17. *Cell Culture Test Methods*; Brown, S.A., Ed.; ASTM STP 810, ASTM: Philadelphia, PA, 1983.
18. Funk, W.D.; MacGillivray, R.T.A.; Mason, A.B.; Brown, S.A.; Woodworth, R.C. *Biochemistry* **1990**, *29*, 1654-1660.
19. McAuslan, B.R.; Johnson, G.; Hannan, G.N.; Norris, W.D. *J. Biomed. Matr. Res.* **1988**, *22*, 963-976.
20. Steele, J.G.; Johnson, G.; Norris, W.D.; Underwood, P.A. *Biomaterials* **1991**, *12*, 531-539.
21. Pendse, G.J.; Karkare, S.; Bailey, J.E. *Biotech. Bioeng.* **1992**, *40*, 119-129.
22. Glacken, M.W.; Adema, E.; Sinskey, AJ. *Biotech. Bioeng.* **1988**, *32*, 491-506.
23. *Heat Shock and Development*; Hightower, L.; Nover, L., Eds.; Results and Problems in Cell Differentiation 17, Springer-Verlag: Berlin, Germany, 1991.
24. *Heat Shock Response of Eukaryotic Cells*; Nover, L., Ed.; Springer-Verlag: Berlin, Germany, 1984.

RECEIVED July 19, 1993

Chapter 18

Nondegradable and Biodegradable Polymeric Particles

Preparation and Some Selected Biomedical Applications

E. Pişkin[1], A. Tuncel, A. Denizli, E. B. Denkbaş, H. Ayhan, H. Çiçek, and K. T. Xu

Department of Chemical Engineering, Bioengineering Division, University of Hacettepe, 06530 Ankara, Turkey

This paper summarizes our recent studies related to production of nondegradable and biodegradable polymeric particles and their use in diverse biomedical applications. Nondegradable monosize polystyrene based particles were prepared in micron-size range by a phase inversion polymerization. Surfaces of these particles were then coated with styrene-acrylate copolymer layers in order to include different functional groups. These particles were radiolabelled with ^{99m}Tc and succesfully used in animals and in human subjects for diagnostic imaging of GIT. Phagocytosis of these particles and their biologically modified forms by leucocytes and also macrophages were investigated. The polystyrene particles were coated with a cross-linked polyvinylalcohol layer, and then a dye (i.e.,Cibacron blue) was attached for specific protein adsorption. The extents of nonspecific and specific protein (i.e., albumin) adsorption on these particles were presented. Biodegradable monosize polyethylcyano-acrylate particles were prepared by a dispersion polymerization. The effects of several parameters on degradation of these particles were discussed. Biodegradable polylactic acid particles with narrow size distribution were produced by solvent evaporation techniques. A model drug was also loaded in these particles, and its release from these particles were investigated.

Monosize polymeric particles have attracted much attention as carrier matrices in a wide variety of biomedical applications (*1*). These polymeric particles have been used in several immunoassays (e.g., latex agglutination, immunoenzymometric assays, etc.) (*2-5*). Various cells (e.g., red cells, B and T lymphocytes, bone marrow cells, etc.) have been separated successfully by using polymeric particles carrying specific ligands on their surfaces (*1,6-9*). A variety of colloidal delivery systems in the form of microspheres and microcapsules have been developed for controlled release of bioactive substances and for targeting therapeutic agents to their site of action (*10-13*). Colloidal particles labelled with radionuclides have been used in nuclear medicine for *in vivo* visualizing of several diseases (*14,15*). Monosize polymeric particles with different size and surface properties have evaluated in the study of phagocytosis process (*16-19*). Polymeric particles and their biologically modified forms (carrying a wide variety of ligands) have been widely used in affinity separation of many important biologicals (*20-22*).

[1]Current address: P.K. 716, Kizilay, 06420 Ankara, Turkey

0097–6156/94/0556–0222$08.00/0

Suspension and emulsion polymerization are two classical polymerization techniques to produce spherical polymeric particles. Larger particles (usually larger than 50 μm) with an appreciable size distribution are produced by suspension polymerization. Submicron polymeric particles (usually smaller than 0.1 μm) with extremely uniform in size are obtained by conventional emulsion polymerization processes. Recent techniques, such as swollen emulsion polymerization, dispersion polymerization, etc. give micron-size (usually between 1-50 μm) monosize polymeric particles (*23*).

Recently, we have also produced monosize nondegradable and biodegradable polymeric particles and used these particles in diverse biomedical applications. In this paper, we reviewed our recent related studies. In each section, we briefly decribed the polymerization procedure, and then gave examples of their biomedical and related applications.

Nondegradable Polystyrene Based Particles

Preparation of PS Particles. We produced monosize polystyrene (PS) particles by following a "phase inversion polymerization" technique which was described in detail elsewhere (*23-27*). In order to obtain PS particles with different size ranges, we studied a wide variety of solvent systems with different polarities (e.g., ethanol/water, isopropanol/ water, and ethanol/2-methoxyethanol). We also changed concentrations of the stabilizer (i.e., polyacrylic acid, PAA), the initiator (i.e., 2,2'-azobisisobutyronitrile), and the monomer to control the size and the monodispersity of these particles.

In a typical polymerization system a glass, jacketed, magnetic drive, sealed cylindrical reactor was used. The dispersion medium was prepared by mixing the relevant alcohol and water in the desired ratio, and the stabilizer was then dissolved in this medium. The initiator was dissolved within the monomer, and this phase was added to the dispersion medium. This polymerization medium was then charged to the reactor. Polymerizations were conducted for 24 h at 75°C with a stirring speed of 150 rpm. A representative picture of the PS particles is given in Figure 1.

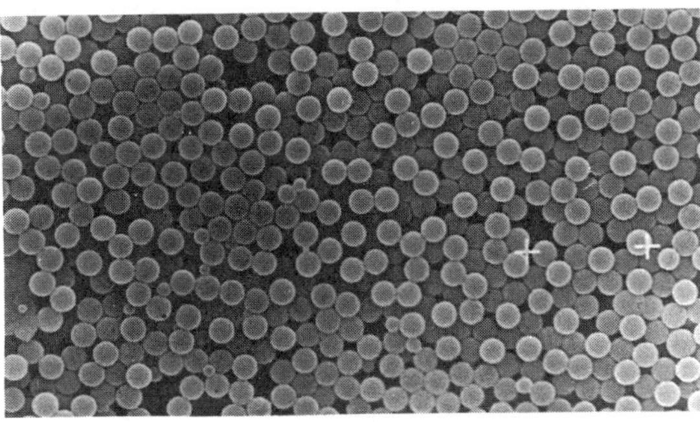

Figure 1. Scanning electron micrograph of monosize PS particles.

Details of our studies related to the variation of the monomer conversion, and particle size and monodispersity with the polymerization parameters were given elsewhere (*26-28*). The following important points elaborated from these studies should be noted.

- Polarity: Polarity of the dispersion medium is one of the most important parameters which controls the average size and monodispersity of the PS particles. The average particle size increases with decreasing polarity of the dispersion medium. Water content is important to achieve phase inversion process and to control monodispersity.
- Initiator concentration: The average particle size increases with increasing initiator concentration.
- Stabilizer Concentration: The average particle size decreases with increasing stabilizer concentration.
- Monomer Concentration: The average particle size and size distribution increase with increasing the monomer concentration.

Surface Modification of PS Particles. In addition to uniformity in size, surface chemistry of the polymeric particles is another main consideration for biomedical applications. Monosize polymeric particles having different surface chemical groups were also synthesized from various other monomers (*29-32*).

Recently, we were able to coat our monosize PS particles described above with styrene/acrylate copolymers. As acrylate monomers, 2-hydroxyethylmethacrylate (HEMA), acrylic acid (AA), and dimethylaminoethylmethacrylate (DMAEMA) were used in order to include functional groups, namely hydroxyl (OH, uncharged), carboxyl (COOH, negatively charged), dimethylamino ($N(CH_3)_2$, positively charged) on the surfaces of the PS particles.

A short description of the copolymerization procedure is as follows: Prior to the copolymerization, the relevant acrylate monomer was mixed with styrene, and the initiator was dissolved in this mixture which was then added to PS latex diluted with water. This medium was stirred at room temperature for 24 h to allow the adsorption of monomers on the PS particles. The adsorbed monomer layer at the outer shelf of the particles were then polymerized for 24 h at 85°C with a stirring speed of 200 rpm. Note that the coating procedure did not change the size and monodispersity of the original PS particles. The existance of the functional groups on the PS particle surfaces were confirmed by ESCA, FTIR and electrophoretic mobility measurements (*33*).

Diagnostic Imaging of GIT with PS Based Radiopharmaceuticals. ^{99m}Tc-sulfur and other colloids, ^{99m}Tc or ^{131}In-labelled diethylenetriamine pentaacetic acid, ^{99m}Tc or ^{131}In-cellulose and many other radiopharmaceuticals have been proposed for scintigraphic visualization of gastrointestinal tract (GIT) (*34-37*). All these radiopharmaceuticals proposed and tested so far have been considered as nonideal because of their following limitations: Most of the carriers are not stable enough to survive the drastic pH values and changes, and enzymatic action prevailing in the gastrointestinal tract without either losing the radiolabel, getting metabolized or decomposed. The free radiolabel released from the carrier during its passage through GIT is absorbed and causes an elevated blood background activity. The free label also accumulates in the thyroid, stomach, and urinary bladder, which then interferes the precision of the colon studies. The labelling efficiency of the existing carriers is low, and they unevenly (nonhomogeneous) distribute within the GIT.

Recently we proposed to use our monosize PS based polymeric particles as a carrier matrix to study GIT transit and morphology. Either the dimethylaminoethylmethacrylate (DMAEMA) or acrylic acid (AA) coated PS particles were used as the basic carrier. These particles were labelled with ^{99m}Tc by applying two different techniques: (i) direct labelling by tin reduction, and (ii) ligand exchange with tin pyrophosphate. Details of these labelling methods were discussed elsewhere (*38*). Briefly, about 99% labelling efficiency was achieved at pH values below 5 for the AA coated PS particles with both methods. The optimum pH for labelling was 5 for the DMAEMA coated PS particles. For these particles, about 90% and 99% labelling efficiencies were obtained by following the first and second labelling methods, respectively. Stabilities of these radiolabelled particles were examined at different time intervals for 48 h after storage at room temperature by Impregnated Thin Layer Chromatography. The stability studies were also repeated at pH 1 and pH 8. Labelling efficiencies were higher than 98% even after 24 h for both types of particles obtained by both methods.

We extensively investigated these particles in *in vivo* studies (*39-42*). These studies were performed by using adult Chincilla rabbits. After overnight fast, ^{99m}Tc-labelled particles were given orally. Whole body scintigrams were obtained immediately and at selected intervals for about 48 h after administration by a gamma camera. The scintigrams were drawn over the GIT and the whole body. The feces were also collected at the end of 24 h and 48 h and were counted.

In scintigraphic studies in rabbits, the label stayed intact during 48 h of observation as there was no dedectable radioactivity outside of GIT ($>90\%$ in GIT). The absence of thyroid, stomach (at late scintigrams) and urinary bladder images reinforced the *in vivo* stability of the labelled particles. The amount of activity recovered in feces was $8.6 \pm 6.7\%$ and $21.3 \pm 13.8\%$ (mean and standard deviations of 5 observations) after 24 and 48 h, respectively.

These PS based pharmaceuticals were also tested clinically, after obtaining informed consents of the volenteers and patients. These studies in man confirmed the findings obtained in rabbits. A scintigram obtained in a human subject is given as an example in Figure 2. Notice that only GI tract was visualized. Almost no radioactivity accumulation was detected outside of GIT. The ratio of GI tract/whole body was higher than 99% .

According to these encouraging results we concluded that currently available pharmaccuticals may bc replaced by these novel pharmaceuticals to study the colon transit time and morphology, because of the following reasons: (i) the labelling is very simple, effective, and reproducible; (ii) the labelled particles are very stable and remain in the GIT until excretion by feces without dissociation; (iii) they allow to monitore the passage of radioactivity through the GIT without any interference from background and other organs.

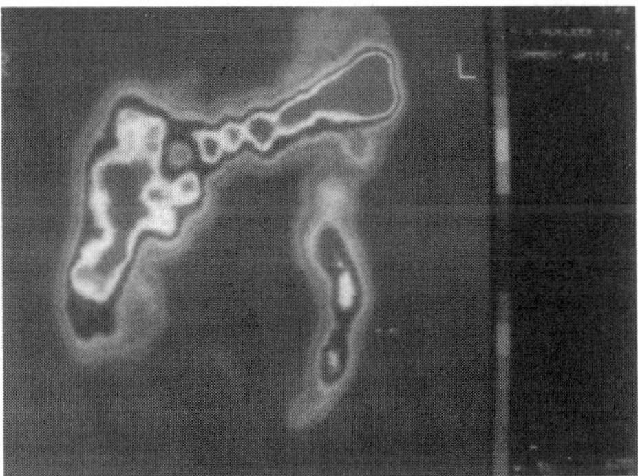

Figure 2. A scintigram obtained in a human object by using ^{99m}Tc-labelled monosize PS particles.

Phagocytosis of PS Based Particles One of the main defense mechanisms of the living system against foreign bodies (e.g. viruses and bacteria) is "phagocytosis", which is usually described as "the internalization of micron-size solid foreign materials by cells". In early studies, vital dyes, colloidal suspensions (i.e. carbon, iron oxide), etc. were utilized for visualization and quantitation of the phagocytic process (*43*).

However, low colloidal stability, lack of size uniformity, aggregation and other problems associated with these model solid particles were noted as important limitations (*43,44*). More recently, synthetic polymeric particles, which can be readily manufactured in desired size range and with different surface properties, have been utilized extensively in studying phagocytosis (*44-47*).

Recently, we investigated phagocytosis of our monosize PS based particles by blood cells and also by mouse peritoneal macrophages. The monosize PS particles with different sizes (0.9-6.0 μm), their modified forms carrying functional groups on their surfaces (i.e., PS/PHEMA, PS/PAA and PS/PDMAEMA), and the PS particles pretreated within bovine serum albumin (BSA) and fibronectin (Fn) aqueous solutions were utilized in these studies (*48*).

Blood samples from healthy volunteers were incubated in heparinised tubes 5% (w/v) in water (10 ml PS latex of a 0.25 ml blood sample) for 20 minutes at 37°C. After fixing in methanol and dying with a proper dye (i.e., Giemsa or Jenner, Merck, Germany), leucocytes were observed by a phase-contrast microscope. In each test at least 200 leucocytes were examined and the average number of particles internalized by one leucocyte was determined. Experiments were repeated three times for each type of particle. Phagocytosis behavior of the peritoneal macrophages freshly isolated from mouse were also studied in Hanks' balanced salt solution and phospahate-buffered saline solution at pH: 7.4, by following the same technique used for leucocytes.

Figure 3 shows some representative pictures of phagocytosis. The following important results should be noted:

- Size of Particles: The particle uptake dropped significantly with the particle size. Both cell types investigated in this study were able to uptake only very few particles having a diameter larger than 4 μm.
- Surface Properties: Table I shows the effects of surface chemistries on the number of particles phagocytosed by different cells. Notice that the uptake of the more hydrophobic PS particles was higher than the less hydrophobic PS/PHEMA particles. Amino groups (i.e., positively charged) on the PS/PDMAEMA particles significantly increased the particle uptake. In contrast to amino groups, carboxyl groups (i.e., negatively charged) on the PS/PAA particles caused a pronounced drop in the number of particles phagocytosed. Biological modification of particle surfaces dramatically changed the behavior of phagocytosis of particles.The Albumin molecules preadsorbed on the PS particle surfaces drastically reduced the number of particles internalized by both cell groups while preadsorbed fibronectin, known as "biological glue", significantly enhanced the phagocytosis.

a

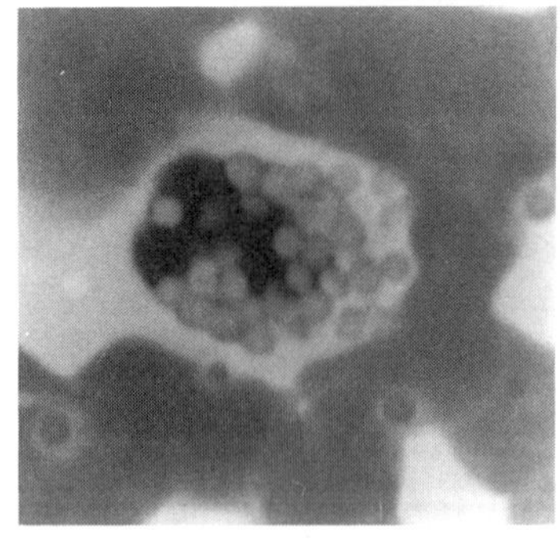

b

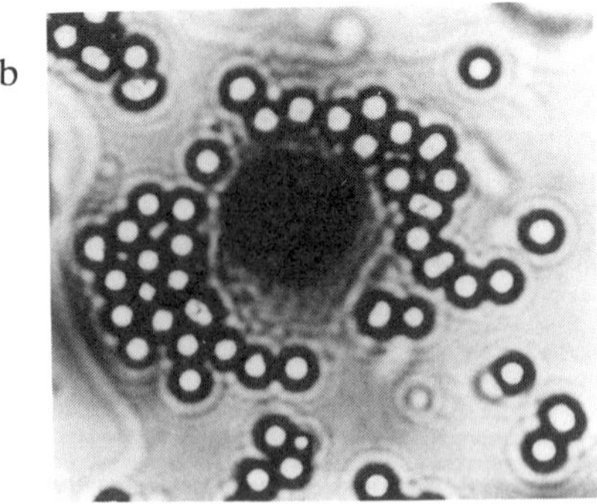

Figure 3. Phagocytosis of PS based particles by leucocytes: (a) plain PS/monocyte; (b) BSA pretreated PS/macrophage.

Table I. Phagocytosis of PS Particles by Different Cells.

PS Particle*	PS Particle No / Cell		
	Neutrophils	*Monocytes*	*Macrophages*
PS	13 ± 4	15 ± 5	42 ± 7
PS/PHEMA	4 ± 2	6 ± 1	10 ± 2
PS/PDMAEMA	20 ± 3	21 ± 4	55 ± 9
PS/PAA	9 ± 3	10 ± 3	18 ± 4
PS/BSA**	0	0	9 ± 6
PS/Fn***	22 ± 8	24 ± 7	54 ± 8

* Particle Diameter: 2.5 μm
** PS particles were soaked in bovine serum albumin (BSA) solution (4 mg BSA /ml) for 24 h.
*** PS particles were soaked in fibronectin (Fn) solution (1 mg Fn/ml) for 24 h.

PS Based Particles for Affinity Chromatography. The interest in and demand for proteins in biotechnology, biochemistry and medicine have contributed to an increased exploitation of affinity chromatography. Unlike other forms of protein separation, affinity chromatography relies on the phenomenon of biological recognition, which enables biopolymers to recognize specifically, and bind reversibly, their complementary ligands (e.g., enzymes) and their substrates, hormones and their receptors, antibodies and their antigens (*20-22, 49-51*).

Unfortunately, preparation of sorbents carrying biological ligands is usually very expensive because the ligands themselves often require extensive purification and it is difficult to immobilize them on the carrier matrix with retention of their biological activity. As alternative to their natural biological counterparts, the reactive triazinyl dyes, have been investigated as ligands for protein affinity separation (*52-54*). These dyes are able to bind proteins in a remarkably specific manner. They are inexpensive, readily available, biologically and chemically inert, and are easily coupled to support materials. Cibacron Blue F3GA, and many other reactive dyes have been coupled to a variety of supports including agarose, cellulose, polyacrylamide, sephadex, silica and glass (*55-60*). Dye-ligand chromatography has now enabled the purification of a wide range of proteins (e.g., lactate dehydrogenase, alcohol dehydrogenase, hexokinase, carboxyl peptidase, etc.) (*52-60*).

Recently we attempted to use our monosize PS particles as a carrier matrix for affinity purification of proteins. We selected albumin as a potential model protein. We studied both nonspecific albumin adsorption on the PS particles and also specific albumin adsorption on the dye-attached PS particles (*49,61*). Some interesting results of these studies are briefly discussed below.

Albumin Adsorption on PS Particles. Bovine serum albumin (BSA) adsorption on various types of polymeric particles have been investigated by several groups (*62-70*). Note that the particles used in these previous studies had been produced either by conventional emulsion polymerization or seeded copolymerization recipes. Therefore, most of these particles contain charged groups on their surfaces coming from the water soluble initiator used in the polymerization recipes. These charged groups exhibit strongly acidic character which, of course, significantly affects the protein-surface interactions.

Here, in the first group of the experiments we investigated adsorption behavior of BSA onto the PS based particles descibed above, which have quite different surfaces than those produced with conventional recipes. In the equilibrium adsorption studies, 10 mg of BSA was dissolved in 10 ml of buffer solution. Ionic strength of the medium was adjusted by using different amounts of NaCl. The PS particles (0.5 g) were added to this BSA solution. Adsorption experiments were conducted for 2 h at a constant temperature of 25°C and with a stirring rate of 40 rpm. The BSA equilibrium concentrations were measured by means of a spectrophotometer at 280 nm. We studied the effects of pH, ionic strength and coexistent electrolytes on adsorption of BSA both at native and denatured states (*49,61*). Figure 4 gives some representative adsorption data. Some important results of these studies are briefly presented below .

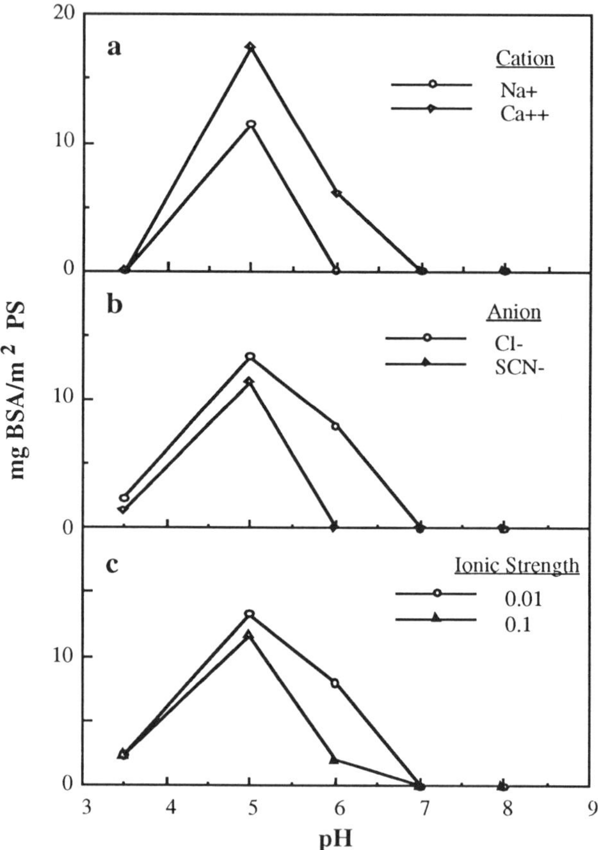

Figure 4. BSA adsorption on PS particles: Effects of (a) cations, (b) anions and (c) ionic strength as a function of pH.

• The maximum adsorption capacities were observed near the isoelectric point (IEP) of BSA (i.e., about 5) in all cases. It has been shown that proteins have no net charge at their isoelectric points (IEP), therefore the maximum amount of adsorption from aqueous solutions are observed at IEP (*62-70*). Below or above their IEP values, proteins are charged positively or negatively, respectively. Therefore, they are more hydrated, which increases their stability and solubility in aqueous phase (means low adsorption). We also obtained significantly lower and nearly zero adsorptions with our particles in both acidic and alkaline regions. We did observe no adsorption of BSA, which is highly hydrated (high solubility in aqueous phase) in alkaline region (pH > 7.0), on our PS based particles in all experimental cases.

• Cations and anions did also take an important role in the adsorption process. In the presence of divalent cation (Ca^{++}), we observed higher albumin adsorption than anions (Cl^- and SCN^-), possibly by stabilizing the protein molecules. The BSA adsorption in the presence of Cl^- anions were lower than those obtained in the presence of SCN^- ions.

• High adsorption values were obtained at lower ionic strengths, as also observed by Shirahama and Suzawa (*68*), which may be due to stabilization of the protein molecules.

Albumin Adsorption on Dye Attached PS Particles. In this part of the study, we prepared dye attached monosize PS particles (4 µm in size) for specific albumin adsorption. In order to attach the dye (i.e., Cibacron Blue F3G-A), firstly, the PS particles were coated with polyvinylalcohol (PVAL) by simple adsorption from aqueous solution containing Na_2SO_4 at 25°C. The adsorbed PVAL layer on the PS surfaces was then chemically cross-linked by using terephatalaldehyde (TPA). For the attachment of dye, the PVAL coated PS particles were incubated with an aqueous solution of Cibacron Blue containing also NaOH in a sealed reactor for 4 h at a 400 rpm of stirring rate and a constant temperature of 80°C. Details of these procedure was given elsewhere (*61*).

The plain PS, PVAL coated PS and Cibacron Blue F3GA attached PS particles were used in the BSA adsorption experiments performed in a buffer medium containing NaCl at 0.01 ionic strength. Figure 5 shows the adsorption isotherms derived at pH: 5.0 and at a constant temperature of 20°C. As it is expected BSA adsorption increased with the initial concentration of BSA. There was a pronounced BSA adsorption on the PS particles (up to 17 mg BSA/m^2) because of the hydrophobic interactions between albumin and the hydrophobic plain PS surfaces. The PVAL coating significantly decreased BSA adsorption on because of the hydrophilic PVAL layer on the polymer surface. High BSA adsorption capacities (up to 40 mg/m^2 PS) were obtained with the Cibacron Blue F3GA attached PS, as it was aimed.

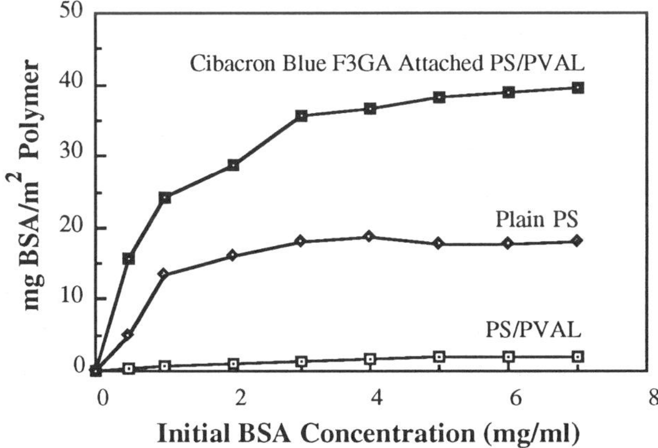

Figure 5. BSA adsorption isotherms.

Biodegradable Monosize Polycyanoacrylate Particles

Biodegradable polymer colloids and suspensions in submicron and in micron-size ranges have been investigated as potential carriers for sustained drug delivery and targeting. Both natural (e.g., albumin) and synthetic (e.g., polylactic acid, polycaprolactone, polyacrylates and polyalkylcyanoacrylates) polymers have been used in these applications.

Polyalkylcyanoacrylates is one of the biodegradable polymer families used for the production of nanoparticles or nanocapsules, which have attracted a great attention due to their ease of preparation, ability to sorb efficiently a large quantities of drugs, their in vivo stability, biodegradability, and biocompatibility (*71-75*).They have been utilized in different applications, such as occular drug carriers to enchance corneal penetration of polar drugs, as carries (with monoclonal antibodies) for targeting of cytotoxic agents to tumor cells (*76-78*).

Recently, we produced monosize polyalkylcyanoacrylate based polymeric particles in micron-size range by a dispersion polymerization method, as potential carriers for diagnosis and therapy. Details of polymerization and degradation kinetics were discussed in detail elsewhere (*79*). Here, the preparation procedure and degradation behaviour of the polyethylcyanoacrylate (PECA) particles are briefly presented.

The PECA particles in the size range of 0.5-2.5 μm were prepared by the polymerization of 2-ethylcyanoacrylate (ECA, Sigma Chemical Co., USA) in an acidic aqueous medium. Polymerizations were performed by using a stabilizer system consisting of a copolymer of polyethyleneoxide/polypropylene oxide (PEO/PPO) (Pluronic Polyol, F-88, BASF Chemical Co., USA) and dextran with a molecular weight of 175 000 (Sigma Chem. Co.,USA). The dispersion medium was an aqueous solution of HCl (37.0% w/w, Merck AG, Germany). Phosphoric acid (Merck AG, Germany) was utilized as the catalyst.

In a typical polymerization, 0.2 ml of ECA was added dropwise to 20 ml of magnetically stirred aqueous solution containing 2.0 mg/ml PEO/PPO, 5.0 mg/ml dextran, 1.0% (v/v) H_3PO_4 and 0.25 N HCl. Polymerizations were performed in sealed glass vials placed in a constant temperature chamber at 20°C. The polymerization period was 48 h. The vials were stirred with a magnetic stirrer at a stirring rate of 1000 rpm. By this particular recipe, monosize spherical PECA particles with a diameter of 1.30 μm were produced. In order to produce monosize PECA particles with different sizes, the PEO/PPO, dextran, HCl, H_3PO_4, and monomer concentrations were varied between 0.0-24.0 mg/ml, 0.0-50 mg/ml, 0.0025-1.00 N, 0.0-5.0 % (v/v), and 0.5-6.0% (v/v), respectively.

Degradation of the PECA particles in buffer solutions at different pH (1.2, 5, 6, 7.4, 7.5, 8.0) and at three different temperatures (4°, 25° and 37°C) was studied.

Figure 6 gives some representative optical micrographs of the PECA particles. Figure 7 shows the degradation of the PECA particles prepared with different polymerization recipes. The following important results should be noted.

- The stabilizer concentration (i.e., PEO/PPO) is one of the most important parameters which effects the size of the PECA particles, and therefore the degradation rate.The PECA particles obtained with higher amount of stabilizer were smaller in size, and degrade much faster.
- The concentration of dextran did not significantly affect the average size of the PECA particles. However, the polymeric particles produced with the polymerization recipe containing no dextran showed an aggregation behaviour.
- The average size and monodispersity did not change significantly with the catalyst (i.e., H_3PO_4) concentration. The degradation rate increased slighlty with the particles produced with higher catalyst concentration, which may be attributed to the decrease in the average molecular weight of PECA with increasing catalyst concentration.

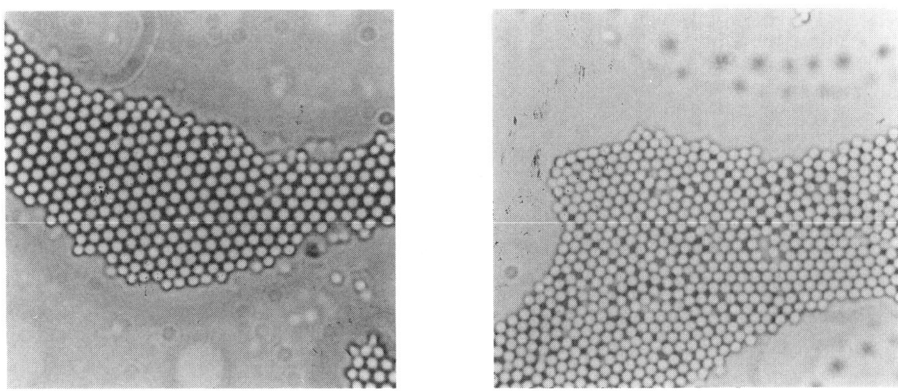

Figure 6. Monosize biodegradable PECA particles.

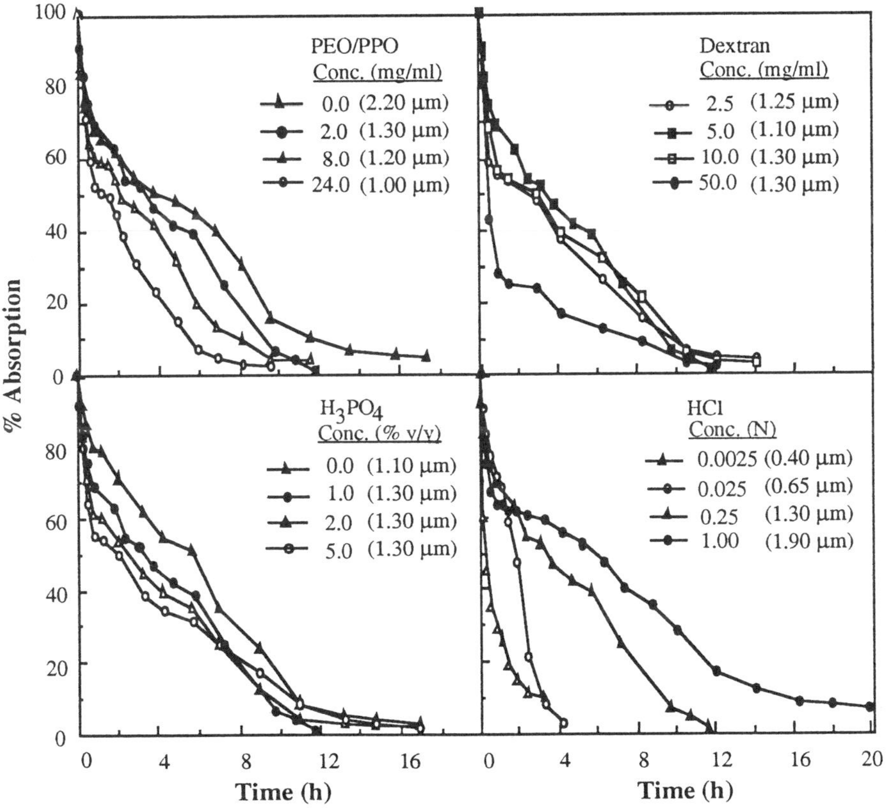

Figure 7. Degradation of PECA particles.

• HCl was used to promote the polymerization rate and to control the microsphere formation by changing the dispersion medium pH (*79*). The average size of the particles decreased higher degradation rate was observed with decreasing HCl concentration.

• The medium pH is known as one of the most effective parameters on the degradation of polyalkylcyanoacrylates (*80-82*). As seen in Figure 8, degradation rate very significantly increased with increasing medium pH, as also reported by others (*80-82*). No significant degradation was observed at acidic pH of 1.2. However, the degradation rate was very high at neutral and alkaline region.

• The degradation rate increased by increasing the temperature of the degradation medium as expected (Figure 8).

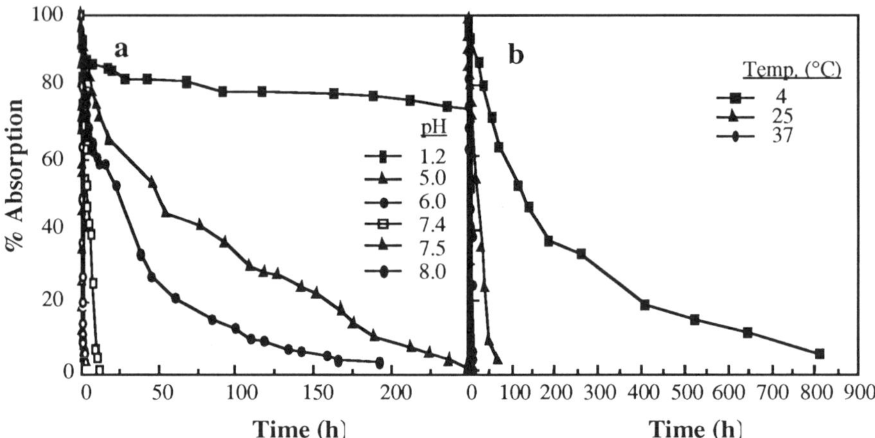

Figure 8. Degradation of PECA particles: (a) at different pH; (b) at different temperatures.

Our studies related to biomedical applications of monosize polyalkylcyanoacrylate particles, such as in sustain release of pH sensetive drugs in GIT, diagnosis and therapy of inflammation, sintigraphy of lymphatic system with radiolabelled particles, are under investigation.

Biodegradable Polylactic Acid Microspheres

Injectable colloidal drug delivery systems have gained great interest in the last years for targeting of potential drugs in order to have more effective local therapy, to preserve the activity of the drug and also to eliminate the possible side effects of the drug on the other parts of the body (*83-87*). Several biodegradable natural and synthetic polymers were employed as carriers in these formulations. Polylactic acid (PLA) was considered as one of the most suitable biodegradable polymers for colloidal drug delivery systems because of its excellent biocompatibility (*88,89*).

Recently we attempted to prepare monosize PLA particles in micron size range as carriers for potential colloidal drug delivery formulations. We prepared PLA particles in a wide size range of 1-50 μm by a conventional solvent evaporation method and its modified form. A tuberculostatic drug (i.e., rifampicin) was loaded in these particles. Here, the preparation procedure, the results of drug loading and release experiments of PLA are briefly presented. Details of these studies can be found elsewhere (*90,91*).

A series of PLA polymers with different average molecular weights (from 5 000 to 100 000) were produced by a condensation polymerization of the respective dimers (i.e., l-lactide and d-l-lactide) (*92*). PLA particles carrying the drug were then prepared by conventional and modified solvent evaporation methods. In the conventional solvent evaporation method, 0.1 g PLA was dissolved in 4 ml of chloroform (Merck AG, Germany) and a proper amount of the drug (i.e., rifampicin) was mixed with this solution. This mixture was then added dropwise to a dispersion medium consist of a 50 ml of aqueous HCl solution (1.4 % v/v) (Merck AG, Germany) containing also 20 mg/ml of the stabilizer (i.e., sodium dodecyl sulfate, Sigma Chemical Co., USA) while the medium was being stirred mechanically at a stirring rate of 2 000 rpm at room temperature. The solvent was evaporated under these conditions in about 4 h. In the modified method, the dispersion medium was an aqueous solution of methyl cellulose (6.5 mg/ml) (BDH, UK), and sonication was also applied in addition to mechanical strirring (*90,91*). The drug release from these particles were followed in pH 7.4 (phosphate buffer) at a temperature of 37°C.

Figure 9 gives the scanning electron micrographs of the PLA particles prepared by conventional and modified solvent evaporation methods. The following important results should be noted.

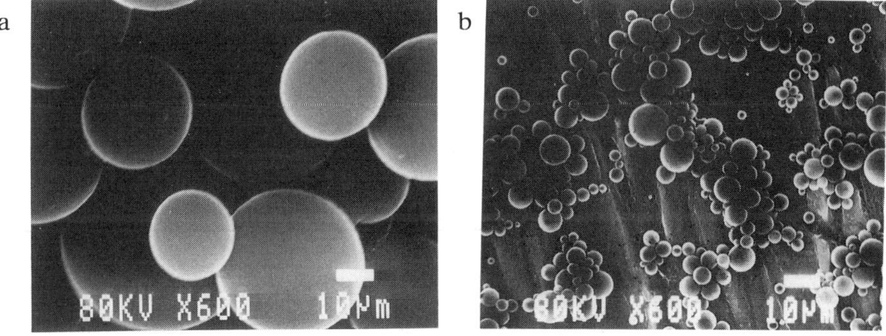

Figure 9. Scanning electron micrographs of PLA particles prepared by: (a) conventional; and (b) modified solvent evaporation methods.

- The average size increased but the size distribution decreased with increasing the PLA molecular weight in the conventional solvent evaporation method. In the modified solvent evaporation method the average size and size distribution did not change with the PLA molecular weight, possibly due to sonication.
- The drug loading did not change significantly with the PLA molecular weight in none of the preparation methods applied. But, the drug loading achieved with the modified solvent evaporation method were significantly higher (i.e., 11 %) relative to those obtained with the conventional solvent evaporation method (i.e., 4 %). Drug loading up to 25% was achieved by adding proper amounts of drug in the dispersion medium. When the drug/PLA weight ratio in the initial mixture was increased the drug loading ratio was decreased.
- The most important parameter affecting the drug release rate was the PLA molecular weight. All of the drug was released in few hours from the particles prepared by using low molecular weight PLA (Figure 10 and 11). However, drug release was very slow in the case of those particles prepared with high molecular weight PLA.

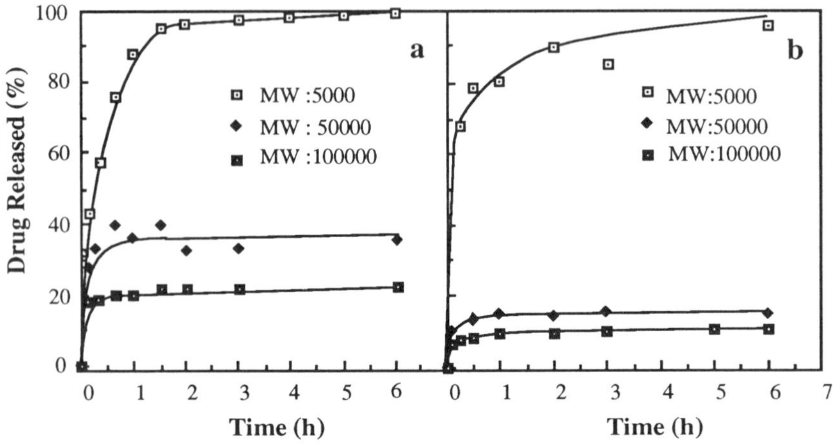

Figure 10. The effect of PLA molecular weight on drug release rate from the PLA particles prepared by: (a) conventional; and (b) modified solvent evaporation methods.

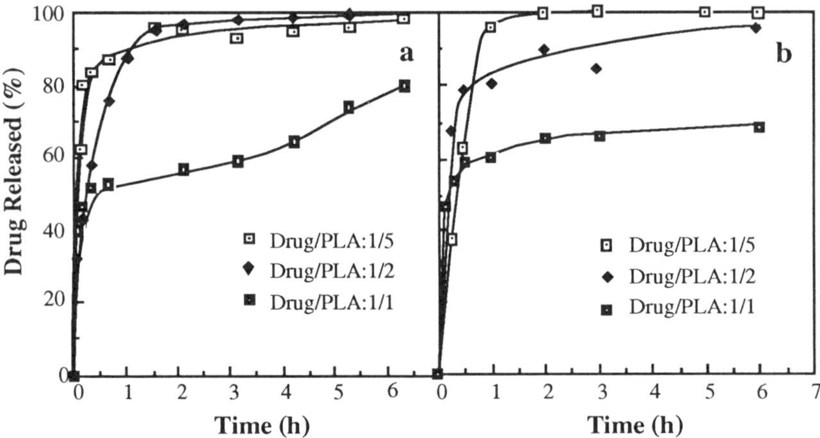

Figure 11. The effect of drug/PLA weight ratio on drug release rate from the PLA particles prepared by; (a) conventional; and (b) modified solvent vaporation methods.

Literature Cited

1. *Microspheres: Medical and Biological Application;* Rembaum, A.; Tokes, Z.A.; Eds., CRC Press, Boca Raton, Florida, 1988.
2. Bangs, L.B. *Uniform Latex Particles*, 3.Edition, Seradyn Inc., Indianapolis, 1987.
3. Millan, J.L.; Nustad, K.; Pederson, B.N. *Clin. Chem.* **1985**, 31, 54.

4. Rembaum, A.; Yen, R.; Kempner, D.; Ugestad, J. *J.Immunol. Meth.*, **1982**, 52, 341.
5. Bormer, O. *Clin. Biochem.*, **1982**, 15, 128.
6. Kumar, R.K.; Lykke, A.W.J. *Pathology,* **1984**, 16, 53.
7. Rembaum, A.; Dreyer, W.J. *Science*, **1980**, 208, 364.
8. Kronick, P.L.; Campbell, G.L.; Joseph, K. *Science.* 200, 1074, 1978.
9. Margel, S.; Beitler, U.; Ofarim, M. *J. Cell. Sci,* **1982**, 56, 157.
10. *Targeted Drugs*; Goldberg, E.P.; Ed., John Wiley, New York, 1983.
11. *Drug Delivery Systems*; Juliano, R.; Ed., Oxford Univ. Press, Oxford, 1980.
12. *Drug Desig;* Ariens, J.J.; Ed., Academic Press, New York, 1980.
13. *Microspheres and Drug Therapy;* Davis, S.S.; Illum, L.; McVie, J.G.; Tomlinson, E.; Eds., Elsevier, Amsterdam, 1984.
14. Strand, S.E.; Anderson, L.; Bergqvist, L. In *Microspheres: Medical and Biological Applications,* Rembaum, A., Tokes, Z.A., Eds., CRC Press, Boca Raton, Florida, 1988, p.193.
15. Froelich, J.W. In *Nuclear Medicine Annual 1985*, Freeman, L.M.; Weismann, H.S.; Eds., Raven Press, New York, 1985, p.23.
16. Walter, H.E.; Krob, E.J.; Garza, R. *Biochim. Biophys. Acta.* **1968**, 165, 507.
17. van Oss, C.J. *Annu. Rev. Microbial.* **1978**, 32, 19.
18. Kawaguchi, H.; Koiwai, N.; Ohtsuka, Y.; Miyamoto, M.; Sasakawa, S. *Biomaterials.* **1986**, 7, 61.
19. Tabata, Y. and Ikada, Y. *Biomaterials.* **1988**, 9, 356.
20. *Affinity Chromatography;* Dean, P.D.G.; Johnson, W.S., Middle, F.A.; Eds., IRL Press, Oxford, 1985.
21. *Affinity Chromatography.* Scouten, W.H.; Ed., John Wiley, New York, 1981.
22. *Immobilized Biochemicals and Affinity Chromatography.* Dunlap, B.R.; Ed., Plenum Press, New York, 1974.
23. Pişkin, E.; Tuncel, A. *Turkish Patent, No: 24125* , 1991.
24. Pişkin, E.; Tuncel, A. *Turkish Patent, No: 24126* , 1991.
25. Kahraman, R. *M. Sc. Thesis*, Hacettepe University, Ankara, Turkey, 1991.
26. Tuncel, A.; Kahraman, R.; Pişkin, E. *J. Appl. Polym. Sci.* **1993** (in press).
27. Tuncel, A.; Kahraman, R. and Pişkin, E. *J. Appl. Polym. Sci.* **1993** (accepted for publication).
28. Tuncel, A.; Denizli, A.; Abdelaziz, M.; Ayhan, H.; Pişkin, E. *Clinical Materials.* **1992**, 11,139.
29. Ugelstad, J.; El-Aasser, M.S.; Vanderhoff, J.W. *J.Polymer Sci.: Polymer Letters Ed.* **1973**, 11, 503.
30. Ugelstad, J.; Mork, P.C.; Berge, A.; Ellingsen, T.; Khan, A.A. In *Emulsion Polymerization_* I. Piirma, Ed., Academic Press, New York, 1982.
31. Okubo, M.; Ikegami, K.; Yamamoto, Y. *Coll.Polym.Sci.* **1989**, 267, 193..
32. Molday, R.S.; Dreyer, W.J.; Rembaum, A.; Yen, S.P.S. *J. Cell. Biol.* **1975**, 64, 75.
33. Pişkin, E.; Tuncel, A.; Denizli, A.; Ayhan, H. *J. Biomat.Sci.: Polym.Ed.,* **1993** (in press).
34. *Radiopharmaceuticals.* Subramanian, G.; Rhodes, B.A.; Cooper, J.F.; Sodd, V.J.; Eds., Soceity of Nuclear Medicine, New York, 1975.
35. Caride, V.J.; Prokop, E.K.; Troncale, F.J.; Buddoura, W.; Winchenbach, K., McCallum, R.W. *Gastroenterology.* **1984**, 86, 714.
36. Isolauri, J.; Koskinen, M.O. and Markula, H. *J.Thorac.Cardiovasc.Surg.* **1987**, 94, 521.
37. Maublant, J.C.; Sournac, M.; Aiache, J.M. and Veyre, A. *J. Nucl. Med.* **1987**, 28, 1199.
38. Ercan, M.T.; Tuncel, S.A.; Caner, B.E.; Pişkin, E. *J. Microencapsulation,* **1993**, 10, 67.

39. Ercan, M.T.; Tuncel, A.; Caner, B.E.; Mutlu, M.; Pişkin, E. *Nuc. Med.Biol,* **1991**, 18, 253..
40. Ercan, M.T.; Tuncel, A.; Caner, B.E.; Mutlu, M.; Pişkin E. In *Proc. Eur. Assoc. Nucl. Med.Cong.* Amsterdam, 1990, p.136.
41. Caner, B.E.; Ercan, M.T.; Kapucu, L.O.; Gezici, A.; Tuncel, A.; Bekdik, C.F.; Erbengi, G.F.; Pişkin, E. In *Proc. Eur. Assoc. Nucl. Med.Cong.* Amsterdam, 1990, p.277.
42. Caner, B.E.; Ercan, M.T.; Kapucu, L.O.; Tuncel, A.; Bekdik, C.F.; Erbengi, G.F.; Pişkin, E. *Nucl. Med. Commun.* **1991**, 12, 539.
43. Kanet, R.I.; Brain, J.D. *RES J. Reticuloendothel. Soc.* **1980**, 27, 201.
44. *Phagocytic Engulfment and Cell Adhesivenes.* van Oss, C.J., Ed., Marcel Dekker, New York, 1975.
45. *Mononuclear Phagocytes;* van Furth, R.; Ed., Blackwell Scientific,Oxford, 1970.
46. Repollet, E.F.; Schwartz, A.; In *Microspheres: Medical and Biological Applications;* Rembaum, A.; Tokes, Z.A.; Eds., CRC Press, Boca Raton, Florida, 1988, p.139.
47. Tabata, Y.; Ikada, Y. In *High Performance Biomaterials*; Szycher, M.; Ed., Technomic Publ.Comp., Basel, 1991, p.621.
48. Ayhan, H. *M. Sc. Thesis*, Hacettepe University, Ankara, Turkey, 1992.
49. Denizli, A. *Ph. D. Thesis*, Hacettepe University, Ankara, Turkey, 1992.
50. Denizli, A.; Tuncel, A.; Olcay, M.; Sarnatskaya, V.; Sergeev, V.; Nicolaev, V.G.; Pişkin, E. *Clinical Materials.* **1992**, 11, 129.
51. Denizli, A.; Kiremitçi, M.; Pişkin, E. *Biomataterial Artificial Cells & Immobilized Biotechnology.* **1993**, 21, 183.
52. *Advances in Biochemical Engineering;* Fiechter, A.; Ed., Spring-Verlag, Berlin, 1982.
53. *Reactive Dyes in Protein and Enzyme Technology;* Clonis, Y.D.; Atkinson, A.A.; Bruton, C.J.; Lowe, C.R.; Eds., Mac Millan, Basingtoke, 1987.
54. Clonis, Y.D. *CRC Crit. Rev. Biotechnol.* **1988**, 7, 263.
55. Lowe, C.R.; Hans, M.; Spibey, N. and Drabble, W.T. *Anal. Biochem.* **1980**, 104, 23.
56. Angal,S.; Dean, P.D.G. *Biochem J.* **1977**, 167, 301.
57. Meldolesi, M.F.; Macchia, V.; Laccetti, P. *J. Biol. Chem.* **1976**, 251, 6.
58. Easterday, R.L.; Easterday, I.M. *Adv. Exp. Med. Biol.* **1974**, 2, 123.
59. Lowe, C.R.; Glad, M.; Larsson, P.O.; Ohlson, S.; Small, D.A.P.; Atkinson, T.; Mosbach, K. *J.Chromatogr.* **1975**, 299, 175.
60. Thresher, W.C. and Swaisgood, H.E. *Biochem. Biophys. Acta.* **1983**, 749, 214.
61. Tuncel, A.; Denizli, A.; Purvis, D.; Lowe, C.R.; Pişkin, E. *J. Chromatography.,* **1983**, 34, 161.
62. Suzawa, T.; Shirahama, H.; Fujimoto, T. *J. Colloid Interface Sci.* **1982**, 86, 144 .
63. Suzawa, T.; Shirahama, H.; Fujimoto, T. *J. Colloid Interface Sci.* **1983**, 3, 49.
64. Shirahama, H.; Suzawa, T. *Colloid Polym. Sci.* **1985**, 263, 141.
65. Shirahama, H.; Suzawa, T. *J. Colloid Interface Sci.* **1985**, 104, 416.
66. Suzawa, T.; Murakami, T.J. *J. Colloid Interface Sci.* **1980**, 78, 266.
67. Zsom, L.J. *Colloid Interface Sci.* **1986**, 111, 434.
68. Shirahama, H.; Takeda, K.; Suzawa, T. *J. Colloid Interface Sci.* **1986**, 109, 552.
69. Denizli, A.; Olcay, M.; Tuncel, A.; Pişkin, E. *Doğa Tr. J. of Chemistry* **1992**, 16, 135.
70. Okubo, M.; Azume, T.; Yamamoto, Y. *Colloid Polym. Sci.* **1990**, 268, 598.

71. Couvreur, P.; Kante, B.; Lenaerts, V.; Scailteur, V.; Roland, M.; Speiser, P. *J. Pharm. Sci.* **1980**, 69, 199.
72. Grislain, L.; Couvreur, P.; Lenaerts, V.; Roland, M.; Deprez-Decampeneere, D.; Speiser, P. *Int. J. Pharm.* **1983**, 15, 335.
73. Kante, B.; Couvreur, P.; Dubois-Krack, G.; De Meester, C.; Guiot, P.; Roland, M.; Mercier, M.; Speiser, P. *J. Pharm. Sci.* **1982**, 1, 786.
74. Ratcliffe, J.H.; Hunneyball, M.; Smith, A.; Wilson, C.G.; Davis., S.S. *J. Pharm. Pharmacol.* **1984**, 36, 431.
75. Couvreur, P.; Kante, B.; Roland, M.; Speiser, P. *J. Pharm. Sci.* **1979**, 68, 12.
76. Kreuter, J. Poly (alkyl acrylate) Nanoparticles: Methods in Enzymology, **1985**, 112, 132.
77. Couvreur, P.; Kante, B.; Roland, M.; Guiot, P.; Baudhuin, P.; Speiser, P. *J. Pharm. Pharmacol.* **1979**, 31, 331.
78. Illum, L.; Jones, P.D.E.; Jones, Baldwin, R.W. and Davis, S.S. *J. Pharm. and Experimental Therapeutics* **1984**, 230, 3.
79. Çiçek, H.; Tuncel, A.; Hayran, M.; Pişkin, E. In *Biodegradable Polymeric Biomaterials* ; Pişkin, E., Ed., Marcel Dekker, New York, 1993.
80. Leonard, F.; Kulkarni, R.K.; Brandes, G.; Nelson, J. and Cameron, J.J. *J. Appl. Polym. Sci.* **1966**, 10, 259.
81. Lenaerts, V.; Couvreur, P.; Christiaens-Leyh, D.; Joiris, E.; Roland, M.; Rollman, B.; Speiser P. *Biomaterials* **1984**, 5, 65.
82. Rainer, H.M.; Lherm, C.; Herbort J. and Couvreur P. *Biomaterials* **1990**, 11, 590.
83. Widder, K.J.; Senyei, A. and Ranney, D.F. *Cancer Res.* **1980**, 40, 3512.
84. Couvreur, P.; Kante, B.; Roland, M.; Guiot, P.; Bauduin, P. and Spesier, P. *J. Pharm. Sci.* **1979**, 68, 1521.
85. Kreuter, J.; Tauber, U. and Illi, V. *J. Pharm. Sci.* **1979**, 68, 1443.
86. Yoshioka, T.; Hashida, M.; Muranishi, S. and Sezaki, H. *Int. J. of Pharm.* **1981**, 81,131.
87. Oppenheim, R.C.; Marty, J.J.; Steward, N.F. *Aust. J. Pharm. Sci.* **1978**, 7, 113.
88. Brady, J.M.; Cutright, D.E.; Miller, R.A.; Battistone, G.C.; Hunsuck, E.E. *J. Biomed. Mater. Res* **1973**, 7, 155.
89. Kulkarni, R.K.; Pani, K.C.; Neuman, C.; Leonard, F. *Arch. Surg.* **1966**, 93, 839.
90. Denkbaş, E.B.; Tuncel, S.A; Pişkin, E. In *Biodegradation and Biodegradable Polymeric Biomaterials* ; Pişkin, E., Ed.; Marcel Dekker, New York , 1993.
91. Denkbaş, E.B., *Ph.D. Thesis*, Hacettepe University, Ankara, Turkey, 1993.
92. Xu, K.T.; Tuncel, S.A; Pişkin, E. In *Biodegradation and Biodegradable Polymeric Biomaterials* ; Pişkin, E., Ed.; Marcel Dekker, New York , 1993.

RECEIVED September 17, 1993

Chapter 19

Semisynthetic Macromolecular Conjugates for Biomimetic Sensors

Mizuo Maeda[1], Koji Nakano[2], and Makoto Takagi[1]

[1]Department of Chemical Science and Technology, Faculty of Engineering, Kyushu University, Hakozaki, Fukuoka 812, Japan
[2]College of General Education, Kyushu University, 4–2–1, Ropponmatsu, Fukuoka 810, Japan

Since the membrane-bound macromolecules provide the most general and powerful mechanisms for biological ion-sensing and -channeling, a new type of polymer-modified electrode was designed based on semi-synthetic macromolecular conjugates. A multiphase material consisting of poly(styrene-*co*-acrylonitrile) and poly(L-glutamate) was synthesized as the block copolymer. The material has made it possible to immobilize the polypeptide onto electrodes just by dip coating. Electrochemical responses towards Ca^{2+} and urea were obtained based on the permeability change of the redox active species as the indicator. The modified electrode has a remarkable stability due to the multiphase structure; vinyl-made membrane supporting the artifical "ion channels". As an extension of the "polymer-acting artificial ion-channel" concept, double-helical DNA was immobilized on an electrode in order to monitor DNA-binding substances.

Signal transduction in living cell surface has been known to take place typically by one of the following three basic mechanisms; 1) ion channel system in which messenger coming from outside the cell causes opening of ion channel, 2) second messenger system where messenger is coupled to a second messenger production indirectly *via* other proteins (*e.g.*, G protein), and 3) the third system where the receptor has a messenger-activated enzyme as its cytoplasmic domain (*1*). The membrane-bound macromolecules (enzymes and/or receptor molecules) thus provide the general and powerful mechanisms for signaling processes.

These unique mechanisms in biological systems would be a good example when one designs a new principle in chemical sensor technology. In this context, the concept of "ion-channel sensors" proposed by Umezawa and coworkers is very promising since the sophisticated mechanism occurring in cell membranes is successfully mimicked by using a lipid-made Langmuir-Blodgett (LB) membrane coated on a solid electrode (*2*). The LB membrane functions as an "ion channel" which can be "opened" specifically by the analyte ion, leading to the permeability change of the electrochemically-detectable ions as the indicator (*e.g.*, ferrocyanide). The electrode has responded to 10^{-3} M of Ca^{2+} ion. Nakashima *et al.* have reported an "ion-gate" lipid monolayer membrane which showed electrochemical responses to pH based on the similar mechanism (*3*). Lipid derivatives thus have been used as receptive entity.

0097–6156/94/0556–0238$08.00/0

In cell membranes, in contrast, lipid molecules form a permeation barrier and thereby establish cell compartments, while membrane proteins are responsible for most of the dynamic processes including ion-sensing and -channeling as described above. This should encourage one to use macromolecules, whichever artificial or biological, as a receptive component of chemical sensors. In fact, most of biosensors reported so far rely on macromolecular biocatalysts such as enzyme proteins. However, low stability of biocomponents is still a serious problem. Stable synthetic polymers are widely used in sensor technology, though few of them play a key role in sensing (*4-6*). Polymeric materials have been rather a supporting player such as matrice, media, insulators, coverings, *etc.*

We describe in this chapter three types of biomembrane-mimetic sensor in which macromolecules serve as chemoreceptive entity based on the "ion-channel concept"; we utilized a certain property of some macromolecules which are not a biocatalytic molecule. 1) A stable polymer membrane bearing synthetic polypeptide was coated on an electrode. It showed an electrochemical response toward 10^{-6} M of Ca^{2+} ion based on the permeability change of the redox active couple ions (*7*). 2) The same electrode was found to give an electrochemical response towards urea, without relying on any enzymatic transducer (*8*). The modified electrode has a remarkable stability due to the vinyl polymer chains as the building blocks of the membrane (Figure 1). 3) As an extension of the 'polymer-acting artificial ion-channel' concept, DNA double strand was immobilized on an electrode in order to monitor DNA-binding substances. The DNA-modified electrode showed a response to a DNA-binding drug (*9*).

Experimental

Preparation of Polypeptide-Immobilized Electrode. 3 g of poly(styrene-*co*-acrylonitrile) (Mn=6000, styrene:acrylonitrile=5:1) having a terminal carboxyl group, which was kindly supplied by Toagosei Chemical Industry Co., Ltd., was dissolved in 8 g of thionyl chloride (Caution!; strongly irritating) and the mixture was heated at 80 °C for 7 h. Then the mixture was dried *in vacuo*, and the resulting solid was

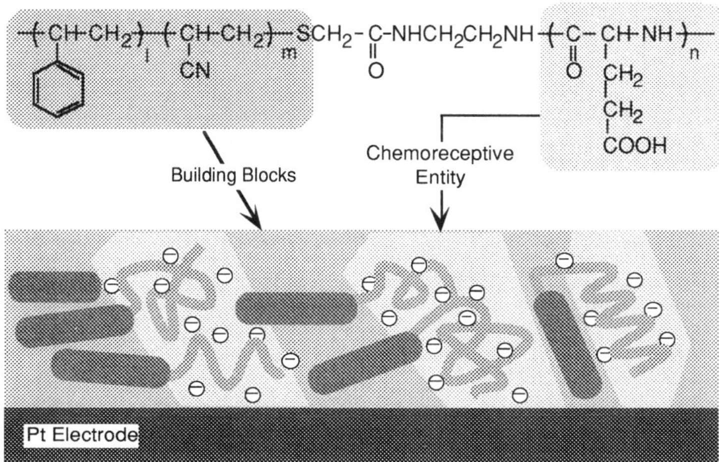

Figure 1. Structure of the polyvinyl-polypeptide block copolymer and schematic illustration for the biomembrane-mimetic sensor; formation of chemoreceprive, channel-like microdomains from poly(L-glutamate) chains.

dissolved in 5 cm^3 of dry benzene (Caution!; a carcinogen). The benzene solution was poured slowly into 100 cm^3 of dry benzene containing 3.5 g of ethylenediamine with stirring. The mixture was stirred for one night, and concentrated to *ca.* 10 cm^3. The solution was poured dropwise into 300 cm^3 of methanol with stirring at room temperature. The precipitate was dissolved in benzene and reprecipitated into methanol; this procedure was repeated three times in total. The precipitate was dried to give white powder (1.6 g). IR spectrum of the polymer showed the absorption at 1660 cm^{-1} due to amide bond; the absorption at 1705 cm^{-1} due to the terminal carboxyl group has disappeared.

The primary amino group-initiated polymerization of γ-methyl L-glutamate N-carboxyanhydride (NCA, a gift from Ajinomoto Co., Ltd.) was carried out as follows, to give a block copolymer; poly(styrene-*co*-acrylonitrile)-*b*-poly(γ-methyl L-glutamate) (Figure 1). A THF solution (5 cm^3) of NCA (0.3 g) was introduced in 1,2-dichloroethane (25 cm^3) of the polymer (0.3 g) and stirred at room temperature for 12 h. IR spectrum of the reaction mixture showed the disappearance of the peaks due to NCA; the spectrum showed some typical absorption bands assigned to poly(γ-methyl L-glutamate) as well as polystyrene and polyacrylonitrile. DPn of the polypeptide segment was 30 as estimated by ^{1}H-NMR in trifluoroacetic acid (Caution!; a strong acid) as solvent. No further purification was made for a use in the electrode modification; the reaction mixture was diluted with 1,2-dichloroethane for the dip coating as follows.

A Pt wire (diameter, 0.5 mm) sealed into the end of a glass tube (diameter, 5 mm) to leave 1 cm of the wire exposed was dipped into 1,2-dichloroethane solution of the block copolymer (1 wt-%) and dried to give the polymer-coated electrode. The electrode was then treated for 12 h with a ternary solvent of water, methanol, and 2-propanol (1:2:2, by volume) containing 0.5 wt-% KOH in order to hydrolyze the methyl ester group of the polypeptide, being converted to the carboxylate group.

Preparation of DNA-Immobilized Electrode. DNA double strands were immobilized on a Au electrode *via* chemisorption, *i.e.*, a co-ordination to Au with a sulfur-containing moiety (*10*). Sonicated calf thymus DNA (30-230 bp, 10 mg) was reacted with 2-hydroxyethyl disulfide (HEDS, 4.6 mg) in the presence of 1-cyclohexyl-3-(2-morpholinoethyl) carbodiimide metho-p-toluenesulfonate (Aldrich, 0.4 g) in 0.4 cm^3 of 0.04 M MES buffer (pH 6.0) for 24 h at 25 °C to give a phosphodiester linkage between the terminal monophosphate ends of the DNA and the hydroxyl group of HEDS (*11*). The reaction mixture was gel-filtrated on an NAP-10 column (Pharmacia), and the macromolecular fractions which displayed the characteristic absorption at 260 nm of the nucleic bases were collected. In the combined fractions containing the disulfide-modified DNA (0.3 mM in base-pair concentration), a polished Au disk electrode (1.6 mm diameter, Bioanalytical Systems) was immersed for 24 h at 5 °C. The modified electrode was then washed with and stored in TE buffer (10 mM Tris-HCl, 1 mM EDTA; pH 7.2) at 5 °C before and between uses.

Electrochemical Measurements. Cyclic voltammetric measurements were performed at 25 °C by using a Solartron Co. Model 1286 potentiostat with a conventional design of a three-electrodes system. A Pt plate (10 x 10 mm) and a standard Ag/AgCl (saturated KCl) electrode were used as a counter and a reference electrode, respectively.

FTIR Measurements. As to the polypeptide-containing block copolymer, the polymer film was formed on a Pt mesh (10 x 10 mm, 60 mesh) by dip coating and treated similarly to that on the electrode. Transmission spectra of the treated film were taken on a Nicolet 510M Fourier transform spectrophotometer. The infrared spectra were measured at 2 cm^{-1} resolution and signal-averaged over 1024 scans.

On the other hand, the disulfide-modified DNA was developed on a Au substrate by immersing it into a TE solution of the modified DNA for 24 h at 5 ℃ and rinsing carefully with TE buffer. The dried substrate was subjected to FTIR measurements in the RAS (reflection absorption spectroscopy) mode. For RAS measurements, *p*-polarized radiation was introduced on the sample at 85° off the surface normal and data was collected at a spectral resolution of 4 cm^{-1} with 2025 scans.

Results and Discussion

Calcium Ion Sensor. Cyclic voltammograms (CV) of ferrocyanide/ferricyanide redox couple with the modified electrode were measured. The peak currents due to the reversible electrode reaction of a $Fe(CN)_6^{4-}/Fe(CN)_6^{3-}$ system on a bare Pt electrode were almost completely suppressed by the coating with the polyvinyl-polypeptide block copolymer. This indicates that the electrode was covered with the hydrophobic polymer and was insulated from redox active species.

When the polymer coated electrode was treated for 12 h with a ternary solvent of water, methanol, and 2-propanol (1:2:2, by volume) containing 0.5 wt-% KOH, the CV peaks have appeared at almost the same potential as, but of much smaller (*ca.* 30 %) current values than, those on the bare one. The alkaline treatment was made in order to hydrolyze the methyl ester group of the polypeptide to the carboxylate group. Thus, with the hydrolysis, the polypeptide segment of the block copolymer would be converted to hydrophilic poly(L-glutamate) (PLG), which should allow the indicator ions to penetrate into the polymer layer.

We assumed polypeptide microdomains would be formed in the matrix from vinyl polymer chains on the electrode (Figure 1), since multiphase materials including block and graft copolymers have been known in general to form microdomain structures in the solid state (*12*). In fact, some of those containing polypeptide segments were found to have microdomains composed of polypeptide (*12-16*). In the present case, the segments from polypeptide was considered to form hydrophilic microdomains after hydrolysis.

The microdomains containing flexible polyanionic chains would function as "ion channels" which allow the electrochemical communication of the redox species with the electrode, so that the CV peaks have reappeared. The electrode surface should be, however, still covered to some extent with hydrophobic vinyl polymer segments which serve as the building blocks of the membrane. In addition, the polyanion network would reduce the local concentration of the anionic redox couple near the electrode surface, mainly due to the electrostatic repulsion. These two can account for the smaller values of peak currents than those on the bare electrode.

When adding Ca^{2+} ions (as $CaCl_2$), the peak currents increased significantly with increasing concentration of Ca^{2+} as seen in Figure 2. It should be noted that the sensitivity to Ca^{2+} depends upon the ionic strength of the supporting electrolyte; 10^{-4} M of Ca^{2+} was the detection limit when using 0.1 M KCl (*7*), whereas the sensor became susceptible to 10^{-6} M of Ca^{2+} with using 10 mM KCl. The value of anodic peak current (i_{pa}) showed almost a linear relationship with the logarithmic concentration of Ca^{2+} in the range of 10^{-6} - 10^{-4} M (Figure 3). Such ion concentration-dependent changes in CV profiles were seen also for Mg^{2+}, but not for Na^+, K^+, and tetrabutylammonium ions. With the addition of Ca^{2+}, the ionic cross-linking of carboxylate groups should occur, resulting in the contracted conformation of poly(L-glutamate) chains. The relatively "open" feature of the "ion channels" as well as the reduced number of anionic sites in the channel was considered to account for the increase in the local concentration of the redox species, leading to the enhancement of the peak currents.

Some experiments were made in order to assure the above discussion on the mechanism of the electrochemical responses. The results with three types of the indicators, namely anionic, neutral, and cationic ones, are listed in Table I. The value α represents a degree of suppression of the peak current by the modification based on the

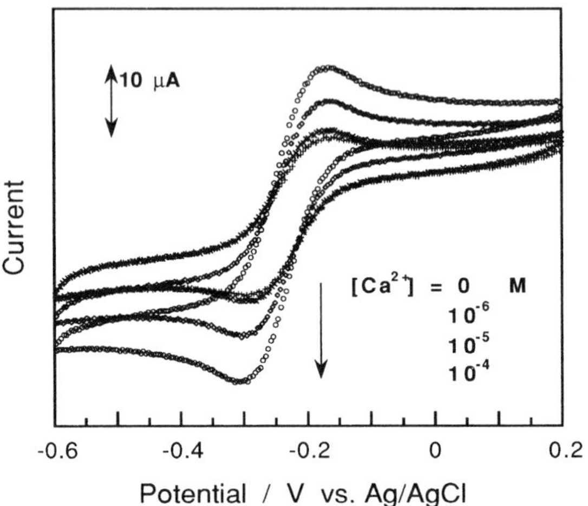

Figure 2. Effects of Ca^{2+} on current-potential curves for the modified electrode; at 25 °C; scan rate, 25 mV/s; $[K_4[Fe(CN)_6]] = [K_3[Fe(CN)_6]] = 1$ mM, [KCl] = 10 mM.

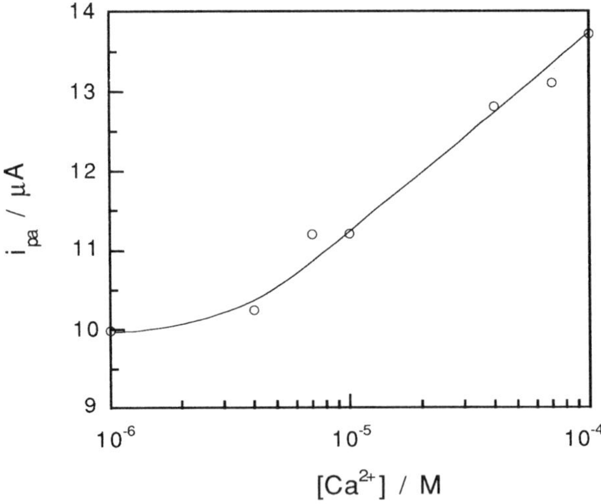

Figure 3. Ca^{2+}-dependent changes in the anodic peak current (*i*pa) of cyclic voltammograms taken similarly to those in Figure 2.

unmodified (bare) as 100. The anionic redox couple showed a large suppression, whereas the cationic one did quite a small. The electrostatic repulsion between the polyanion on the electrode and the anionic indicator ions were thus found to be operative. On the other hand, the value β demonstrates the effect of charge of the indicator on the response toward Ca^{2+} ion. The electrostatic repulsion was again seen to be a major factor; the decrease in the number of the anionic sites brought about by the binding with Ca^{2+} should reduce the repulsion between the modified layer and the anionic markers.

Table I. Effect of charge of redox-active ion as the indicator on the peak suppression in CV profiles (α) and the susceptibility to Ca^{2+} (β)

Indicator	*Charge*	α^a	β^b
$Fe(CN)_6^{3-}$	negative	37.1	71.3
p-quinone	neutral	66.1	17.8
$Ru(bpy)_3^{2+}$	positive	82.6	6.5

[a] α = *i*pa(modified) x 100 / *i*pa(bare)
[b] β = {*i*pa($[Ca^{2+}]$=10 mM) - *i*pa($[Ca^{2+}]$=0 M)} x 100 / *i*pa($[Ca^{2+}]$=0 M) ; the values were determined for the modified electrode.
[indicator] = 5 mM, [KCl] = 0.1 M, at 25 °C.

The nonionic marker, however, showed the substantial response; the mechanism cannot be explained only by the charge. The conformational change induced by the ionic cross-linking on the addition of Ca^{2+} would be an additional important factor for the sensor response.

In order to obtain information as to the structure of PLG, FTIR measurements were made for the film from the polypeptide-containing block copolymer. Figure 4 shows the FTIR spectra of the polymer film. The intact film showed typical absorption for ester at 1740 cm^{-1}, which disappeared almost thoroughly with the alkaline hydrolysis. Absorption band assigned to the Amide I became rather broad after the hydrolysis, suggesting that the polypeptide took random coil conformation. Ca^{2+} treatment made a small change which involved a slight shift of the absorption at 1394 cm^{-1} to 1397 cm^{-1} and a sharpening of the Amide I band. The former change would be due to the binding of carboxylate with Ca^{2+}, whereas the latter may be ascribable to the conformational change of PLG chain.

It is known that a conformational change of polypeptide takes place on applying various stimuli including H^+, ions, chemical substances, solvent composition, and so on. By taking these advantages, the multiphase polymer membranes containing polypeptide would serve as a key material for chemical modification of electrodes. In fact, the PLG-modified electrode has responded electrochemically towards urea as shown in the following section; the concentration of urea can be determined directly, without relying on enzymes.

Our newly-designed ion sensor based on the multiphase material thus demonstrated important features of "ion channel"-dependent biological sensing. The sensor is prepared quite simply, just by dip-coating and hydrolysis. In addition, the modified electrode has a remarkable stability; the electrode has exhibited a response to Ca^{2+} repeatedly for more than two months, when stored in 0.1 M KCl at room temperature. The facile modification as well as the excellent stability should be ascribed to the multiphase structure; vinyl-made membrane supporting "ion channels". Synthetic polypeptides have been studied extensively as models of proteins, though their use in

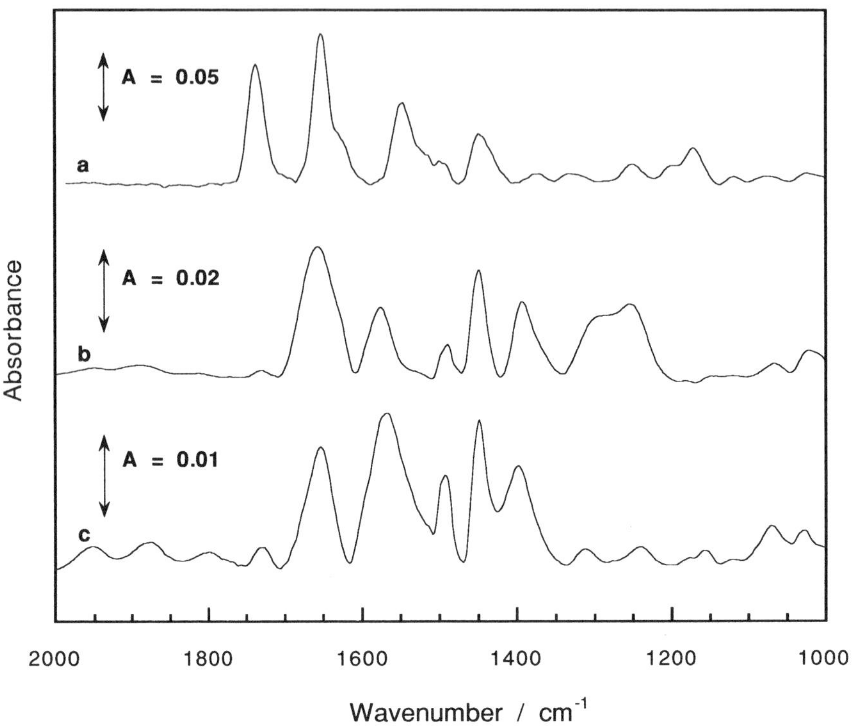

Figure 4. FTIR spectra of the polymer film from the polyvinyl-polypeptide block copolymer; (a) the intact, (b) that treated with the alkaline (KOH) solution, and (c) the hydrolyzed film treated with Ca^{2+}.

practical purposes has been rather limited. The strategy described here would facilitate the use of synthetic polypeptides for advanced, intelligent materials.

Non-Enzymatic Urea Sensor. One of the important features of the "ion channel-mimetic sensor" is that the sensor may respond electrochemically to uncharged, non-dissociable and/or redox-inactive molecules.

We describe in this section an enzyme-free electrode responsive to urea. A sensitizer used was again a synthetic polypeptide, poly(α-L-glutamate) (PLG), which undergoes conformational changes depending on the concentration of urea. PLG was easily immobilized on a Pt wire with the aid of the multiphase polymer material, poly(styrene-*co*-acrylonitrile)-PLG block copolymer, as described above. Electrochemical response towards urea was obtained based on the permeability change of the redox-active couple ions (the "ion-channel" mechanism).

Cyclic voltammograms of ferrocyanide/ferricyanide redox couple with the modified electrode are shown in Figure 5. The peaks due to the reversible electrode reaction of a $Fe(CN)_6^{4-}/Fe(CN)_6^{3-}$ system appeared at almost the same potential as those on the bare Pt electrode, though the current values were considerably small. When adding aqueous urea, however, the peak currents increased significantly with increasing concentration of urea as seen in Figure 5. The cathodic peak current (*i*pc) showed almost a linear relationship with the logarithmic concentration of urea in the range of $5x10^{-3}$ - $6x10^{-1}$ M (Figure 6; correlation coefficient, 0.987; n = 9). The dynamic range bears comparison with the conventional enzyme electrodes, as discussed later. Such urea-dependent changes in CV profiles were not seen at all on a bare Pt electrode within this concentration range.

Urea is widely known as denaturant for proteins. The denaturation process has been considered to be due to the breaking of hydrogen bonds and of hydrophobic bonds (*17*), resulting in the conformational change from ordered to random-coil form. Urea exerts its denaturing action also on synthetic polypeptides including poly(L-glutamic acid) (*17,18*). The effect has been utilized to regulate the permeability of a polymer membrane incorporating unionized poly(L-glutamic acid) chains (*19*).

The urea-induced enhancement observed here of the peak current can be explained by the permeability change of the redox species through the PLG-modified layers. However, the mechanism for the permeability change would not be simply ascribable to the conformation of the electrode-bound PLG chains as in the previous work (*19*), since the polypeptides were believed to be in the fully ionized, randomly coiled state through the present measurements. In fact, the intrinsic viscosity of PLG random coils was reported to be insensitive to urea at ionic strength of 0.1 (*18*). Thus, the drastic change observed in the present CV profiles is rather surprising.

An explanation could be possible that a strong, attractive interaction of urea with the peptide groups (and/or the side chain carboxyl groups) would cross-link PLG chains to be contracted, resulting in the relatively "open" feature of the "ion channel". Although the putative favorable interaction of urea with those groups has been found too small, if any, to induce conformational changes of fully-ionized PLG in aqueous solution (*18*), the PLG chains in the present case are densely anchored on the electrode and are sorrounded by the matrix from the hydrophobic vinyl polymer chains: the electrode surface bristling with PLG chains could be a peculiar environment for the action of urea as well as for the electrochemical reaction of ferrocyanide/ferricyanide redox couple. Another well-demonstrated effect of urea on PLG is to raise the pKa of the side chain carboxyl groups (*17*). If this is the case on the electrode, the reduced number of anionic sites brought about by the addition of urea would account for the increase in the local concentration of the anionic redox species, leading to the enhancement of the peak currents.

A urea sensor has conventionally consisted of a urease-immobilized membrane and electrode for the determination of the metabolites; electrodes for ammonium ion (*20*), pH (*21*), ammonia gas (*22*), and carbon dioxide (*23*) have been used. These sensors were all potentiometric ones to give Nernstian responses, the linear concentration range of which was typically between 10^{-4} and 10^{-1} M of urea. If one wishes to detect urea with using the ion- or gas-selective electrodes, urease is indispensable not only for the specific recognition based on the bioaffinity interaction but also for chemical conversion of urea to detectable species, since urea is essentially electro-inactive. On the contrary, the polymer-modified electrode described here does not rely upon enzymes; the concentration of urea can be determined directly.

DNA-Binding Drug Sensor. The "artificial ion-channel" concept was applied to a deoxyribonucleotides (DNA)-immobilized electrode in order to monitor DNA-binding substances. In this section, we describe an electrochemical sensor, which comprises a DNA double helix as a receptive component. Though most of biosensors rely on proteins including enzymes and antibodies, DNA is also an important host molecule in biological affinity reactions (*24*). For example, some polycyclic aromatic molecules bind to DNA by inserting themselves between the base pairs of the double helix. This process is known as intercalation. Metal ions also bind to DNA with some selectivity. DNA prefers magnesium ion to other alkali and alkaline earth metal ions. Some proteins recognize a certain sequence of nucleic bases and bind to that site. Thus an electrode modified with DNA double strands would be useful for detecting these DNA-binding molecules.

DNA double strands were immobilized on a gold electrode *via* chemisorption, that is, a co-ordination to gold with a sulfur-containing moiety. In order to clarify the

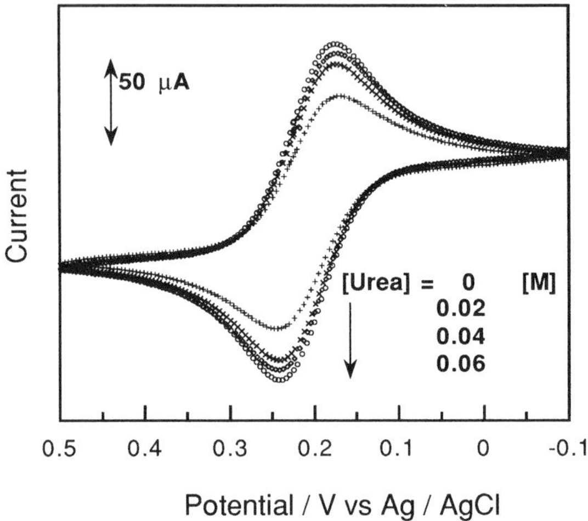

Figure 5. Effects of urea on current-potential curves for the poly(L-glutamate)-modified Pt electrode at room temperature; scan rate, 25 mV/s; $[K_4[Fe(CN)_6]] = [K_3[Fe(CN)_6]] = 5$ mM, [KCl] = 0.1 M.

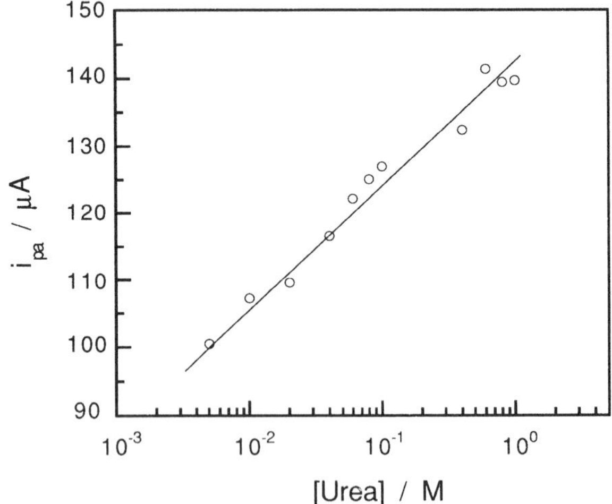

Figure 6. Urea-dependent changes in the cathodic peak current (*i*pc) of cyclic voltammograms taken similarly to those in Figure 5.

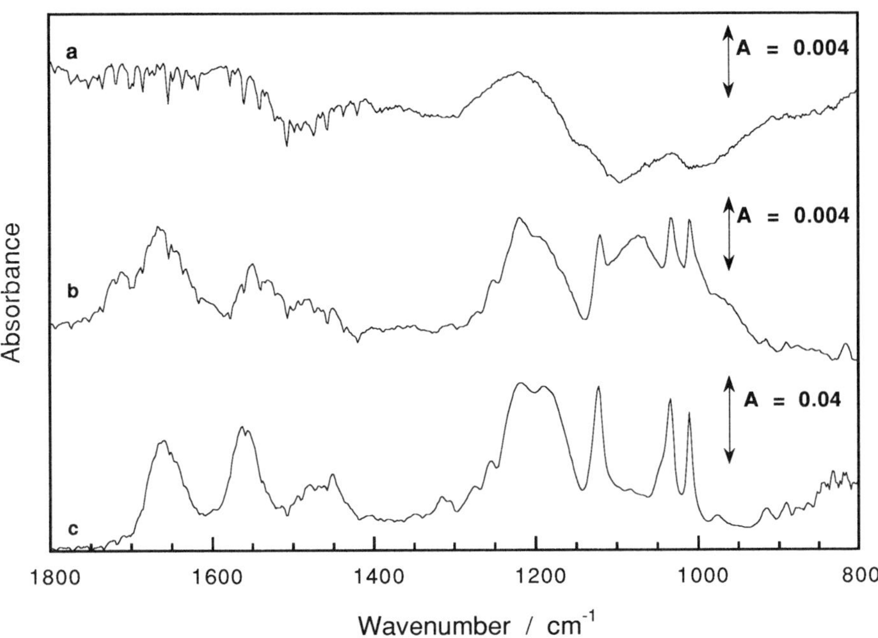

Figure 7. FTIR spectra of DNA layer formed on a Au surface; a) reflection absorption (RA) spectrum of Au treated with native DNA; b) RA spectrum of Au treated with disulfide-terminated DNA; c) transmission spectrum of DNA (cast film on CaF_2).

modification of gold surface with disulfied-terminated DNA, we carried out FTIR spectral measurements on a reflection absorption mode. Figure 7b is reflection absorption spectrum of gold surface treated with the disulfide-modified DNA, and Figure 7c is transmission spectrum of DNA pellet as a reference. They are almost identical with each other. On the other hand, the gold surface treated with native DNA did not show any typical peak (Figure 7a). Thus we can conclude that DNA was immobilized on the gold surface with the aid of sulfur-containing group of the DNA.

Cyclic voltammograms of ferrocyanide/ferricyanide redox couple with the bare and the modified electrodes are shown in Figure 8. The peak currents due to the reversible electrode reaction of a $Fe(CN)_6^{4-}/Fe(CN)_6^{3-}$ system on the bare Au electrode were significantly suppressed by the treatment with the disulfide-modified DNA. In contrast, the treatment with unmodified DNA made no suppression, and that with 2-hydroxyethyl disulfide (HEDS) did only a slight as seen in Figure 8. These results indicate that the surface-anchored DNA blocks the electrochemical reaction of $Fe(CN)_6^{3-/4-}$ with the underlying Au electrode, due to the electrostatic repulsion between the polyanionic DNA and the anionic redox couple ions.

When adding quinacrine (*25*), the well-known antimalarial drug, the peak currents increased with increasing concentration of the drug (Figure 9). The anodic peak current (i_{pa}) showed almost a linear relationship with the concentration of quinacrine in the range of 10^{-7} - $5x10^{-7}$ M and then saturated over the concentration of 8 x 10^{-7} M (Figure 10). Neither bare Au electrode nor that treated with HEDS showed response to quinacrine. The binding of the cationic drug to the immobilized polyanionic host (*i.e.*, DNA) should reduce the negative charge on the electrode surface, resulting in the enhancement of the current intensity (Figure 11).

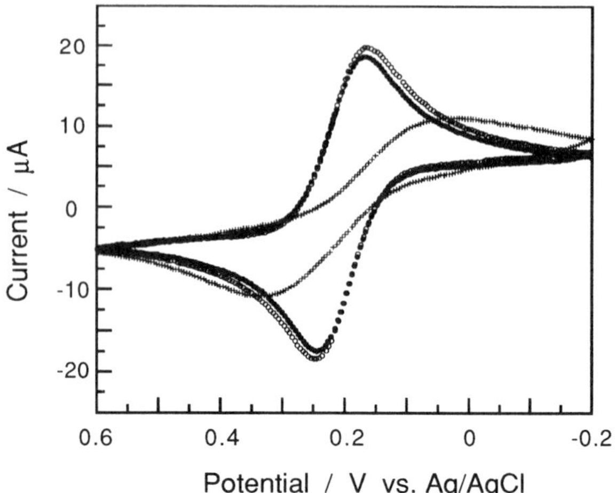

Figure 8. Cyclic voltammograms of the bare and the modified Au electrodes at 25 °C; scan rate, 25 mV/s; $[K_4Fe(CN)_6]] = [K_3[Fe(CN)_6]] = 5$ mM, [KCl] = 10 mM; (○) bare, (●) treated with 2-hydroxyethyl disulfide, (+) treated with the disulfide-modified DNA.

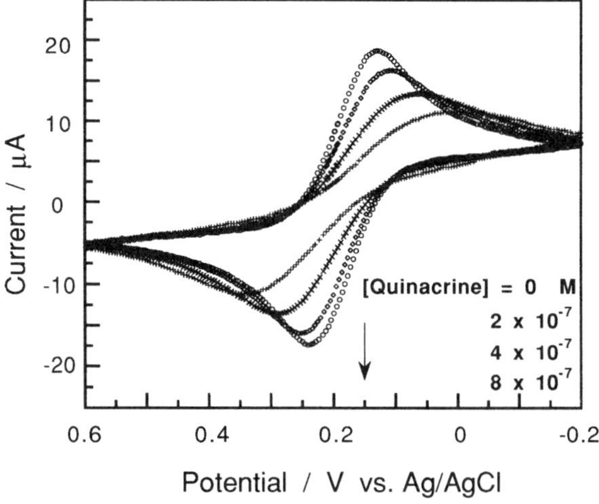

Figure 9. The quinacrine-dependent changes in cyclic voltammograms of the DNA-immobilized Au electrode; experimental conditions are the same as those in Figure 8.

Such concentration-dependent changes in CV profiles on the DNA-immobilized electrode were not seen for Na^+ ion in the concentration range of 10^{-7} - 10^{-3} M. The effect of quinacrine upon the peak recovery was thus much stronger than that of NaCl: 8×10^{-7} M of the drug seems sufficient to cancel the electrostatic repulsion, whereas 10^{-3} M of NaCl does not at all. This can be attributed to the specific and strong interaction between DNA double helix and quinacrine (*25*).

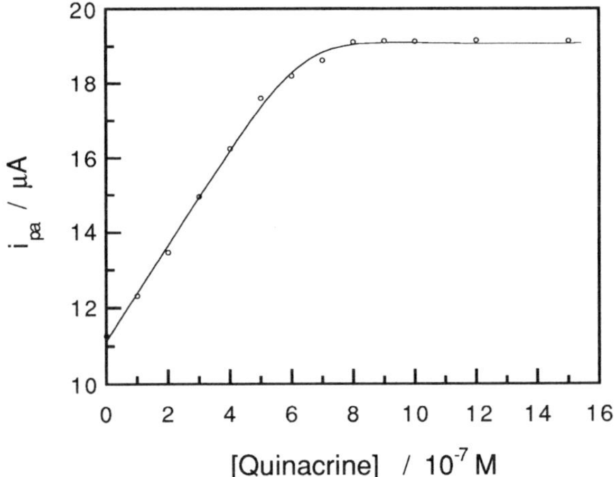

Figure 10. [Quinacrine] vs. anodic peak current (*i*pa) on the DNA-immobilized Au electrode; experimental conditions are the same as those in Figure 8.

The intercalation binding of Quinacrine with DNA double helix can be expressed by the equation;

$$\text{DNA(bp)} + \text{Q} \rightleftharpoons \text{DNA(bp)} - \text{Q},$$

where DNA(bp) is a base pair as a potential binding site and Q is Quinacrine. Then the binding constant can be defined as;

$$K = [\text{DNA(bp)-Q}] / [\text{DNA(bp)}][\text{Q}].$$

It has been reported that intercalation binding of quinacrine has maximum saturation at one bound molecule per two base pairs (*25*). In other words, after one site is occupied, binding at neighboring site is excluded. So at saturation of binding, the concentration of the occupied site should be the same as that of the free site.

$$[\text{DNA(bp)-Q}]_{\text{sat.}} = [\text{DNA(bp)}]_{\text{sat.}}$$

Now the binding constant can be simply expressed by the reciprocal of concentration of free quinacrine at saturation.

$$K = 1 / [\text{Q}]_{\text{sat.}}$$

If we assume the saturation of the peak current would correspond to the saturation of quinacrine binding to the DNA, the concentration of free quinacrine at saturation point can be read from Figure 10 as 6.2 x 10^{-7} M. Then we can obtain apparent binding constant to be 1.6 x 10^{6} M^{-1}. According to the literature, quinacrine has binding constant of 1.5 x 10^{6} M^{-1} in the similar conditions to the present ones in terms of temperature and concentration of coexisting salt (*25*). This fair agreement between two values may support our discussion on the mechanism of response. In addition, this result would encourage us to use this DNA-modified electrode for

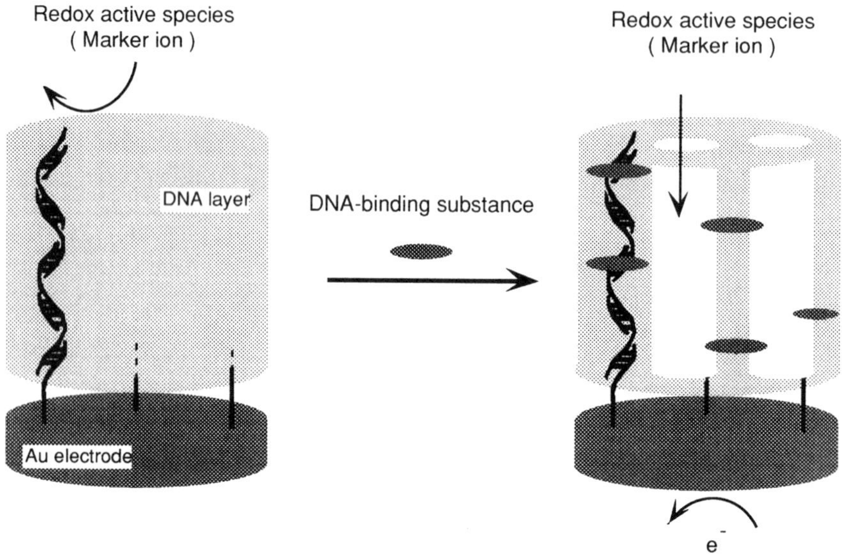

Figure 11. Schematic illustration of DNA-dependent ion-channel mechanism.

determining binding constant of a series of DNA-binding drugs. A certain kind of polycyclic aromatic compounds are known to be mutagenic and/or carcinogenic (*26*). The sensor comprising DNA as a receptive component would be useful for evaluation of interaction between DNA and such family of compounds.

Acknowledgments

Important contributions to the work described herein have been made by Yasunori Tsuzaki, Hiroaki Taira, Yuji Fujita, and Yoshiharu Mitsuhashi. We thank Mr. H. Horiuchi for technical support in the preparation of glass materials including electrical cells and electrodes. This work was supported in part by the Kawasaki Steel 21st Century Foundation. Financial support by a Grant-in-Aid for Scientific Research from Ministry of Education, Science and Culture of Japan is also acknowledged.

Literature Cited

1. Hardie, D.G. *Biochemical Messengers*, Chapman and Hall, London, 1991.
2. a) Sugawara, M.; Kojima, K.; Sazawa, H.; Umezawa, Y. *Anal. Chem.*, **1987**, *59*, 2842. b) Nagase, S.; Kataoka, M.; Naganawa, R.; Komatsu, R.; Odashima, K; Umezawa, Y. *Anal. Chem.*, **1990**, *62*, 1252.
3. Nakashima, N.; Takada, Y.; Kunitake, M.; Manabe, O. *J. Chem. Soc., Chem. Commun.*, **1990**, 845.
4. Christian, L.M.; Seitz, W.R. *Talanta*, **1988**, *35*, 119.
5. Ho, M.Y.K.; Rechnitz, G.A. *Anal. Chem.*, **1987**, *59*, 536.
6. Kurauchi, Y.; Yanai, T.; Ohga, K. *Chem. Lett.*, **1991**, 1411.
7. a) Maeda, M.; Tsuzaki, Y.; Nakano, K.; Takagi, M. *J., Chem. Soc., Chem. Commun.*, **1990**, 1529. b) Nakano, K.; Fujita, Y.; Maeda, M.; Takagi, M. *Polymer*, **1992**, *33*, 3397.

8. Maeda, M.; Fujita, Y.; Nakano, K.; Takagi, M. *J. Chem. Soc., Chem. Commun.*, **1991**, 1724.
9. Maeda, M.; Mitsuhashi, Y.; Nakano, K.; Takagi, M. *Anal. Sci.*, **1992**, *8*, 83.
10. a) Laibinis, P.E.; Whitesides, G.M.; Allara, D.L.; Tao, Y.; Parikh, A.N.; Nuzzo, R.G. *J. Am. Chem. Soc.*, **1991**, *113*, 7175. b) Nakano, K.; Taira, H.; Maeda, M; Takagi, M. *Anal. Sci.*, **1993**, *9*, 133.
11. Nicola, N.A.; Kristjansson, J.KR.; Fasman, G.D. *Arch. Biochem. Biophys.*, **1979**, *193*, 204.
12. Gallot, B.R.M. *Adv. Polym. Sci.*, **1978**, *29*, 85.
13. Ballot, J.P.; Douy, A.; Gallot, B. *Makromol. Chem.*, **1978**, *177*, 1889.
14. Nakajima, A.; Kugo, K.; Hayashi, T. *Macromolecules*, **1979**, *12*, 844.
15. Maeda, M.; Inoue, S. *Makromol. Chem., Rapid Commun.*, **1981**, *2*, 537.
16. Higuchi, S.; Mozawa, T.; Maeda, M.; Inoue, S. *Macromolecules*, **1986**, *19*, 2263.
17. Hermans, J. *J. Am. Chem. Soc.*, **1966**, *88*, 2418.
18. Skolnick, J.; Holtzer, A. *Macromolecules*, **1980**, *13*, 1311.
19. Chung, D.-w.; Maeda, M.; Inoue, S. *Makromol. Chem.*, **1988**, *189*, 1635.
20. Guilbault, G.G.; Nagy, G.; Kuan, S.S. *Anal. Chim. Acta*, **1973**, *67*, 195.
21. Masinl, M.; Guilbault, G.G. *Anal. Chem.*, **1977**, *49*, 795.
22. Paastathopoulos, D.S.; Rechnitz, G.A., *Anal. Chim. Acta*, **1975**, *79*, 17.
23. Guilbault, G.G.; Shu, F.R., *Anal. Chem.*, **1972**, *44*, 2161.
24. Waring, M.J. *Annu. Rev. Biochem.*, **1981**, *50*, 159.
25. Wilson, W.D.; Lopp, I.G., *Biopolymers*, **1979**, *18*, 3025.
26. Lee, M.L.; Novotny, M.V.; Bartle, K.D. *Analytical Chemistry of Polycyclic Aromatic Compounds*, Academic Press, New York, NY, 1981.

RECEIVED August 30, 1993

Chapter 20

Enzyme Immobilization on Polymerizable Phospholipid Assemblies

Alok Singh[1], Michael A. Markowitz[1], Li-I Tsao[2], and Jeffrey Deschamps[3]

[1]Center for Bio/Molecular Science and Engineering, Code 6900, Naval Research Laboratory, Washington, DC 20375–5348
[2]Department of Biochemistry, Georgetown University, Washington, DC 20040
[3]Laboratory for Structure of Matters, Code 6030, Naval Research Laboratory, Washington, DC 20375

Two phospholipids, one saturated and one polymerizable, containing iminodiacetic acid functionality linked to their phosphate headgroup are synthesized to build metal chelated lipid assemblies. These lipids were non-ideally miscible with analogous phosphatidylcholine as determined by monolayer studies at air/water interface. The metal chelating lipids produced vesicles when mixed with polymerizable phosphatidylcholine. These vesicles upon binding with Cu^{2+} ions were used for binding enzyme - bovine carbonic anhydrase II (EC 4.2.1.1) - utilizing their affinity towards histidine present on the surface of carbonic anhydrase. The catalytic activity of surface-bound protein molecule on polymerized and non-polymerized vesicles was measured. Only the polymerized vesicles showed improved stability and sustained activity of enzymes.

The versatility of phospholipids in biological membranes has inspired researchers to explore the potential of synthetic phospholipids to build functionalized self-organized microstructures (*1*). Phospholipids provide a modular approach to rationally build mono- and bimolecular assemblies and to control the density of reactive functionalities accessible from the dispersion medium. The nature and number of these sites on membrane surface influence the interactions of organic, inorganic, or biological molecules at membrane interface (*2,3*). The formation of supramolecular assemblies from synthetic materials and their use in diverse application areas including molecular recognition, biosensors, and controlled release technology have been the area of active research (*1,4-7*).

Polymerizable phospholipids, on the other hand, have been used in the stabilization of molecular assemblies and in the development of strategies to expand

0097–6156/94/0556–0252$08.00/0

the usefulness of the lipid assemblies (*8,9*). These strategies include the altering of surface properties of polymer films (*10*) and construction of polymerized lipid-membrane surface equipped with reactive sites suitable for chemical interactions (*11,12*).

Spontaneous formation of organized structures from lipids combined with the preservation of morphologies and functional features of the supramolecular assemblies by polymerization can lead to the development of a hybrid approach to recognizing and immobilizing macromolecules on their surface. Instead of depending on covalent bonds for immobilization, the present work relies on the hydrogen bonds or electrostatic forces in the immobilization scheme. The idea is to immobilize a macromolecule by coordinating its binding sites to a metal ion which is already complexed with ligand. This could be accomplished by incorporating a metal-chelating phospholipid in the membrane formed from phosphatidylcholine. We have focussed on iminodiacetic acid (IDA) moiety as chelating site on phospholipid headgroup because of its universal use in affinity chromatography for the separation and purification of proteins (*13*).

The binding constant of Cu^{II}-iminodiacetate with imidazole in aqueous solutions are quite high, $10^{3.5}M^{-1}$ (*14*). In principle, comparable binding constants can be expected for the IDA linked to phospholipid headgroups. A two fold increase in binding constants due to the presence of two histidines may provide a ground for preference in binding of macromolecules based on the number of histidines available for making complex with iminodiacetate. Thus, proteins or macromolecules can be effectively immobilized on a vesicle surface consisting of metal ions ligated to IDA functionality. Randomly distributed IDA-phospholipids in lipid membranes offer such an opportunity for binding with proteins containing multiple histidine sites. Utilizing the affinity of Cu-IDA complex with substituted bis-imidazole, Arnold and coworkers (*15,16*) have demonstrated specificity for bis-imidazole by matching the spatial distribution of surface coordinating group in bulk polymers with iminodiacete. In other reports, materials prepared by following similar approach have shown promise as selective supports for chromatographic separations of proteins (*17*) and as antibody mimics for radioassay of small molecules (*18*). The concept of preorganization of the lipid monomers to produce a polymer network retaining vesicular shape has been demonstrated previously by polymerizing monomer counterions present on vesicle surface (*19,20*).

We planned to extend the utility of polymerizable lipid vesicles to protein immobilization and molecular recognition schemes by taking advantage of Cu-iminodiacetate affinity towards histidines. The chemical structure of the lipids used in this work is shown in figure - 1. Vesicles are made by mixing polymerizable lipid 1,2 bis (tricosa-10,12-diynoyl)-sn-glycero-3-phosphocholine (**1**, $DC_{8,9}PC$) with nonpolymerizable 1,2 dipalmitoyl-sn-glycero-3-phospho-(N,N-bis carboxymethyl)-2-aminoethanol (**2a**, DPPIDA) or polymerizable 1,2 bis (tricosa-10,12-diynoyl)-sn-glycero-3-phospho-(N,N-bis carboxymethyl)-2-aminoethanol (**2b**, $DC_{8,9}PIDA$). The first step involves the demonstration of the proof of principle that enzyme can be immobilized on the surface of vesicles and can maintain its catalytic activity. The binding of an enzyme to the vesicles involving more than one available binding sites will require a lipid system which facilitates the lateral mobility of lipids in the assemblies (*21*). The next step involves the selection of an enzyme which consist several surface-exposed histidine residues. Bovine Carbonic anhydrase II (EC 4.2.1.1) was found to be an ideal choice, since it contains six histidine residues, four of them are available within a distance of 6 A. The four available sites may improve binding affinity by about four fold to the metal chelating sites on vesicles surface. Once the binding is achieved, the bilayers can be fixed in its morphology by cross-linking through photo-polymerization.

This study includes two types of mixed lipid systems; a) non-polymerizable IDA lipid in polymerized vesicles (**1** mixed with **2a**) and b) polymerizable IDA lipid (**2b**) copolymerized with **1** in vesicles. Enzyme immobilization was carried out on copper chelated vesicles before and after polymerization following a protocol shown in Scheme-1.

The binding scheme provides a straightforward route to enzyme immobilization without involving enzyme in chemical reactions. Enzyme immobilization on surfaces usually results in substantial loss of their activity (*22,23*).

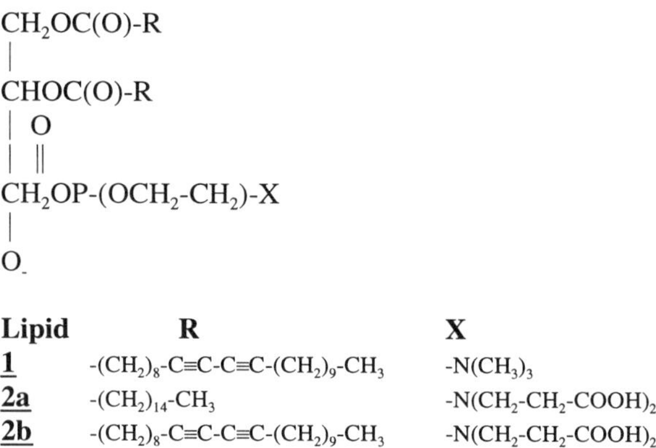

Lipid	R	X
1	-$(CH_2)_8$-C≡C-C≡C-$(CH_2)_9$-CH_3	-$N(CH_3)_3$
2a	-$(CH_2)_{14}$-CH_3	-N(CH_2-CH_2-COOH$)_2$
2b	-$(CH_2)_8$-C≡C-C≡C-$(CH_2)_9$-CH_3	-N(CH_2-CH_2-COOH$)_2$

Figure 1. Structure of phospholipids.

Materials and methods

General. Chloroform, methanol, ether, and acetone were obtained from Burdick and Jackson. All other chemicals and solvents were purchased from Aldrich Chemical Company. Triple distilled water was used in the lipid characterization studies. Polymerizable diacetylenic lipid 1,2 bis (tricosa-10,12-diynoyl)-sn-glycero-3-phosphocholine was synthesized following the procedure developed in our lab (*24*). Synthetic intermediates and lipids were characterized by IR (Perkin-Elmer 1800 FT-IR), and NMR (Brucker MSL-360 300 MHz) Spectrometer. Purity of all the compounds was monitored by thin layer chromatography on silica gel (E.M. Merck) using chloroform:methanol:water (65:25:4, Solvent A) for lipid **1** and

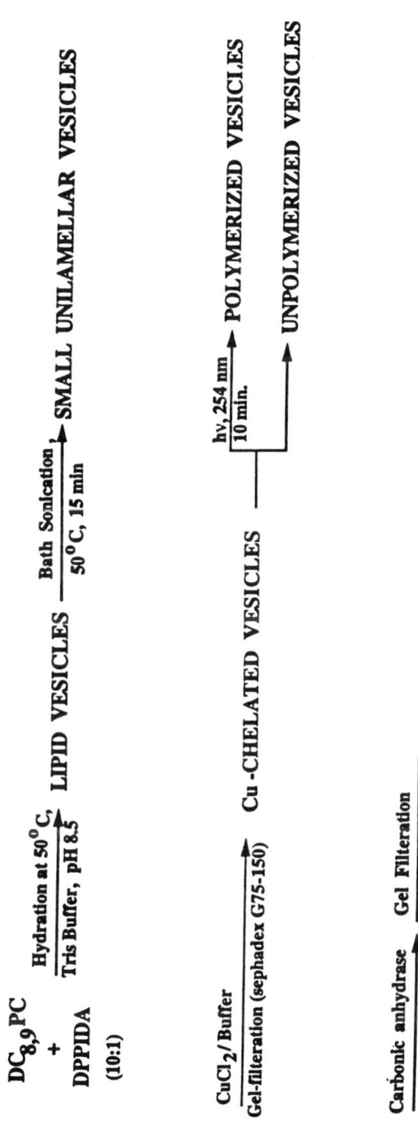

Scheme 1. Protocol for enzyme immobilization

chloroform:methanol:water:acetic acid (65:25:4:6, Solvent B) for lipids **2a** and **b**. Bovine carbonic anhydrase II (EC 4.2.1.1)and p-nitrophenyl acetate were obtained from Sigma Chemical Company (St. Louis, MO).An 80 μg/mL enzyme stock solution was prepared in 100 mM Tris -HCl (pH 8.5) buffer and stored refrigerated.

Synthesis of 1,2 dipalmitoyl- and 1,2 bis(tricosa-10,12-diynoyl)-*sn*-glycero-3-phospho-(N,N-bis carboxymethyl)-2-aminoethanol (2a,b). The phospholipids were synthesized following the synthetic route depicted in scheme-2. The 1,2 dihydroxy -*sn*- glycero-3-phosphatidyl-N,N-biscarboxymethyl ethanolamine (GPIDA) was prepared in high yield starting from commercially available solketal. Acylation of the GP.IDA was performed following the similar protocol described for phosphatidylcholine (*24*). The final product (final yield ranged 31 - 35% for both **2a,b**) was purified on a cation exchange resin to insure the absence of metal ions. The proton NMR (300 MHz) of the lipid in chloroform-*d* showed the following chemical shifts: δ_{ppm} 0.88 (t, 6H, CH_3); 1.2 (s,48H, -(CH_2)-); 1.6 (m center,2H -CH_2-CH_2-COO); 2.3 (m center,4H, -CH_2-COO); 3.78 (m center,6H, -CH_2-N-(CH_2)$_2$-); 4.8 (m center,6H, -OCH_2-); and 5.3 (m, 1H, -CHOCO-). In lipid **2b** chemical shift for protons due to -CH_2-C≡C- were found merged with that of -CH_2-COO protons.

Thermal Behavior of Lipid Dispersion. Differential scanning calorimetric (DSC) studies were performed with a Perkin-Elmer DSC-7 differential scanning calorimeter.Samples were prepared by weighing 1. 2 mg lipid into the DSC pan, injecting 60 μL H_2O, and leaving at 60°C for 3 hrs in sealed pan to ensure complete hydration. Samples were scanned following a cooling and heating cycle at a rate of 1°C/min.

Miscibility Behavior of Ligand Phospholipid on Langmuir Film. The lipids **1** and **2b** were separately dissolved in $CHCl_3$. Monolayer of the pure lipids and of the mixtures were spread with a Hamilton syringe on subphase of 10 mM aq. $CuCl_2$ (23 °C) and compressed 5 min after spreading. The mixtures of the two lipids were made from the two stock solutions. Monolayer were compressed at a rate of 50 cm^2/min up to their collapse pressures and were reproducible to within 1 A^2/molecule. Compression data was obtained using a film balance which measured surface pressure using a Wilhelmy plate sensing device (NIMA).

Preparation of Enzyme Immobilized Vesicles. Polymerizable lipid **1** (9 mg) was mixed with nonpolymerizable metal chelating lipid **2a** (0.8 mg) and polymerizable metal chelating lipid **2b** (1.0 mg) in 90:10 (mole/mole) ratio using chloroform as solvent. The solvent was removed to form a thin lipid film on the walls of the glass tubes. The vacuum dried (4 hrs.) film was dispersed in 2 mL, 0.05M Tris-HCl (pH 8.5) by incubating at 55°C for two hours followed by sonication at 50°C for 12 minutes in a bath sonicator.

The vesicle dispersions were then treated with 0.01mM $CuCl_2$ in the Tris buffer. A small precipitation was observed which dispersed upon vortex mixing. The unbound copper salt was removed by gel filtration on a sephadex G 75-125 column. Copper bound vesicles were then divided into two equal volumes. One portion was polymerized at 4°C by irradiating with 254 nm light for 10 minutes.

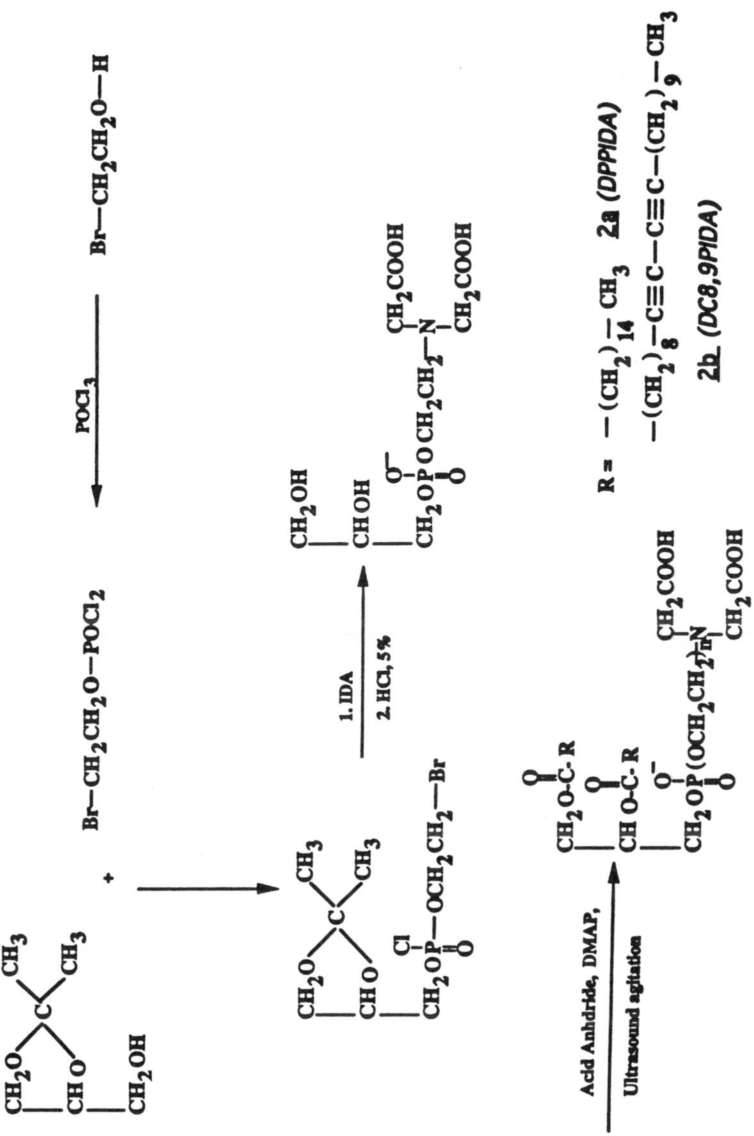

Scheme 2. Scheme for the synthesis of Metal chelating lipids

Both, polymerized and unpolymerized samples were incubated with enzyme at room temperature for 3-4 hours, and gel filtered to remove unbound enzyme from enzyme bound to vesicles.

General Procedure for Enzyme Assay. Enzyme assays were performed using a Durrum stopped-flow spectrophotometer equipped with an A/D converter and the OLIS stopped-flow operating system from Online Instrument Systems Inc. (Jefferson, GA).One syringe of the stopped-flow spectrophotometer was loaded with substrate (5mM p-nitrophenyl acetate in 30% acetone/water), and the other syringe was loaded with the preparation to be tested in 100mM Tris (pH 8.5). Data was collected at the rate of 200 points in 120 seconds. Absorbance was monitored at 402 nm with a slit width set to 1.0. Data was analyzed using the "DATAFIT" module of the stopped-flow operating system. All data analyzed were fit with a linear regression.

Results and Discussion

The metal chelating IDA lipids were passed through cation exchange resin before the preparation of vesicles or initiating monolayer studies. These lipids show a streaking pattern on the thin layer chromatographic plates when the plates are developed by conventional lipid solvent systems. These patterns leave an impression that the lipid is contaminated. Use of solvent system B ceased the streaking pattern. Additionally, pre-development of TLC plates with methanol: 10% aq. ammonium acetate (2:1) also removed this artifact. Saturated DPPIDA lipid **2a** showed a chain melting transition peak at 55.1 °C in DSC thermogram. Lipid **1** has chain melting transition at 43.1°C and **2b** should have melting transition lower to that of **2a**. Using these transition numbers as guides, we prepared vesicles at 55°C which is above their chain melting transition temperatures of the mixed system.

The two sets of vesicles were prepared by mixing lipid **1** with **2a**, and **1** with **2b**. Sonication for about 10 minutes provided dispersion with constant turbidity at 400 nm. Each set of vesicles showed an increased turbidity upon mixing with copper chloride solution, which diminished in minutes upon vortex mixing. This effect can be due to an abrupt change in ionic strength of the medium as well as to electrostatic attractions between the copper bound vesicles. The copper bound vesicles came in the void volume during gel-filtration (7.5 ml) while the free copper eluted in fractions collected after 20 mL. The fractions with free copper salt were visible due to their intense blue color. The copper bound vesicles were divided into two parts, one of which was polymerized. Polymerization produced an orange to reddish color dispersion. Polymerization was confirmed by TLC analysis. Most of the lipid was found at the origin of the TLC (solvent A).

The mixing behavior of polymerizable lipids was studied by plotting surface pressure-area isotherms of mixed $DC_{8,9}$PIda/$DC_{8,9}$PC (**1/2b**) monolayer at 23°C. Figure-2 illustrates the representative pressure-area isotherm for **1**, **2b**, and 3:1 mixture of **1** and **2b**. Since the acyl chain lengths of the two lipids are the same, differences in monolayer properties and mixing behavior can be attributed to the different headgroup shapes. On 10 mM $CuCl_2$ subphase, metal chelating lipid has a limiting area of 58.2 A^2/molecule, liquid expanded to liquid condensed (LE/LC) transition at 41.8 mN/m, and collapses at 55.5 mN/m.In contrast, **1** exhibits the LE/LC transition at 5 mN/m, collapses at 38.0 mN/m and has a limiting area of 60.5 A^2/molecule (as opposed to an area of 52 A^2/molecule and collapse pressure 36.5 mN/m on pure water (*25*).

The plot of molecular area of the lipid mixtures versus mole fraction of $DC_{8,9}$PIda demonstrate that the lipids are miscible in all proportions (Figure-3). However, the mixing behavior is complex, particularly with mixtures containing more than 50% $DC_{8,9}$PIda. The mixtures containing 50% or less $DC_{8,9}$PIda exhibit more ideal behavior than the other mixtures. An ideal miscibility of lipids should favor the lateral motion of the lipids which is essential for binding more than one histidine moieties of a protein to the surface of lipid assemblies.

The enzyme carbonic anhydrase was bound to the surface of a vesicle utilizing both polymerized and non-polymerized vesicles. The surface available histidine moieties due to their affinity towards cu-iminodiacete were utilized for immobilization (*26*). Vesicles prepared by mixing diacetylenic phosphocholine with polymerizable or saturated metal chelating lipids were eluted on a sephadex column. Vesicles come in the void volume (7 mL). Each type of vesicles were eluted in a total of four fractions (average fraction was about 1 mL). The fractions were monitored at 400 nm for the presence of vesicles. Average vesicle size of Cu-bound vesicles before polymerization was found to be 1300 A. We did not observe any changes in vesicle size during polymerization. On the similar column, enzyme carbonic anhydrase was collected soon after the elution of vesicles. Moreover, enzyme eluted slowly spreading to about 8 one mL fractions.

The results are summarized in Figure-4. Three sets of experiments were performed employing; a) unpolymerized vesicles, b) polymerized vesicles made from polymerizable lipids, and c) polymerized vesicles containing non-polymerizable metal chelating lipid and polymerizable phosphocholine. In the case of unpolymerized vesicles, no enzyme activity was observed (figure-4C). The non-polymerizable **2a** in polymerized vesicles showed an enzyme activity which was equivalent to enzyme concentration between 0.4-0.8 μg/ml (as shown in figure-4B). Maximum activity was observed in polymerized vesicles made from polymerizable phosphocholine and metal chelating lipid. The graph depicted in figure-4A illustrates that the estimated activity is equivalent to an enzyme concentration between 1.6-3.2 μg/mL carbonic anhydrase.

Polymerization is found assisting the immobilization of carbonic anhydrase on the surface of vesicles. In these examples, polymerization is stabilizing the vesicular structures as well as eliminating the motional properties of lipids i.e., lateral mobility and rotational motion. This explains the poor immobilization of enzyme on polymerized vesicles containing non-polymerizable metal chelating lipid. Detailed studies on the kinetics of enzyme binding with non-polymerizable metal-chelated lipid inserted in polymer constrained environment of polymerized vesicles may provide an insight about the role of polymerization on enzyme immobilization.

Conclusions

The activity of immobilized carbonic anhydrase on fully polymerized vesicles and no enzyme activity on unpolymerized vesicles clearly leads to the conclusion that cross-linking of the bilayers is a necessary step in the immobilization of enzymes on vesicle surface. Polymerization seemed to help in persevering the architecture and functionalities of organized assemblies. The ligand bound metal ions not only offer

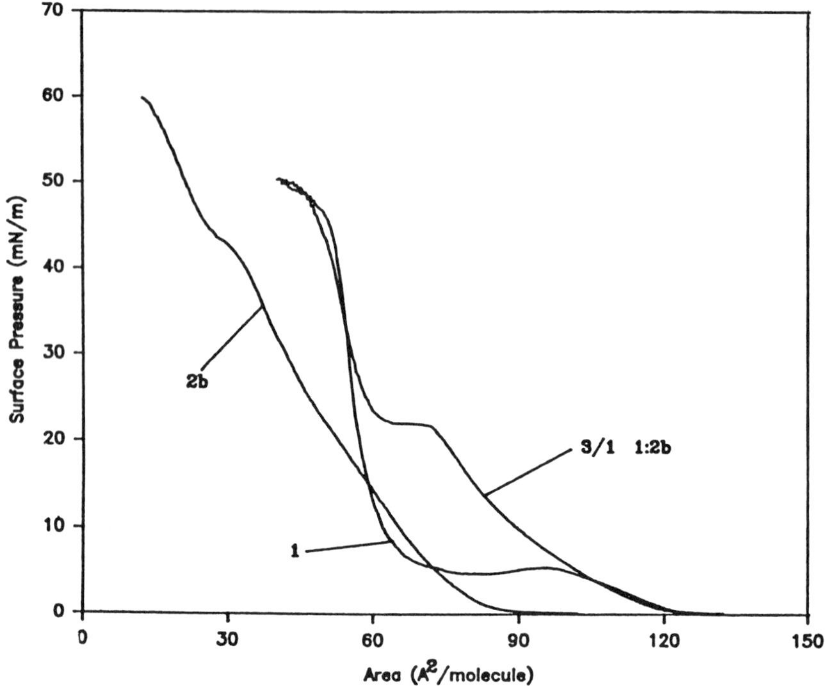

Figure 2. Surface pressure-area isotherms of lipid **1**, **2b**, and mixture of **1 and 2b** at 23°C

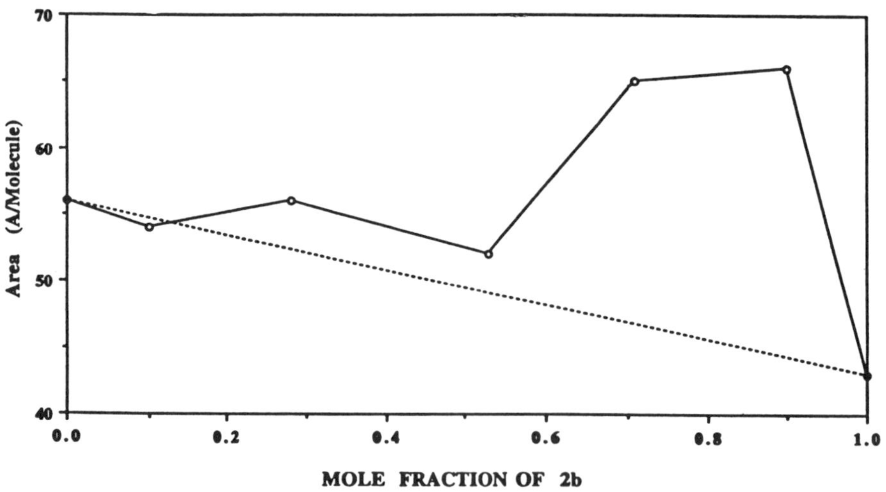

Figure 3. Molecular area of **1** and **2b** mixtures at 30mN/m

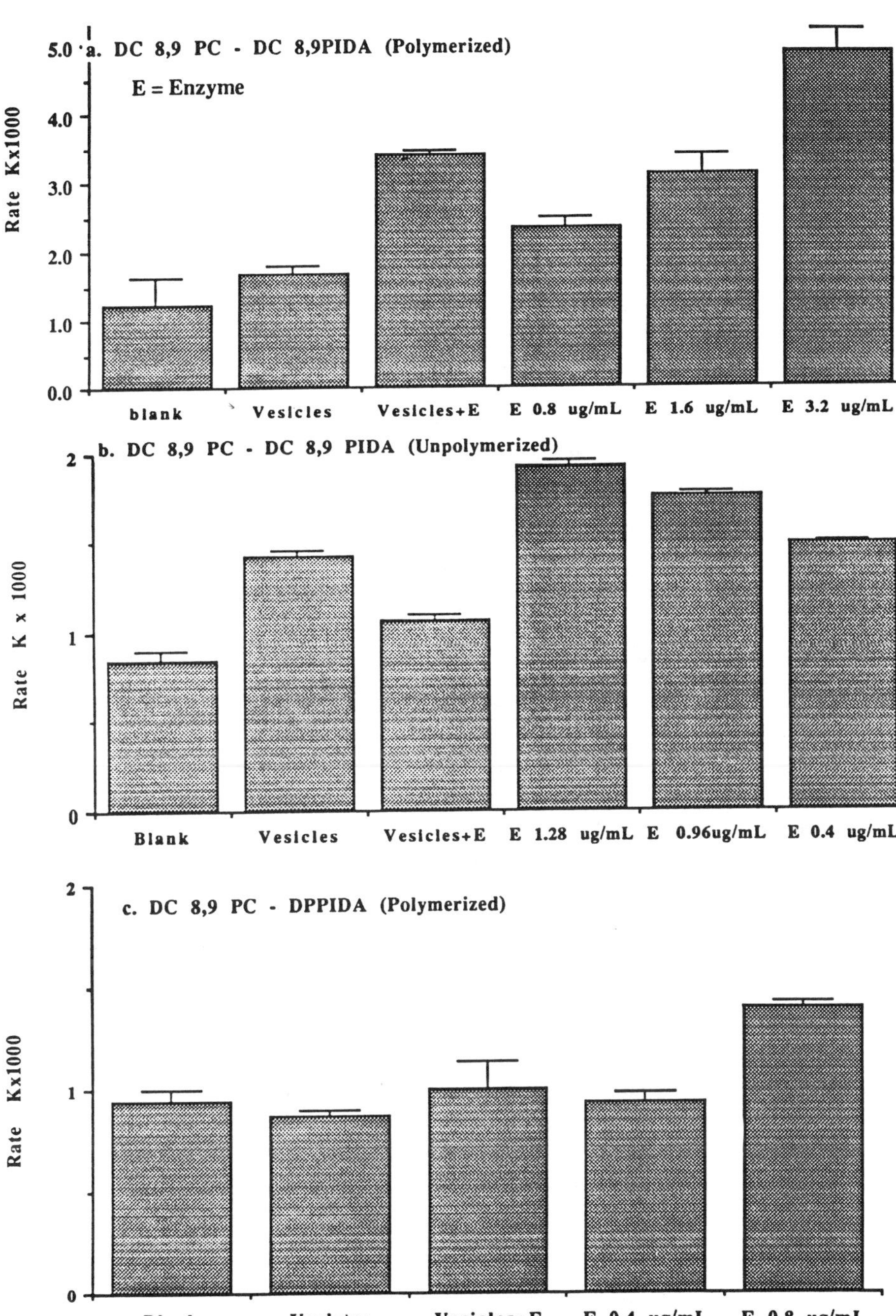

Figure 4. Activity of Carbonic anhydrase bound to the vesicles

an attractive alternative for immobilization of macromolecules, but also open the way for making patterns of chelated metal ions that match to the metal binding sites of a macromolecule. Monolayer studies on the mixing behavior of lipids indicate a non-ideal mixing of lipids takes place in lipid assemblies. Other useful feature of the approach is the capability to replace the metal ions to alter the selectivity and binding affinities of macromolecules to ligand bound metal.

Acknowledgement

We thank to our collaborator Dr. Frances H. Arnold (Caltech, Pasadena) for comments and suggestions. This work was supported by the Office of the Naval Research.

References

1. Eibl, H. *Angew. Chem. Int. Ed. Engl.* **1984**, *23*, 257.
2. *Membrane structure and functions- The state of the art*; Gaber, B. P., Easwaran, K.R. K., Eds.; Adenine Press, New York, **1992**; pp 1-19; 37-51.
3. *Synthetic Microstructures in Biological Research*; Schnur, J. M., Peckerar, M., Eds.; Plenum Press, New York, **1992**.
4. Prime, K. L.; Whitesides, G. M. *Science*, **1991**, *252*, 1164-1167.
5. Whitesides, G. M., Biebuyck, H. in *Molecular Recognition: Chemical and Biological Problems*; Roberts, S.M., Ed., Royal Society of Chemistry, Cambridge, Great Britain, **1989**, pp.270-285.
6. Grainger, D. W., Sunamoto, J., Akiyoshi, K., Goto, M., Knutson, K. *Langmuir*, **1992**, *8*, 2479.
7. Lehn, J-M. *Angew. Chem. Int. Ed. Engl.*, **1988**, *27*, 89; **1990**, *29*, 1304)
8. Singh, A.; Schnur, J. M. in *Polymerizable Phospholipids in Phospholipid Handbook*; Cevc, G., Ed.; Marcel Dekker Inc., New York, **1993**; pp 233-293.
9. Ringsdorf, H., Schlarb, B., Venzmer, J. *Angew. Chem. Int. Ed. Engl.*; **1988**, *27*, 113.
10. Regen, S. L.; Kirszenstejn, P.; Singh, A. *Macromolecules*, **1983**, 16, 335.
11. Markowitz, M. A.; Schnur, J. M.; Singh, A. *Chem. Phys. Lipids* **1992**, *62*, 193.
12. Markowitz, M.; Baral, S.; Brandow, S.; Singh, A. *Thin Solid Films* **1993**,*224*, 242.
13. For compilation of articles on metal-affinity separations, see *Methods: A companion to methods in Enzymology* **1992**, *4*, 1-108.
14. Sinha, P. C.; Saxena, P. K.; Nigam, N. B.; Srivastava, M. N. *Ind. J. Chem.* **1989**, *28A*, 33515.
15. Dhal, P. K.; Arnold, F. H. *Macromolecules* **1992**, *25*, 7051.
16. Dahl, P. K.; Arnold, F. H. *J. Amer. Chem. Soc.* **1991**,*113*, 7417.
17. Wulf, G. in *Polymeric Reagents and catalysis*; Ford, W. Ed.; ACS Symposium Series 308: American Chemical Society, Washington, DC, 1986; pp. 186-230.
18. Viatakis, G.; Anderson, L. I.; Muller, R.; Mosbach, K. *Nature* **1993**, *361*, 645
19. Regen, S. L.; Shin, J. S.; Hainfeld, J. F.; Wall, J. S.*J. Amer. Chem. Soc.* **1984**, *106*, 5756.
20. Alieve, K. V.; Ringsdorf, H.; Schlarb, B.; Liester, K. H. *Makromol. Chem. Rapid Commun.* **1984**, *5*, 345.

21. Yvonne, L. In *The Physical Chemistry of Lipids: From alkanes to Phospholipids*; Small D. M. Ed., Plenum Press, NY, **1986**, pp.
22. Bhatia, S. K.; Cooney, M. J., Shriver-Lake, L. C.; Fare, T. L.; Ligler, F. L. *Sensors and Actuators* **1991**, *B 3*, 311.
23. Hua, J. D.; Liu, L.; Qiu, Y-M. *J. Macromol. Sci - Chem* **1989**, *A26*, 495.
24. Singh, A. *J. Lipid Res.* **1990**, *31*, 1522.
25. Singh, A.; Markowitz, M. A.; Tsao, Li-I *Chem. Phys. Lipids* **1993**, *63*, 191.
26. Sulkowski, E. in *Protein purification: Micro to macro,* Burgess, R., ed; Alan R. Liss Inc. Pub., **1987**, pp 149-162.

RECEIVED March 7, 1994

Chapter 21

Controlled Release from Liposomes of Long-Chain Polymerizable Diacetylenic Phosphocholine and a Short-Chain Saturated Phospholipid

Michael A. Markowitz, Eddie L. Chang, and Alok Singh

Center for Bio/Molecular Science and Engineering, Code 6900, Naval Research Laboratory, Washington, DC 20375–5348

The controlled release properties of polymerized and unpolymerized liposomes composed of mixtures of 1,2-bis(tricosa-10,12-diynoyl)-sn-glycero-3-phosphocholine ($DC_{8,9}PC$) and 1,2-bis(dinonoyl)-sn-glycero-3-phosphocholine (DNPC) were studied. Usually, the short chain "spacer" lipid enhances polymerization. However, we found that the extent of polymerization of the lipid mixtures depended on the method of dispersing the lipid mixture: Vortexed and extruded liposomes polymerized but extensively sonicated dispersions did not polymerize. The encapsulated fluorescent markers also affected polymerization; liposomes polymerized in the presence of 4-methylumbelliferyl-ß-glucuronide but not in the presence of 6-carboxyfluorescein or Sulphorhodamine 101. Liposomes containing 20 % or 33 % DNPC had smaller diameters, 836 A and 576 A, respectively, than liposomes composed of 100 % $DC_{8,9}PC$ (5540 A). Vortexed lipid mixtures containing 50 % DNPC formed cylindrical microstructures. Leak rates from liposomes increased as DNPC content increased and polymerized liposomes were leakier than nonpolymerized liposomes. Liposomes containing 33 % DNPC were more stable to lysis by Triton X-100 than liposomes composed of 100 % $DC_{8,9}PC$.

For some time, there has been interest in the controlled release properties of stabilized liposomes.(*1-3*) A number of methods for stabilizing liposomes have been developed. Liposomes have been prepared from lipids which contain polymerizable groups such as methacrylate, butadiyne, or diacetylene and subsequently polymerized.(*4-6*) Another strategy is to make liposomes from lipids with amino acid headgroups and then polymerize the liposome by polycondensation.(*7*) Alternatively, stabilized liposomes have been prepared from prepolymerized amphiphiles.(*8*) Recently, the design and preparation of

Published 1994 American Chemical Society

polymerized liposomes which release encapsulated material in response to a specific chemical or enzymatic stimuli has been an area of much interest.(*1*) Some of these systems utilized the construction of "corked" liposomes from a mixture of polymerized and unpolymerized lipids.(*1,9*) Release of encapsulated material can then be triggered by light, change in pH, or temperature.(*9,10-12*) Detergents or organic solvents have been used to uncork by dissolving unpolymerized lipids within the stabilized liposome.(*13*)

Most of the controlled release studies involving polymerized liposomes have used phospholipids with polymerizable groups located in the middle of the acyl chain and at the acyl-chain terminus. The polymerizable groups at the chain terminus can be incorporated with relative ease into the lipid and polymerize efficiently. Phospholipids with polymerizable diacetylene groups in the middle of the acyl chain require both precise alignment of the diacetylenes in neighboring chains and some structural flexibility for polymerization to occur. Because of these structural constraints, the diacetylenic lipids have been less preferred for controlled release studies. However, a study of the permeability of liposomes formed from diacetylenic lipids has shown that the polymerized liposomes are less leaky than the nonpolymerized liposomes.(*14*) Recently, studies have been conducted which demonstrate that the polymerizability of multilamellar vesicles formed from these lipids can be considerably enhanced if short chain spacer lipids (with acyl chains equal in length to the portion of the diacetylenic acyl chain proximal to the diacetylene) are mixed in.(*15*) X-ray crystallographic studies have indicated that the polymerization enhancement is due to increased structural flexibility provided by the addition of the spacer lipids rather than more precise alignment of the diacetylenes due to interdigitation of the lipid acyl chains within the bilayer.[16] Therefore, we decided to study the properties of these mixed lipid vesicles with the goal of developing a controlled-release system which would respond to specific analytes or at a specific temperature. The initial objective was to investigate the effect of encapsulated material and lipid-dispersing conditions on polymerization. Because polymerization of the diacetylenic phospholipids requires precise alignment of the diacetylenes, interactions between the encapsulated material and lipid headgroups, which alter lipid packing, will affect polymerization. By relating the structure of the encapsulated material to its effect on polymerization, some insight can be gained into what types of encapsulants interact with phosphocholine headgroups, and, in turn, should affect the release of the material. A second objective of this study was to examine the effect of the spacer lipid on microstructure morphology and leakage of encapsulated material. A recent study examining protein insertion into these polymerized liposomes indicated that the mixed spacer lipid/diacetylenic lipid system was more stable against detergent effects than the diacetylenic lipid system at reduced temperature.(*17*) Therefore, we wanted to examine more closely the stability of the liposomes to detergent lysis. Since the position of the diacetylene group can be moved within the acyl chains, spacer lipids with different chain lengths could be used. Thus, the controlled release properties of the mixed liposomes could be tailored for a specific application by altering lipid ratio or spacer lipid chain length.

The results of this paper describe our initial efforts to systematically characterize the formation and leakage properties of mixed vesicles formed from 1,2-bis(tricosa-10,12-diynoyl)-sn-glycero-3-phosphocholine (**1**) and the spacer lipid 1,2-dinonoyl-sn-glycero-3-phosphocholine (**2**) (Figure 1). The polymerization of these liposomes in the presence of fluorescent dyes was studied. The effect of lipid ratio on liposome size and morphology was investigated by light scattering and transmission electron microscopy (TEM). The effect of the ratio of diacetylenic lipid to spacer lipid in mixed liposomes (polymerized and unpolymerized) on leakage of an encapsulated fluorescent dye was examined. In addition, the effect of detergent on leak rates from polymerized liposomes as a function of lipid ratio was studied.

Materials and Methods

1,2-Bis(tricosa-10,12-diynoyl)-sn-glycero-3-phosphocholine was synthesized according to literature procedures.(*18,19*) 1,2-Bis(dinonoyl)-sn-glycero-3-phosphocholine was purchased from Avanti Polar Lipids (Birmingham, AL). 6-Carboxyflourescein was purchased from Molecular Probes. 4-Methylumbelliferyl-ß-glucuronide (MUGL) and ß-glucuronidase were purchased from Sigma Chemical Co. Sulforhodamine 101 was obtained from Dr. Anne L. Plant (NIST). Chloroform and Sephadex G-50 gel was purchased from Aldrich Chemical Co. Triton X-100 was purchased from Boehringer Mannheim. Flourimetry was performed with a SLM 8000C spectroflourimeter (SLM Instruments).

Liposome Preparation. Aliquots of $CHCl_3$ solutions of diacetylenic and spacer lipids were added to a test tube and the solvent was evaporated under a steady stream of Ar. The samples were kept under vacuum for a minimum of 6 hrs and then, a known volume of buffer containing the fluorescent marker was added. The lipids were hydrated at 55 °C (above the phase transition of the mixed liposomes) in the presence or absence of fluorescent dye containing buffer for 30 min and then dispersed by vortex mixing for 30 s (forming MLVs). Some of the dispersions were further processed by sonication (Branson cup sonifier, Model 250) at 55 °C or by multiple extrusions through 0.2 m filters (Millipore Co.) at 55 °C. Liposomes with encapsulated dye were separated from excess fluorescent marker by passing the liposomes through successive Sephadex G-50 centrifuge minicolumns (5 mL) within a centrifuge at 1100 rpm for 4 min (23 °C).

Size and Morphology of Liposomes. The size of the particles was determined by light scattering (Coulter particle counter). Morphology of the particles was determined by transmission electron microscopy (TEM) using a Zeiss EM-10 microscope. A drop of each dispersion was dried on a carbon coated copper grid and then stained with 2 % aq. phosphotungstic acid.

Polymerization. Liposomes to be polymerized were exposed to UV irradition (254 nm, Rayonet photochemical reactor) at 5-8 °C for times ranging from 15 sec to 15 min. Polymerization was monitored by thin layer chromatography (65:25:4 $CHCl_3$:CH_3OH:H_2O v/v/v).

Release Studies. An assay developed by Thompson and Gaber was used to study leakage from the liposomes.(*20*) Leakage rates were determined by adding 50 L of the gel filtered liposomes containing encapsulated MUGL (3 mM) to 3 mL of buffer in a cuvette at different time intervals. The buffer was 20 mM aq. Tris-HCl (pH 8.3). The presence of encapsulated MUGL was detected by monitoring its weak fluorescence (excitation at 319 nm, emission at 381 nm). Then 50 L of a 1 mg/mL solution of *E. Coli* ß-glucuronidase was added to the cuvette. Dye leaking from the liposomes was cleaved by the enzyme and the fluorescence of the 4-methylumbelliferyl product was monitored (excitation at 365 nm, emission at 455 nm). Complete release from unpolymerized liposomes was achieved by the addition of 10 uL of an aq. 25 mM solution of Triton X-100. The same amount of detergent was also used to study the effect of detergent on polymerized vesicles.

Results

The effect of different liposomal formation methods on the polymerization of the mixed lipid systems was examined using mixtures of **1**:**2** in ratios of 1:1, 2:1 and 5:1, dispersed separately by vortex mixing, sonication and extrusion. UV polymerization of the vortexed dispersion proceeded smoothly in 15 to 45 seconds (5 °C) as evidenced by thin layer chromatographic (TLC) analysis. No diacetylenic lipid monomer was detected. In addition, the dispersion turned a deep purple color which indicates that a highly conjugated polymer was formed.(*14,17,21-23*) The length of time necessary to polymerize the liposomes depended on the ratio of diacetylenic lipid to spacer lipid. The higher the percentage of diacetylenic lipid present, the longer the time necessary to polymerize the lipids. A dependence of liposome polymerization on sonication time was also observed: For sonication times of less than 15 min, exposure to UV radiation for 5 min at 5 °C was required for polymerization as determined by TLC (complete loss of diacetylenic lipid monomer and formation of product which does not migrate on the TLC plate). After sonication for 15 min and then extended exposure to UV radiation (15 min, 5 °C), only monomer lipid was detected by TLC. Lipid ratio did not have an effect on the polymerization of the sonicated liposomes. Long exposure (15 min, 5°C) to UV radiation was required to polymerize the extruded liposomes. Complete reaction of monomer was determined by TLC analysis. In addition, the irradiated dispersions were yellow indicating that conjugated polymers were formed although not as long as those of the polymerized vortexed dispersions. The effect of encapsulated dye on polymerization was also observed. In the presence of 50 mM 6-carboxyfuorescein or Sulphorhodamine 101, no polymerization of any dispersion, regardless of lipid ratio or the method of dispersing the lipids, was detected by TLC. However, mixed liposomes (vortexed and extruded) did polymerize in the presence of encapsulated MUGL. It may be that the negatively charged 6-carboxyfluorescein and Sulphorhodamine 101 molecules interact with the phosphocholine headgroups sufficiently to disrupt the alignment of the diacetylenes within the acyl chains.

The effect of lipid ratio on the particle size and morphology of vortexed dispersions (20 mM Tris-HCL, pH 8.3) was determined by light scattering and transmission electron microscopy (Table I). As spacer lipid content was increased

Table I. Dependence of Particle Size and Morphology on [DNPC]

% $DC_{8,9}PC$	*% DNPC*	*Hydrodynamic Radius*	*Morphology*
100	0	5540	liposomes, tubule shards
80	20	836	liposomes
66.7	33.3	576	liposomes
50	50	1000-6000[1]	cylinders

1. Diameter measurements varied widely for identical mixtures.

from 0% to 20 %, the particle size decreased significantly. In addition, the 4:1 **1**:**2** mixture contained only spherical liposomes whereas the 100 % diacetylenic lipid dispersion produced liposomes along with a few shards of hollow microcylinders, tubules. Increasing the spacer lipid content to 33 % produced a dispersion of liposomes with a slightly smaller diameter. However, further increasing the spacer lipid content to 50 % resulted in a major change in particle size and morphology. Addition of an aliquot of the 1:1 dispersion to buffer (20 mM Tris-HCl, pH 8.3) produced a heterogeneous dispersion as opposed to the homogeneous mixing obtained with the other lipid mixtures. The dispersion remained heterogeneous after vigorous shaking. As a result of the heterogeneous nature of the diluted dispersion, the values obtained for particle diameter fluctuated over a wide range. TEM revealed that the dispersion of 1:1 **1**:**2** contained only cylindrical microstructures (Figure 2). The diameters of the microstructures ranged from 0.2 m to 2.0 m and their length ranged from 10 m to 70 m. When the vortexed dispersion was extruded through 0.2 m or 0.1 m filters at 55 °C, a mixture of the cylindrical microstructures and liposomes was formed. After 24 hrs., the range of cylinder diameters decreased significantly to 0.01 m to 0.05 m and the cylinders agglomerated.

The permeability of unpolymerized and polymerized liposomes was studied using MUGL. The release of MUGL from extruded liposomes made with various ratios of diacetylenic lipid and spacer lipid was monitored over time. The results of experiments done with unpolymerized liposomes are shown in Figure 3. As the percentage of spacer lipid was increased from 0 % to 33 %, a progressive increase in the amount of released MUGL over time was observed. However the increase in leakage over pure diacetylenic liposomes, as spacer lipid concentration, was only 1.7 fold and 3.0 fold for mixtures of **1** and **2** in lipid ratios of 3:1 and 2:1, respectively. The results of the leakage experiments performed with polymerized liposomes are shown in Figure 4. The overall leak rates were higher for the polymerized liposomes than for the unpolymerized liposomes (1.5-2.0 fold increase). As with the unpolymerized liposomes, the leak rates increased with spacer lipid content. The leak rate increased 2.4 fold for the 3:1 mixture and 3.1 fold for the 2:1 mixture.

The effect of Triton X-100 (10 x CMC) on the release of MUGL from polymerized liposomes is shown in Figure 5. The addition of the detergent reduced the activity of the *E. Coli* ß-glucuronidase by 25 % but did not cause lysis

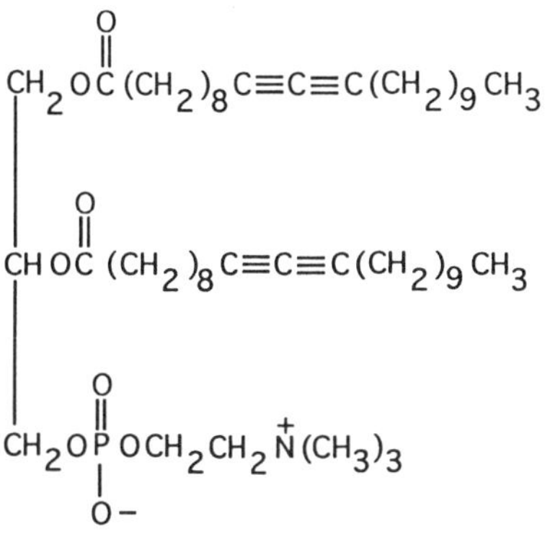

$$\begin{array}{l} CH_2OC(=O)(CH_2)_8C{\equiv}C{\equiv}C(CH_2)_9CH_3 \\ | \\ CHOC(=O)(CH_2)_8C{\equiv}C{\equiv}C(CH_2)_9CH_3 \\ | \\ CH_2OP(=O)(O^-)OCH_2CH_2\overset{+}{N}(CH_3)_3 \end{array}$$

1

$$\begin{array}{l} CH_2OC(=O)(CH_2)_7CH_3 \\ | \\ CHOC(=O)(CH_2)_7CH_3 \\ | \\ CH_2OP(=O)(O^-)OCH_2CH_2\overset{+}{N}(CH_3)_3 \end{array}$$

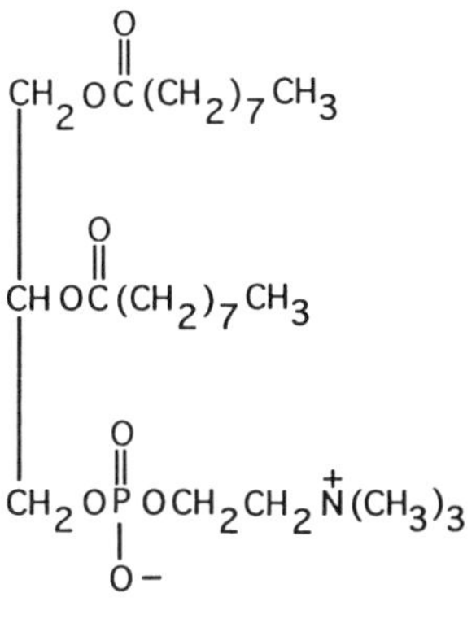

2

Figure 1. Chemical structure of 1,2-*bis*(tricosa-10,12-diynoyl)-*sn*-glycero-3-phosphocholine (**1**) and 1,2-*bis*(nonoyl)-*sn*-glycero-3-phosphocholine (**2**).

Figure 2. TEM of cylindrical microstructures formed from a 1:1 ratio of **1** and **2**.

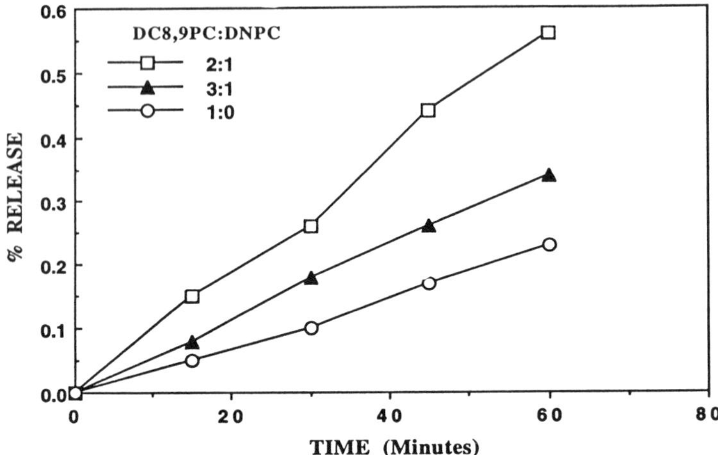

Figure 3. Release of MUGL from unpolymerized liposomes: ○, 100 % diacetylenic lipid; ▲, 3:1 mixture of **1** and **2**; [] , 2:1 mixture of **1** and **2**.

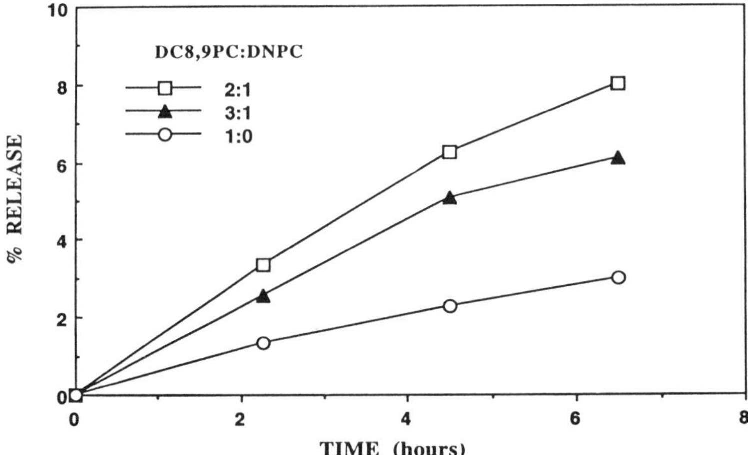

Figure 4. Release of MUGL from polymerized liposomes: ○, 100 % diacetylenic lipid; ▲, 3:1 mixture of **1** and **2**; [], 2:1 mixture of **1** and **2**.

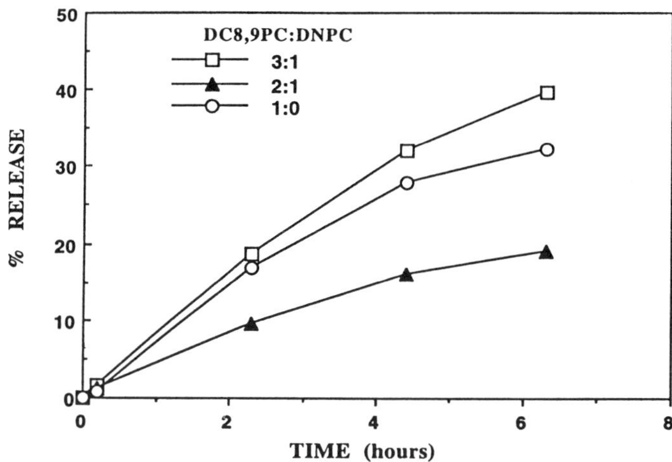

Figure 5. Release of MUGL from polymerized liposomes in the presence of Triton X-100 (10xCMC): ○, 100 % diacetylenic lipid; ▲, 3:1 mixture of **1** and **2**; [], 2:1 mixture of **1** and **2**.

of the mixed lipid vesicles. The release of MUGL was still the rate limiting step in the assay. The effect of lipid ratio on leak rate from polymerized liposomes in the presence of detergent was significantly different from that in the absence of detergent. As the spacer lipid content was increased from 0 % to 25 %, a small increase (1.2 fold) in the leak rate of the polymerized liposomes in the presence of detergent was observed. However, further increasing the spacer lipid content to 33 % resulted in a 1.7 fold decrease in the leak rate from that obtained with 100 % diacetylenic lipid. The increase in leak rate for the 100 % diacetylenic lipid system in the presence of detergent was 13 fold over that in the absence of detergent. As the spacer lipid content was increased, the difference in leak rates in the absence and presence of detergent decreased. For the 3:1 **1:2** liposomes the increase was 6.5 fold while for the 2:1 **1:2** liposomes the increase was 2.4 fold.

Discussion

Polymerized liposomes are being actively investigated for their use in controlled release applications such as drug delivery and biosensors.(*24*) At present, in order to obtain liposomes which have the release characteristics required for a specific need, the amphiphiles used to construct the liposome as well as the precise mixture of amphiphiles required must be designed and synthesized from start each time. This is a time consuming and expensive process. One of our main objectives in this research is to design a series of liposomes constructed from diacetylenic and short chain spacer lipids with predictable leak rates which could be utilized in controlled release applications. In order to accomplish this goal, it was necessary to characterize both the effect of spacer lipid on microstructure formation properties and to determine the methods that could be used to modulate leak rates from the mixed liposomes. The results of our initial study demonstrate that the ratio of diacetylenic phospholipid to spacer phospholipid has a significant effect on the microstructure formation properties of the mixed lipid systems. In addition, the results demonstrated that liposome stability could be modulated by the lipid ratio and by polymerization.

Another goal of this study was to examine the effect of encapsulated material on polymerization of the liposomes. The results indicate that highly charged encapsulants may adversely affect polymerization. No polymerization was observed in the presence of 6-carboxyfluorescein, which has negative charges arising from two carboxylate and one phenolic groups, or Sulphorhodamine 101, which has charges arising from an amine and two sulfonate groups. The liposomes were polymerizable in the presence of 4-methylumbelliferyl-ß-glucuronide (containing one ionizable carboxylate group) which is a considerably less charged molecule. The results suggest that highly charged encapsulants interact with the phosphocholine headgroups of the lipids possibly altering lipid packing sufficiently to disrupt the alignment of the diacetylene groups. In the case of 6-carboxyfluorescein, previous results indicate that while almost all of the dye is encapsulated in the liposome some of the dye may be associated with it.(*25*) In addition to the encapsulant, liposome formation method had an effect on polymerization. Enhanced polymerization was observed with vortexed dispersions; less extensive polymerization was observed with extruded liposomes; little or no polymerization was observed with highly sonicated dispersions. Vortexing lipid dispersions produces multilamellar vesicles (MLV's) while extrusion produces large unilamellar vesicles (LUV's) and sonication produces small unilamellar vesicles (SUVs).

Light scattering and TEM studies of the mixed lipid systems revealed that the addition of spacer lipid to diacetylenic lipid had a major impact on microstructure morphology. While 100 % diacetlenic lipid dispersions produced a mixture of liposomes with traces of a few tubule shards, addition of 20 to 33 % spacer lipid favored the complete formation of enclosed liposomes. In addition, the mixed lipid vesicles were significantly smaller in size than those formed from only diacetylenic lipid. The greatest influence of spacer lipid on microstructure formation was observed for the mixture containing 50 % of **2**. Instead of small diameter liposomes, cylindrical microstructures were formed. The reduction in size of the cylinders over 24 hrs indicates that the large structures initially formed were intermediates which formed relatively quickly and then slowly transformed to the smaller cylinders. The results suggest that between 33 % and 50 % spacer lipid content, microstructures formed are on the edge of stability between liposomes and cylinders.

The MUGL controlled release studies demonstrate that liposomes formed from mixtures of diacetylenic and spacer phospholipids can be used to modulate release rates and liposomal stability. Leak rates were dependent on lipid ratio and polymerization. Leak rate increased with increased spacer lipid content. However, the relatively small increment in leak rate for unpolymerized liposomes as a function of spacer lipid content (ca. 3 fold over that for liposomes formed from 100 % **1**) demonstrates that the addition of spacer lipid did not significantly disrupt the stability of the liposomes. Polymerizing the liposomes made them 1.5

to 2 times leakier than the unpolymerized liposomes. The increased leakage may be due to the presence of polymer boundaries which result from the formation of a series of smaller polymer chains rather than a single large one.(*26*) The effect of polymerization on leakage provides an additional means to tailor a mixed liposome in order to obtain a controlled release system with a particular leak rate.

Finally, the lipid ratio in polymerized liposomes also had a major impact on the stability of the mixed liposomes to detergent. The small increase in leakage for polymerized liposomes in the presence of detergent over that in the absence of detergent when spacer lipid content went from 0 % to 25 % suggests an inherent stability of these mixed liposomes to detergent. Surprisingly, increasing spacer lipid content to 33 % produced liposomes which were more stable to detergent than the liposome formed from only diacetylenic phospholipid. The enhanced stability of the polymerized mixed lipid vesicles to detergent can be further evidenced by comparing the leak rates in the presence and absence of detergent. Detergent had a much greater effect on liposome stability for the liposomes formed from 100 % **1** than for liposomes formed from either 3:1 or 2:1 mixtures of **1**:**2**. Further studies are required to fully characterize these mixed liposome systems. It would be of interest to compare the present data with formation properties and leak rates of mixtures of diacetylenic phospholipids in which the diacetylene group is at a greater distance from the glycerol backbone with longer chain spacer lipids. In addition, the stability of the mixed liposomes to other detergents and analytes can be investigated in order to determine if the liposomes are stable to specific reagents.

Acknowledgements

We gratefully acknowledge funding provided by the Office of Naval Research through an NRL Synthetic Membranes program.

References

1. Bangham, A. D.; Hill, M. W.; Miller, N. G. A. in *Methods in Membrane Biology, Vol. 1*; Korn, E. D., Ed.; Plenum Press: New York, 1985, 1.
2. Ringsdorf, H.; Schlarb, B.; Venzmer, J. *Angew. Chem. International Ed.*, **1988**, *27*, 113-158.
3. Singh, A.; Schnur, J. M. in *Phospholipid Handbook*; Cevc, G., Ed.; Marcel Dekker: New York, 1993, 233-291.
4. Regen, S. L.; Singh, A., Oehme, G.; Singh, M. *J. Am. Chem. Soc.*, **1982**, *104*, 791.
5. Hub, H.-H.; Hubfer, B.; Koch, H.; Ringsdorf, H. *Angew. Chem.*, **1980**, *92*, 962.
6. O'Brien, D. F.; Kilingbiel, R. T.; Specht, D. P.; Tyminski, P. N. *Ann. N. Y. Acad. Sci.*, **1985**, *446*, 282.
7. Neuman, R.; Ringsdorf, H.; Patton, E. V.; O'Brien, D. F. *Biochim. Biophys. Acta*, **1987**, *898*, 338.
8. Kunitake, T.; Nakashima, N.; Takarabe, K.; Nagai, M.; Tsuge, A.; Yanagi, H. *J. Am. Chem. Soc.* **1981**, *103*, 5945.

9. Okahata, Y *Acc. Chem. Res.*, **1986**, *19*, 57.
10. O'Brien, D. F.; *Photochem. Photobiol.*, **1979**, *29*, 679.
11. Fuhrhop, J. H.; Liman, U.; David, H. H. *Angew. Chem. Int. Ed. Engl.*, **1985**, *24*, 339.
12. Sullivan, S. M.; Huang, L. *Biochim. Biophys. Acta*, **1985**, *812*, 116.
13. Ohno, H.; Takeoka, S.; Tsuchida *Polym. Bull.*, **1985**, *14*, 487.
14. Leaver, J.; Alonso, A.; Durrani, A. A.; Chapman, D. *Biochim. Biophys. Acta*, **1983**, *732*, 210.
15. Singh, A.; Gaber, B. in *Applied Bioactive Polymeric Materials*; Geberlein, C. G.; Carraher Jr., E.; Foster, V. R., Eds.; Plenum Publishing Corp.: 1988, 239-249.
16. Rhodes, D. G.; Singh, A. *Chem. Phys. Lipids*, **1991**
17. Ahl, P. L.; Price, R.; Smuda, J.; Gaber, B. P.; Singh, A. *Biophys. Biochim. Acta*, **1990**, *1028*, 141-153.
18. Chapman, D. *Biocompatible Surfaces*, 1982, US Patent No. 4348, 329.
19. Singh, A. *J. Lipid. Chem.*, **1990**, *31*, 1422.
20. Thompson, R.; Gaber, B. P.; *Anal. Lett.*, **1985**, *18 (B15)*, 1847-1863.
21. Bangham, R. H.; Chance, R. R. *J. Polym. Sci., Polym. Phys. Edn.*, **1976**, *14*, 2037.
22. Kuhn, H. *Fortshr. Chem. Org. Naturst.*, **1959**, *17*, 404.
23. Exarhos, G. J.; Risen, W. M.; Baughman, R. H. *J. Am. Chem. Soc.*, **1976**, *98*, 481.
24. *Biotechnical Applications of Lipid Microstructures*: Gaber, B. P.; Schnur, J. M.; Chapman, D., Eds.; Plenum Press: New York, 1988.
25. Weinstein, J. N.; Ralston, E.; Leserman, L. D.; Klausnar, R. D.; Dragsten, P.; Henkart, P.; Blumenthal, R. *Liposome Technology, Vol. III*; Gregoriades, G., Ed.; CRC Press: Boca Raton, FL, 1984, 183-204.
26. Stefely, J.; Markowitz, M. A.; Regen, S. L. *J. Am. Chem. Soc.*, **1988**, *110*, 7463.

RECEIVED January 10, 1994

Immobilization and Stabilization Methods

Chapter 22

Immobilization of Glucose Oxidase on Polyethylene Film Using a Plasma-Induced Graft Copolymerization Process

Chee-Chan Wang[1] and Ging-Ho Hsiue

Department of Chemical Engineering, National Tsing Hua University, Hsinchu 300, Taiwan

Plasma induced graft copolymerization of acrylic acid, which was incorporated onto PE films, was investigated here. Glucose oxidase was immobilized onto this novel grafted polymeric film with and without using poly(ethylene oxide) used as spacer. ESCA was used for analyzing the plasma treated PE, plasma induced polyacrylic acid graft PE and poly(ethylene oxide) modified PAAc graft PE films. The immobilized GOD with spacer (PAAc-PEO-GOD) was observed to record a higher activity than that without spacer (PAAc-GOD). The pH and thermal stabilities of the immobilized GOD without spacer (PAAc-GOD) were higher than those of the immobilized GOD with spacer (PAAc-PEO-GOD) as well as free enzyme.

Plasma treatment and plasma deposition polymerization provides a unique and powerful method for the surface chemical modification of polymeric materials without altering their bulk properties.(*1-3*) These techniques offer the possibility to improve the performance of existing biomaterials and medical devices and for developing new biomaterials·(*4-6*)

Modifying the polymer surface by graft copolymerization has also been made possible by free radicals or peroxides generated by the plasma treatment. A technique which is similar to radiation induced graft copolymerization via preirradiation in air was proposed by Ikada and co-workers.(*7*) This approach was presented for the graft copolymerization of acrylamide onto the surface of polymer films being treated with an oxidative or inert gas plasma. This was then followed by exposure to air. The peroxide formed during or after plasma exposure was utilized for initiating graft copolymerization onto the plasma treated polymers. (*8-9*)

A series of studies regarding surface modification of polymeric films with graft polymerization have been started by Wang and co-workers(*10*) by using a plasma induced graft copolymerization technique. In this examination, plasma induced grafting procedures have been adopted for this work for producing polyacrylic acid (PAAc) grafted PE. This support has also been used here for immobilizing the

[1]Current address: Department of Chemical Engineering, Van Nung Institute of Technology, Chungli 320, Taiwan

0097–6156/94/0556–0276$08.36/0

glucose oxidase. Poly(ethylene oxide) (PEO) is used here for exploring the effect of spacer length on the activity of the immobilized GOD. The pH and thermal stabilities for the immobilized GOD have also been addressed.

Experimental

Materials. The substrate polymer used for enzyme immobilization is a linear low-density polyethylene film (thickness of 20μm), which was obtained from USI Far East Co.. The acrylic acid monomer (R.D.H. 62023) was redistilled in a vacuum (bp.39.0/10 mmHg) before use. Glucose oxidase (E.C.1.1.3.4) from *Aspergillus niger*, type V was purchased from Sigma Chemical Co. (St. Louis, MO). The bisamino PEO samples of molecular weight 4000 were provided as a gift from the Nippon Oil and Fats Co. Ltd., Japan. The coupling agents, glutaraldehyde (25% aqueous solution) and 1-cyclohexyl-3-(2-morpholinoethyl) carbodiimide metho-p-toluene sulfonate (CMC), were obtained from Merck. All other chemicals used in this study were of reagent grade and used as obtained.

Apparatus. The sensor system consisted of a dissolved oxygen electrode (YSI Model 5750, BOD bottle probe), and a D.O. Meter (Suntex Model SD-60) and a recorder (Eyela Model TR-250). (*11-12*) A spectrometer Perkin-Elmer, PHI-560. AMSAM/1650 was employed for carrying out the Electron Spectroscopy for Chemical Analysis (ESCA) measurement of plasma treated films, grafted films and PEO modified PAAc graft PE films at a pass energy of 1253.6 ev with a Mg Kα X-ray source.

The Preparation of Substrate Polymer. The PAAc grafted PE membrane was prepared according to the procedure of Wang and Hsiue.(*10*) A glow discharge reactor Model PD-2 plasma deposition (Figure 1) was used with a bell jar type reaction cell manufactured by Samco Corp., Kyoto, Japan. The frequency applied was 13.56 MHz and an impedance matching unit was required.

Graft Copolymerization of Acrylic Acid onto PE Surface. In a typical reaction, the films exposed to oxygen after Ar plasma treatment (90W, 60sec, 300mtorr) were placed in glass ampoules containing a 50% acrylic acid monomer and Mohr's salt ($1x10^{-3}$ M) aqueous solution. The ampoules were sealed after degassing three times and kept at 80^oC for 24 h. The reaction scheme is illustrated in Figure 2. PAAc grafted films were removed from the monomer solution and washed with deionized water in ultrasonics, which was followed by soaking in deionized water overnight. These films were then stored in deionized water. The amount of grafted PAAc was determined as follows: each PAAc grafted PE membrane was reacted for 2 h, at 60^oC, with 10 ml of 0.01M NaOH, then 5 mL of the supernatant were back titrated with 0.01M HCl using a titrator (METTLER DL21). A 60 $\mu g/cm^2$ grafted sample was selected in this study for the coupling process.

Coupling of PEO Spacer onto PAAc Grafted PE. In a typical reaction, the PAAc grafted PE films were added to 5 mL acetate buffer solution (0.05 M, pH 3.5) containing bisamino PEO (200 mg/mL). The PEO solutions also contained 4 mg/mL of coupling agent CMC. The reaction was carried out for 16 hours at 4^oC with constant shaking. The overall reaction scheme of the preparation of substrate polymer is shown in Figure 3.

Coupling of GOD onto PAAc grafted PE. The GOD that has been covalently immobilized on the PAAc modified PE is shown in Figure 4A. A typical

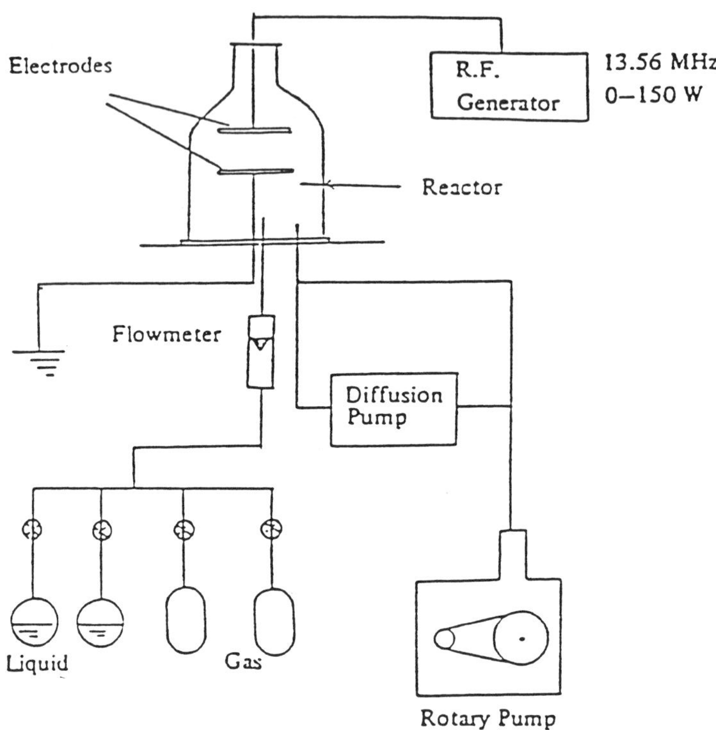

Figure 1. Schematic diagram of the plasma treatment apparatus.

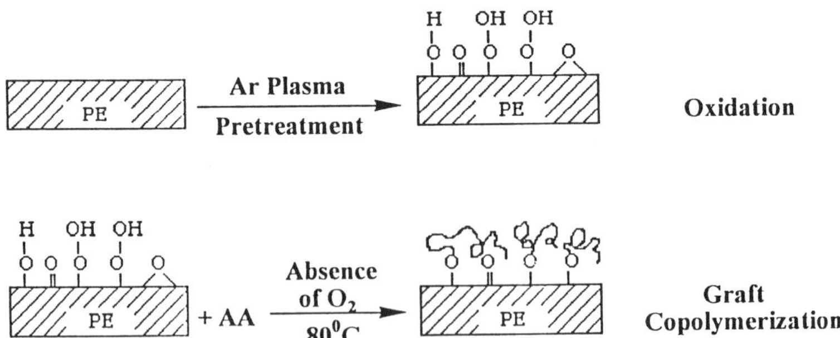

Figure 2. Schematic representation of the surface modification of PE.

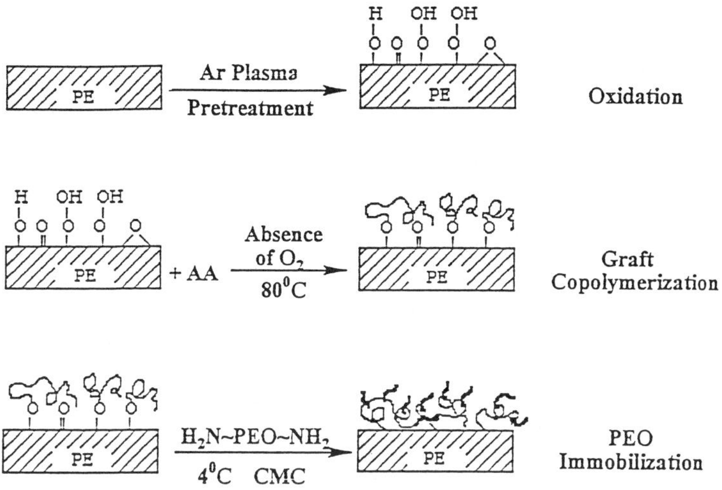

PE — Ar Plasma Pretreatment → PE — Oxidation

PE + AA — Absence of O_2, 80^0C → PE — Graft Copolymerization

PE — H_2N~PEO~NH_2, 4^0C CMC → PE — PEO Immobilization

Figure 3. Schematic diagram of the bisamino PEO immobilization onto PE film functionalized with Ar plasma induced graft copolymerization of acrylic acid.

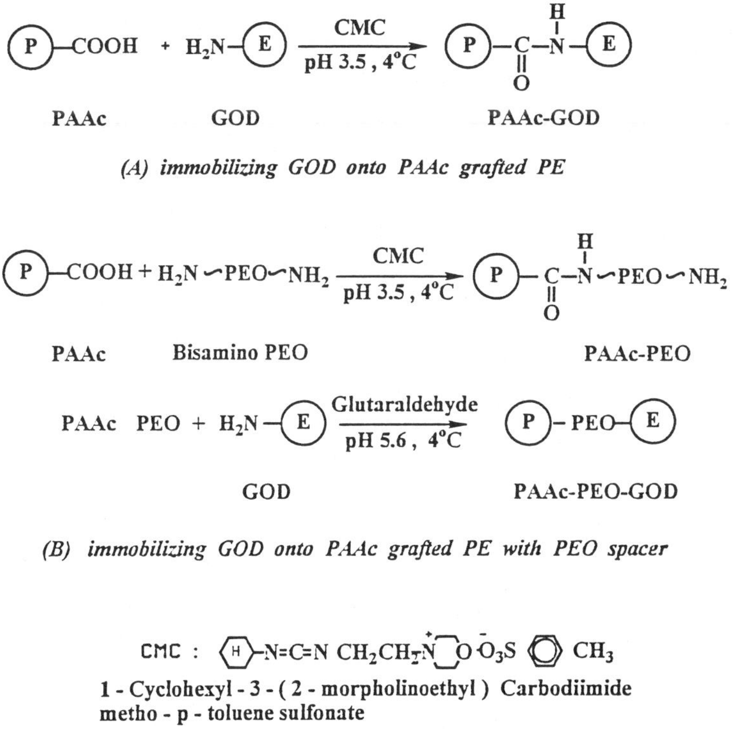

(P)–COOH + H_2N–(E) —CMC, pH 3.5, 4°C→ (P)–C(=O)–N(H)–(E)

PAAc GOD PAAc-GOD

(A) immobilizing GOD onto PAAc grafted PE

(P)–COOH + H_2N~PEO~NH_2 —CMC, pH 3.5, 4°C→ (P)–C(=O)–N(H)~PEO~NH_2

PAAc Bisamino PEO PAAc-PEO

PAAc PEO + H_2N–(E) —Glutaraldehyde, pH 5.6, 4°C→ (P)–PEO–(E)

GOD PAAc-PEO-GOD

(B) immobilizing GOD onto PAAc grafted PE with PEO spacer

CMC : (H)–N=C=N CH_2CH_2–$\overset{+}{N}$(morpholino)O $\overset{-}{O_3}S$–(C₆H₄)–CH_3

1 - Cyclohexyl - 3 - (2 - morpholinoethyl) Carbodiimide metho - p - toluene sulfonate

Figure 4. Schematic diagram of immobilizing GOD onto (A) PAAc grafted PE, and (B) PAAc grafted PE with PEO spacer.

immobilization procedure without spacer is as follows: one piece of 60 $\mu g/cm^2$ PAAc grafted PE was added to 4 mL of 0.05 M acetate buffer, pH 3.5. 20 mg of CMC and 1 mL (5 mg/mL) of glucose oxidase solution (in 0.05 M buffer, pH 3.5) were then added. The mixture was shaken at 4^oC for 16 h. The membrane was then filtered and rinsed with deionized water three times. The enzyme membrane (PAAc-GOD) was soaked further in acetate buffer, pH 5.6, and was used for determining the activity of GOD immobilized membrane.(*12*)

Coupling of GOD onto PEO Spacer Modified PAAc Grafted PE. The immobilization of GOD on the PAAc grafted PE membrane through PEO chains being used as a spacer is shown in Figure 4B. A typical immobilization procedure with spacer is as follows: one piece of PEO modified PAAc grafted PE (PAAc-PEO) was added to 4 mL of 0.05M acetate buffer, pH 5.6. 0.5 mL glutaraldehyde (25% aqueous solution) and 1 mL (5mg/mL) of glucose oxidase solution (in 0.05M acetate buffer, pH 5.6) were then added. The mixture was shaken at 4^oC for 16 h. The membrane was then filtered and rinsed with deionized water three times. The enzyme membrane (PAAc-PEO-GOD) was soaked further in acetate buffer, pH 5.6, and was used for determining the activity of GOD immobilized membrane.

Assay of Enzyme Activity. Throughout all the experiments, the PAAc grafted PE membrane without immobilized GOD or the PAAc grafted PE membrane coupled only with the PEO spaces were used as "control" in the determination of the amount of immobilized GOD. The activity of the immobilized enzymes was expressed as the relative activity in percent based on that of free enzymes.

The amount of immobilized GOD was estimated by subtracting the amount of GOD determined in supernatant after immobilization from the total amount of GOD used for immobilization. The amount of GOD in the supernatant was determined at 455 nm with the Hitachi U2000 spectrophotometer. The activity of the free and immobilized GOD was assayed using the electrochemical procedure.(*12*)

Stability Measurements of Immobilized GOD. The thermal stability of the immobilized GOD was evaluated by measuring the residual activity of GOD exposed to various temperatures in a 0.05M acetate buffer solution of pH 5.6 for 10 minutes. Following heating, the samples were immediately cooled and assayed for enzymatic activity at 30^oC. The remaining activities were expressed as being relative to the original activities assayed at 30^oC without heating. The kinetics of thermoinactivation was investigated by determining the residual activity of the free and immobilized GOD after incubation at 65 ^oC for various periods of time.

The pH stability of the free and immobilized GOD was studied by incubating them in 0.05 M buffer solution at 30^oC and various pH regions for 20 minutes. The activity was also assayed using the above mentioned techniques.

Results and Discussions

ESCA Analysis of the Preparation of Substrate Polymer. Preliminary screening studies were carried out on PE surfaces treated by Ar plasma and then exposed to oxygen and allowed to polymerize with acrylic acid monomer and react with bisamino terminated PEO molecules. This screening process resulted in a selection of the best reaction system for further study. Figure 5 shows the ESCA survey scan and C1s spectra, during each stage of the PEO immobilization reaction onto a PE film which was modified by a plasma induced graft copolymerization of acrylic acid. Bisamino PEO-4000 was used in this reaction at a concentration 200 mg/mL. The survey scan of the control PE demonstrated only carbon present on the films surface, as expected (Figure 5A).

The C1s spectrum of the PE film exposed at an Ar plasma is different from that of control PE. After the PE is treated at an Ar plasma and then exposed to oxygen, the oxygen peaks become evident in the survey spectrum (Figure 5B). The oxygen came from the post-reaction of free radicals on the PE film which was treated at an Ar plasma and then exposed to oxygen. The C1s spectrum of an Ar plasma treated PE can be resolved into three distinct peaks which correspond to the three different carbon atomic environments. These correspond to carbon atoms with a single bond to oxygen at 286.4 eV (e.g. alcohol, epoxy, ether, ester, hydroperoxy, peroxide), carbon atom with two bonds to oxygen at 287.8 eV (e.g. aldehyde or ketone), and carbon atoms with three bonds to oxygen at 289.3 eV (e.g. carboxylic acid or ester). The C1s spectrum of the PAAc grafted PE is quite different from that of Ar plasma treated PE (Figure 5C).

Following the film reaction with bisamino PEO-4000, the C1s spectrum also changed (Figure 5D). The presence of oxygen and nitrogen on the film surface is demonstrated by the survey scan of the bisamino PEO immobilized PE. The C1s spectrum of -C-O-C- peak increases in intensity relative to that of PAAc modified PE. Thus, bisamino PEO can be immobilized on PE films exposed at an Ar plasma and polymerized with acrylic acid. The binding energies assigned to these peaks are based on values previously determined by Clark .*(13)*

Enzyme Immobilization. The amount of immobilized GOD is markedly influenced by the initial GOD concentration, in the low concentration level below about 4.0 mg/mL, but then levels off. (Figure 6) However, the surface concentration of GOD immobilized onto a PAAc grafted PE membrane with PEO spacer is less than that without a spacer. This occurs as a result of the spacer having less functional groups to react with GOD. The mechanism of immobilizing GOD onto PAAc grafted PE membrane is that in which coupling occurs between the carboxyl groups of the PAAc chain, and is mediated by carbodiimide, CMC and amino group of GOD. But, the mechanism of immobilizing GOD onto bisamino PEO modified PAAc grafted PE membrane (PAAc-PEO) is illustrated in Figure 4 in which case coupling occurs for both the amino groups of PEO and GOD with glutaraldehyde. The initial GOD concentration was kept at 5 mg/mL throughout the following experiments. The maximum amounts of immobilized GOD were 0.45 $\mu g/cm^2$ for PAAc grafted PE (PAAc-GOD) and 0.39 $\mu g/cm^2$ for PEO modified PAAc grafted PE (PAAc-PEO-GOD).

The effects of the surface concentration of GOD immobilized onto PAAc and PAAc-PEO are illustrated in Figure 7. The relative activity of the immobilized GOD without spacer (PAAc-GOD) is clearly seen in this figure to decrease gradually with the decreasing surface concentration of the immobilized GOD. However, the immobilized GOD with a PEO spacer (PAAc-PEO-GOD) gave an almost constant relative activity even at low surface concentrations. This result may be accounted for in terms of structural deformation of the immobilized GOD molecules as previously reported by Ikada.*(14,15)*

The immobilized enzyme should undergo strong deformation, especially in the lower surface concentration region without spacer (Figure 8A), whereas the immobilized GOD molecule with spacer (Figure 8B) must be protected from the structure deformation even in the lower surface concentration region owing to the spacer effect.

Determination of Michaelis Constant and Maximum Reaction Velocity. The permeability of GOD immobilized onto a PAAc grafted PE membrane was studied by adding standard concentrations of the substrate, glucose, from 25-1000

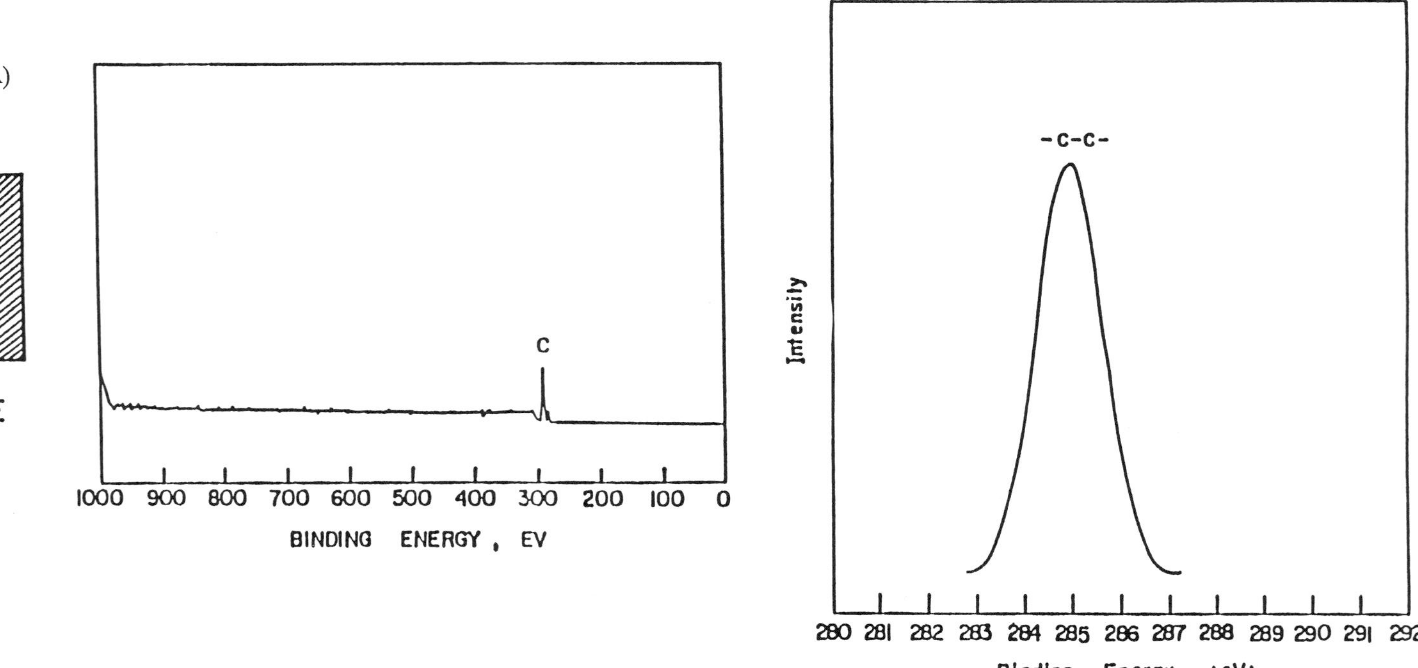

Figure 5A. ESCA survey scan and C1s spectrum during stage of the surface modification of PE.

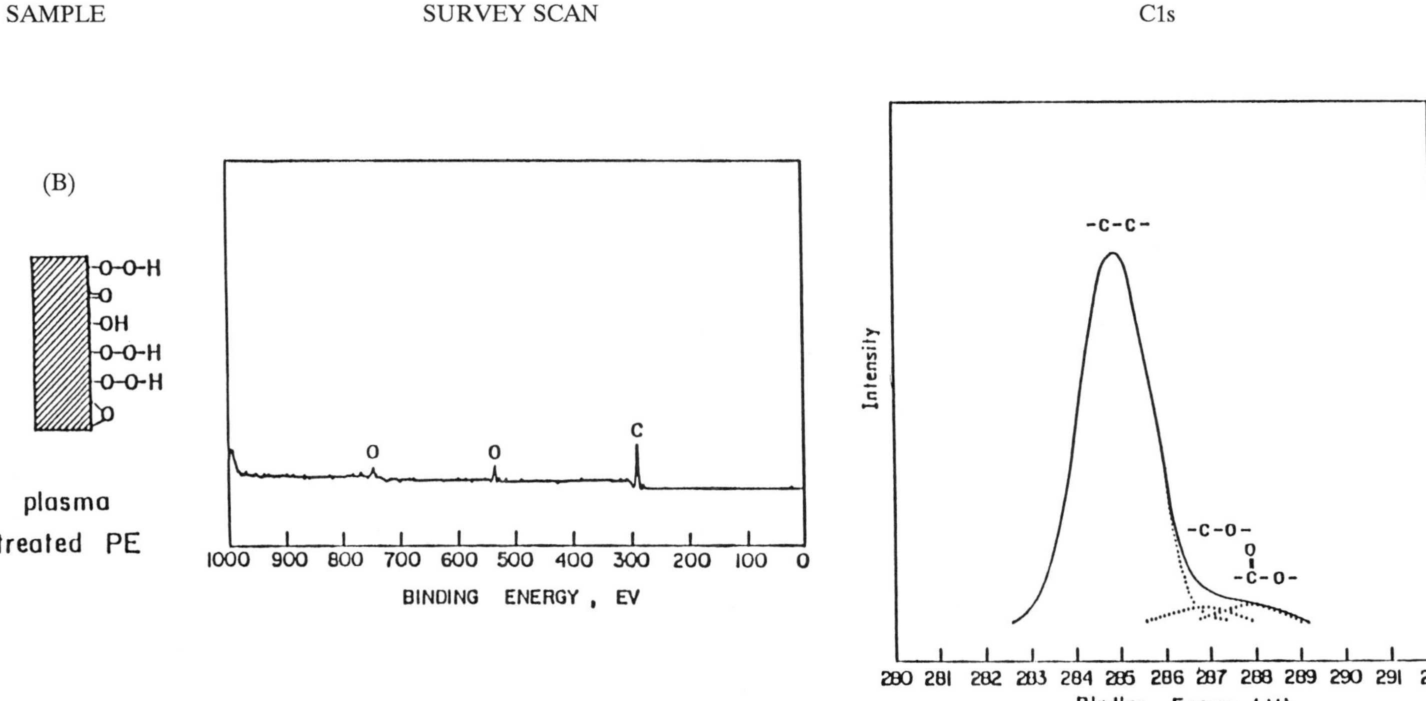

Figure 5B. ESCA survey scan and C1s spectrum during stage of the surface modification of PE.

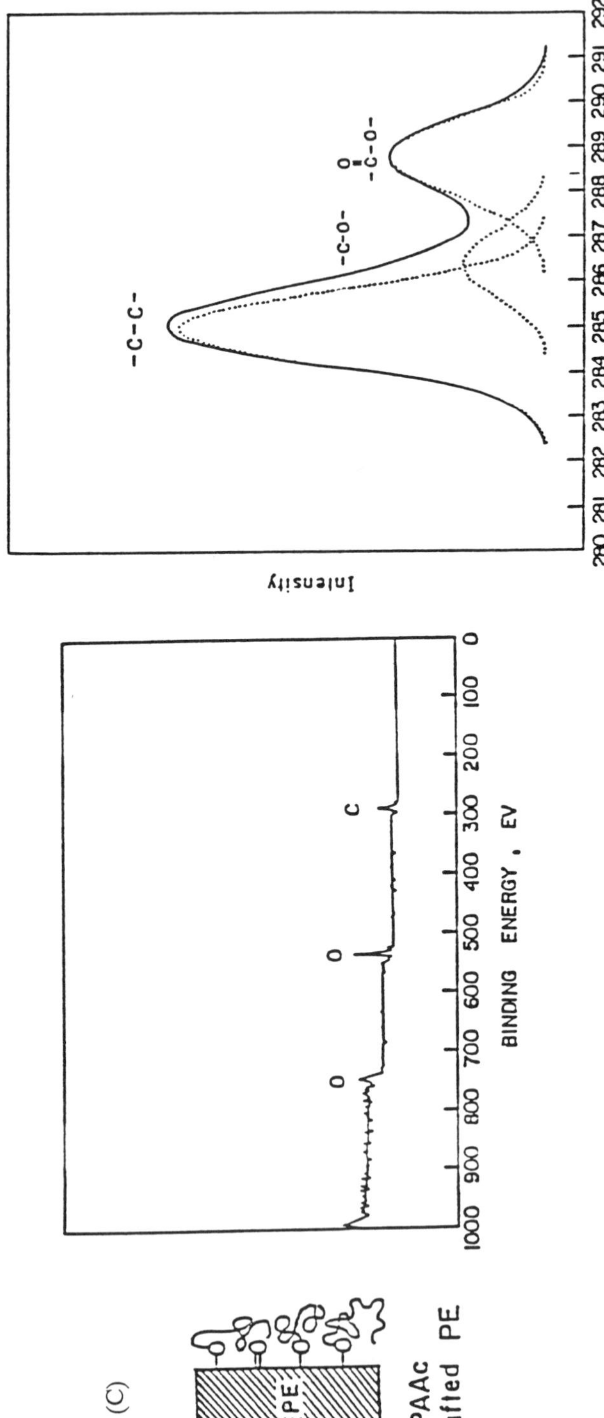

Figure 5C. ESCA survey scan and C1s spectrum during stage of the surface modification of PE.

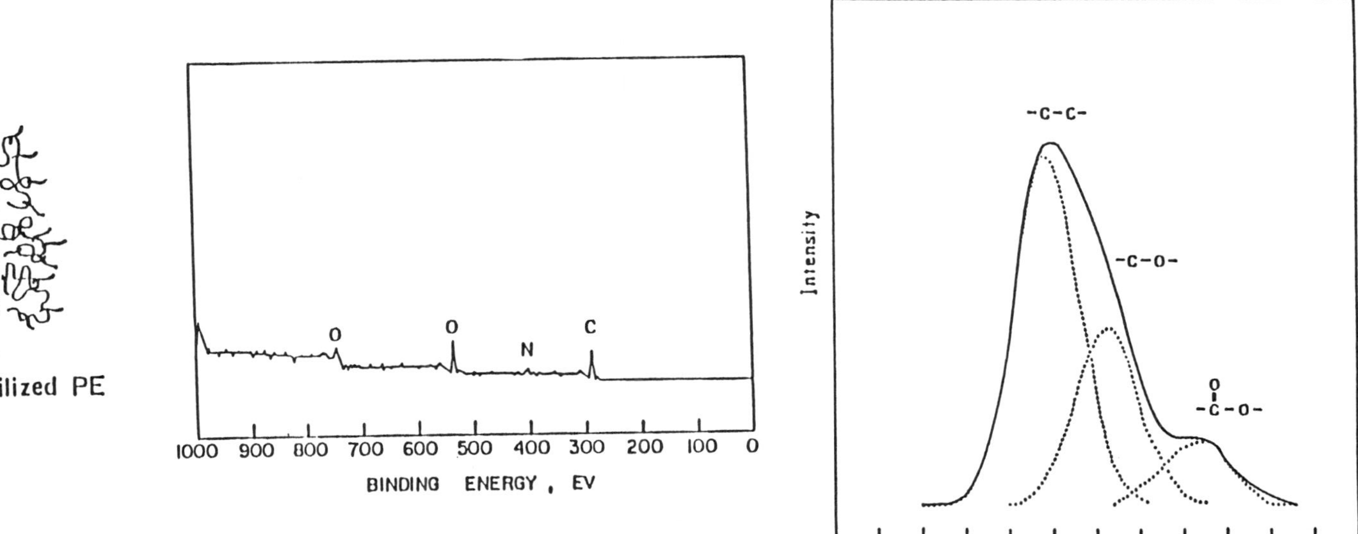

Figure 5D. ESCA survey scan and C1s spectrum during stage of the surface modification of PE.

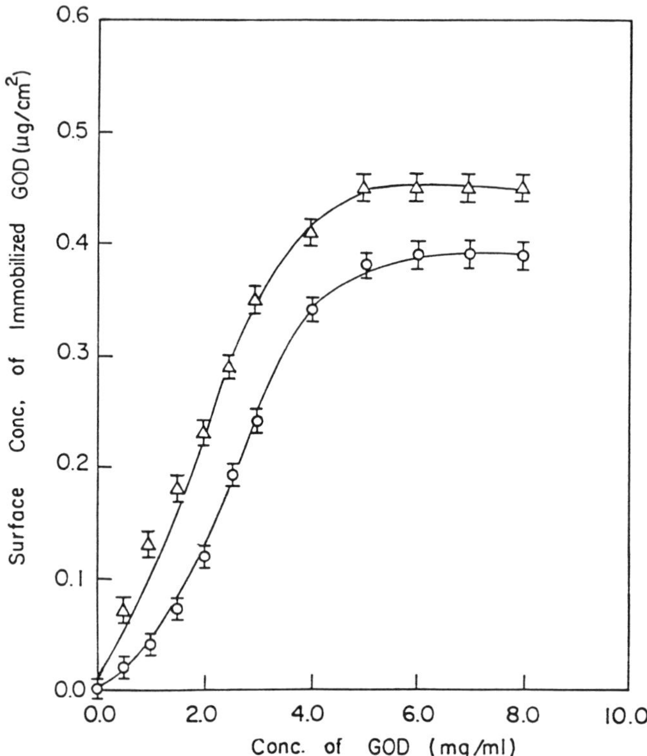

Figure 6. Effect of the GOD concentration on the amount of GOD immobilized onto (Δ) PAAc grafted PE; (O) PEO spacer modified PAAc grafted PE, 4°C and 16 h.

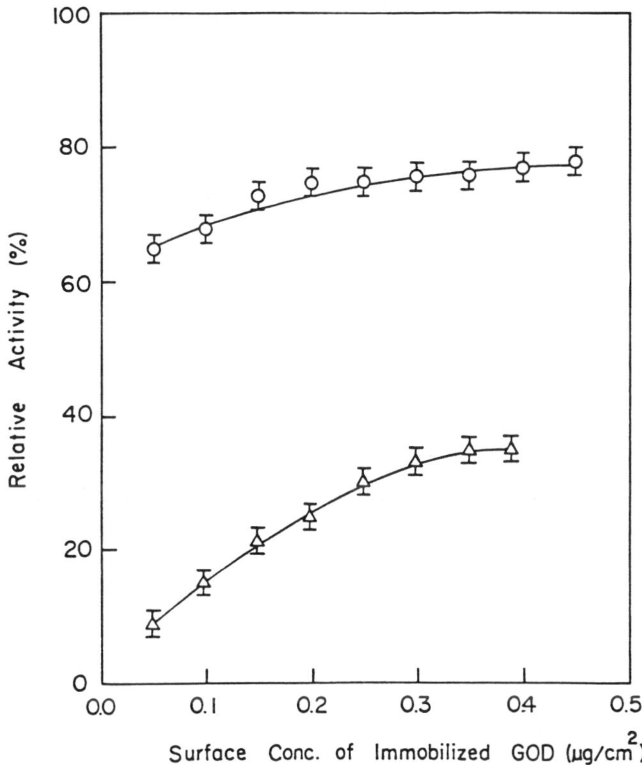

Figure 7. Effect of the surface concentration (SC) of the immobilized GOD on the relative activity (RA). Glucose 500ppm, pH 5.6 and 30°C (Δ) PAAc-GOD, (O) PAAc-PEO-GOD.

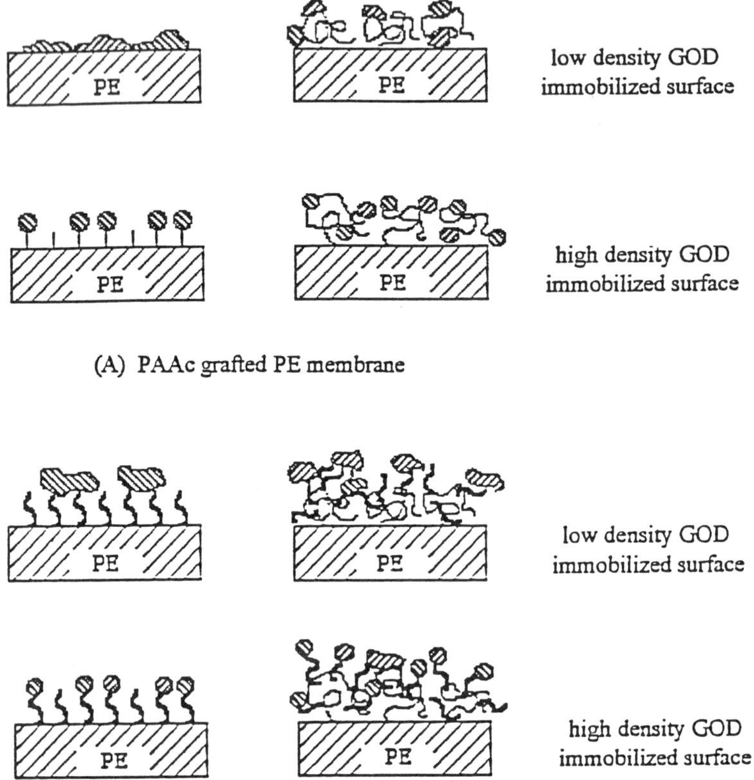

Figure 8. Schematic representation for molecular state of enzyme immobilized onto the surface of PAAc grafted PE (A) GOD immobilization without spacer and (B) GOD immobilization with spacer.

ppm, and measuring the dissolved oxygen obtained at the oxygen sensor. The enzymatic reaction is catalytically controlled since these results provide straight lines in the Lineweaver-Burk plot (Figure 9). The Michaelis constant, K_m, and the maximum reaction velocity V_m for the native GOD, PAAc-GOD, and PAAc-PEO-GOD, estimated from Figure 9, are presented in Table I. The apparent Michaelis constant, K'_m of the immobilized GOD without spacer (PAAc-GOD) is higher than that of the immobilized GOD with spacer (PAAc-PEO-GOD) and the native enzyme. This may be due to the limitation of diffusion resistance. Glucose is apt to decrease for immobilized GOD without spacer owing to the stereohindrance as compared to the case of immobilized GOD with spacer or the native GOD. On the other hand, the maximum reaction velocity, V_m, value of PAAc-GOD offers the lowest value suggesting that the relative activity of immobilized GOD has decreased in the course of covalent fixation without spacer.

Thermal Stability. The thermal stability of immobilized enzymes is one of the most important criteria of their application. The enzyme in most immobilized enzyme preparations, especially in covalently bound systems, is more resistant against heat and other denaturing agents than the soluble form.(*16*) The effect of temperature on the stability of the three forms of GOD in acetate buffer is shown in Figure 10. The temperature maximum of the immobilized GOD is the same as that of the native GOD(30°C), but broader in the range of a higher temperature. The immobilized GOD at 65°C at 10 min is observed to exhibit activity two times that of the soluble enzyme. However, the immobilized GOD without a PEO spacer (PAAc-GOD) is more stable than that with spacer (PAAc-PEO-GOD). The higher stability of the immobilized GOD without spacer than that with a spacer is shown in Figure 8 which probably be ascribed to the stabilization of GOD molecule owing to the multipoint attachment of the GOD molecule to the surface of the PAAc grafted PE membrane. This multipoint attachment consequently leads to a reduction in molecular mobility that is common principle of enzyme stabilization.(*17*)

Comparative studies of thermoinactivation of the native and the immobilized GOD at temperature 65°C are presented in Figure 11. Three forms of the enzyme were incubated at temperatures 65°C and their activity changes were periodically evaluated in presence of 500 ppm of glucose at 30°C, pH5.6. The kinetic curve of thermoinactivation of the immobilized preparation at 65°C reveals a two-stage process characterized by the following constants: k_1= $1.35\text{x}10^{-1}$ min^{-1} and k_2= $1.71\text{x}10^{-2}$ min^{-1} for PAAc-GOD, and k_1= $9.23\text{x}10^{-2}$ min^{-1} and k_2= $2.09\text{x}10^{-2}$ min^{-1} for PAAc-PEO-GOD. These constants are summarized in Table II. These constants represent the rate of transition of the enzyme structure to intermediate forms based on a two stage process.(*18,19*)

$$E_1 \xrightarrow{k_1} E_2 \xrightarrow{k_2} E_3$$

where E_1 is the native form of the enzyme; E_2 is an active intermediate form; and E_3 is the denaturalized form of the enzyme.

The rate of transition from E_1 to E_2 forms for the immobilized enzyme is $1.35\text{x}10^{-1}$ $\text{min}^{-1}$for PAAc-GOD and $9.23\text{x}10^{-2}$ min^{-1} for PAAc-PEO-GOD which are lower than the value of $1.46\text{x}10^{-1}$ min^{-1}retrieved for the native enzyme at 65°C.

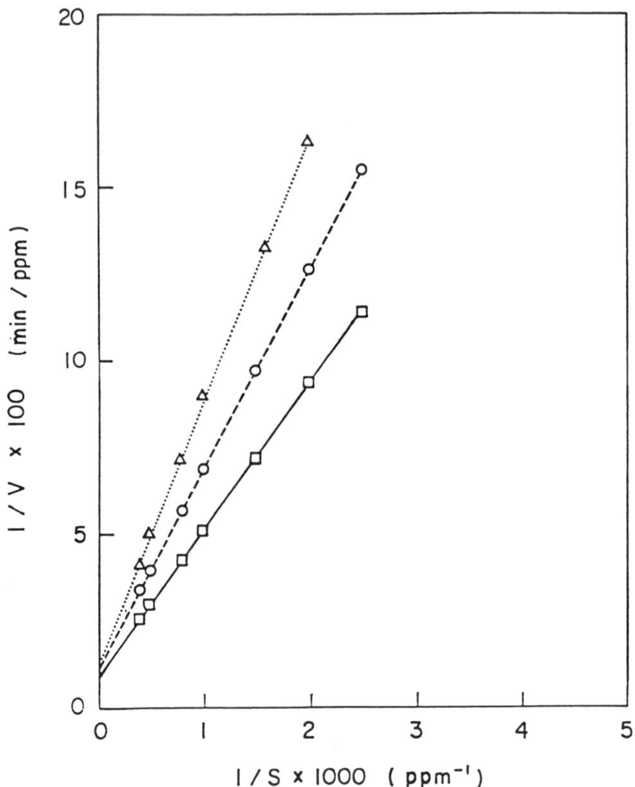

Figure 9. Lineweaver-Burk plots of 1/V vs. 1/S. (□) native GOD, (O) PAAc-PEO-GOD, and (Δ) PAAc-GOD.

Table I. Michaelis Parameters K_m and V_{max} at pH 5.6 and 30°C

Sample	K_m (mM)	V_{max} (mM / min)
GOD	23.58	5.62
PAAc-GOD	31.44	4.17
PAAc-PEO-GOD	27.33	4.75

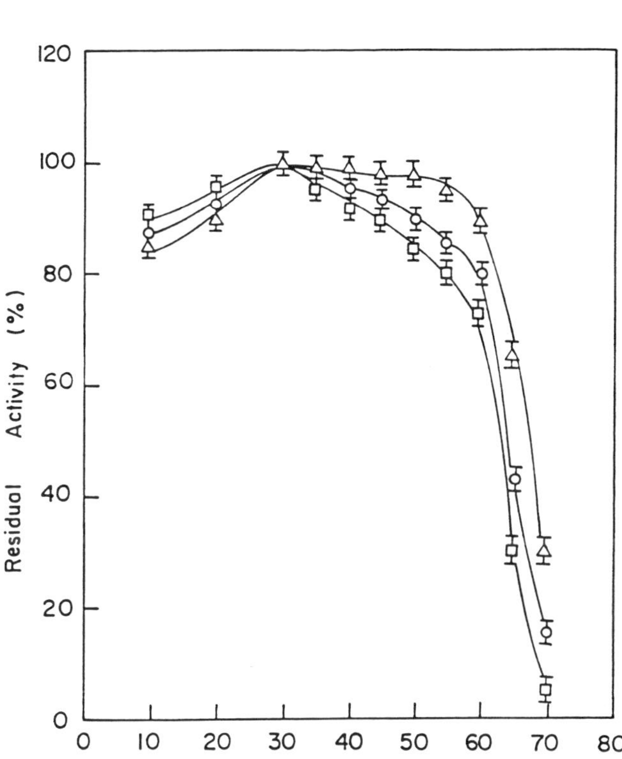

Figure 10. Effect of temperature on the residual activity of glucose oxidation at pH 5.6 and 30°C. (□) native GOD, (O) PAAc-PEO-GOD, and (Δ) PAAc-GOD based on 10 minutes reaction at various temperatures in 0.05M acetate buffer (pH 5.6).

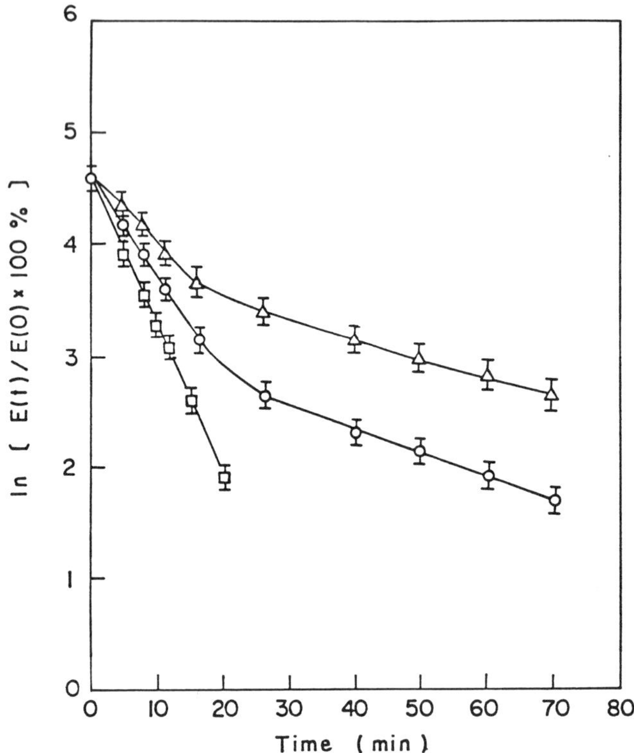

Figure 11. Activity change ln [E(t)/E(o)] by heat treatment at 65°C of glucose oxidation at pH 5.6.(□) native GOD, (○) PAAc-PEO-GOD, and (△) PAAc-GOD

Table II. Deactivation Constants (k_1 and k_2) Values for Soluble and Immobilized GOD Incubated at 65°C

Temperature °C	*Native GOD* k_1 (10^2 min^{-1})	*PAAc-PEO-GOD* k_1 (10^2 min^{-1})	k_2	*PAAc-GOD* k_1 (10^2 min^{-1})	k_2
65	14.6	9.23	2.09	13.5	1.71

On the other hand, the second constant (k_2) representing the rate of transition from the second active form of GOD to an inactivated form is higher for the PAAc-GOD ($1.71x10^{-2}$ min^{-1}) than for the PAAc-PEO-GOD ($2.09x10^{-2}$ min^{-1}). The results on the rate of thermoinactivation suggest that the second transition step is the rate determining step for the denaturation of the immobilized enzyme.

pH Stability. The pH effect on the activity of the immobilized and soluble forms of the enzyme was studied in adequate buffers (pH 2.8-9.6) and is presented in Figure 12. Two forms of immobilized GOD have the same optimum pH value of the native form (pH5.6), but the pH profile is considerably widened owing to diffusional limitations of the immobilized enzyme molecules.(*20*) PAAc-GOD displays the greatest stability at higher pH values than the PAAc-PEO-GOD and native GOD. This is probably due to the stabilization of GOD molecule resulting from multipoint attachment of the GOD molecule to the surface of the PAAc grafted PE membrane.

Durability. The durability of the immobilized GOD at repeated oxidation is also very important in applications. Figure 13 illustrates the effect of repeated use on the residual activity of glucose oxidation by the immobilized GOD, PAAc-GOD and PAAc-PEO-GOD, respectively. A 500 ppm of glucose solution, pH5.6, was used for verifying the GOD immobilized membrane at 30°C. The activity is seen to be retained without any significant loss, irrespective of the spacer interposition.

Storage Stability. The obtained glucose oxidase immobilized membranes (PAAc-GOD and PAAc-PEO-GOD) were kept in acetate buffer solution, pH5.6, at 4°C. The glucose oxidase activity of the membrane was determined at seven-day intervals. The PAAc-PEO-GOD retained 68% glucose oxidase activity after six months of storage (Figure 14). The PAAc-GOD form has a higher activity (86%) than PAAc-PEO-GOD. However, the activity of free GOD lost more than 50% of their initial activity under seven days storage. The high stability of the immobilized GOD can be attributed to the prevention of autodigestion and thermal denaturation as a result of the fixation of GOD molecules on the surface on PAAc grafted PE.

Conclusion

The novel synthesized acrylic enzyme carriers polyacrylic acid grafted PE and poly(ethylene oxide) modified polyacrylic acid grafted PE were observed in this study to be a promising glucose oxidase carrier. The bioactivity of PEO modified PAAc grafted PE membrane (PAAc-PEO-GOD) was higher than that of PAAc grafted PE membrane (PAAc-GOD). The apparent Michaelis constant, K'_m values were larger for immobilized GOD than for free enzyme, while V_m values were smaller for the immobilized GOD. The optimum pH value of glucose oxidase was not affected by the immobilization reaction. However, the activity of immobilized glucose oxidase was observed to be considerably widened. Immobilized enzyme demonstrated a reduced sensitivity to thermoinactivation as compared to the free form. The immobilized GOD without spacer (PAAc-GOD) exhibited a higher thermal stability than that with spacer (PAAc-PEO-GOD). The storage stability of the immobilized GOD was higher than native enzyme. The initial enzymatic activity of the immobilized GOD remained almost unchanged without any elimination and inactivation of GOD, indicating excellent durability.

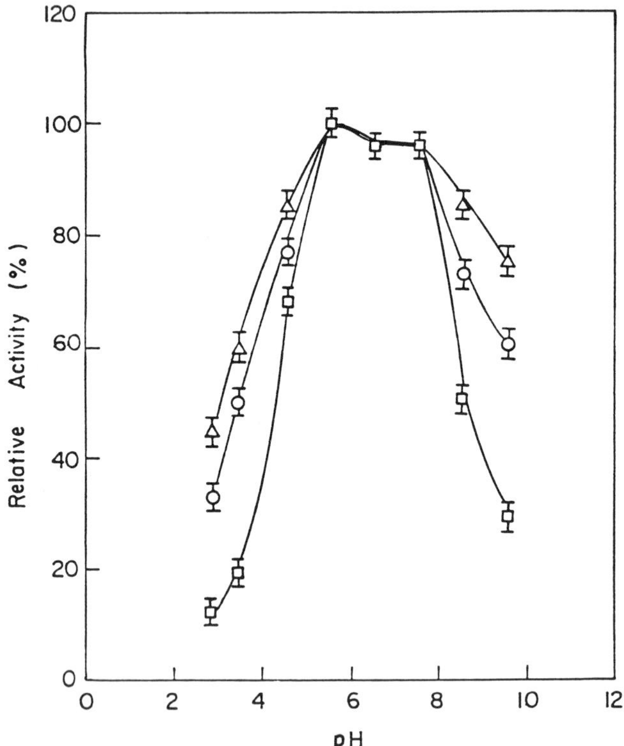

Figure 12. Effect of pH of the reaction medium on the residual activity of glucose oxidation at 30 °C. (□) native GOD, (O) PAAc-PEO-GOD, and (Δ) PAAc-GOD

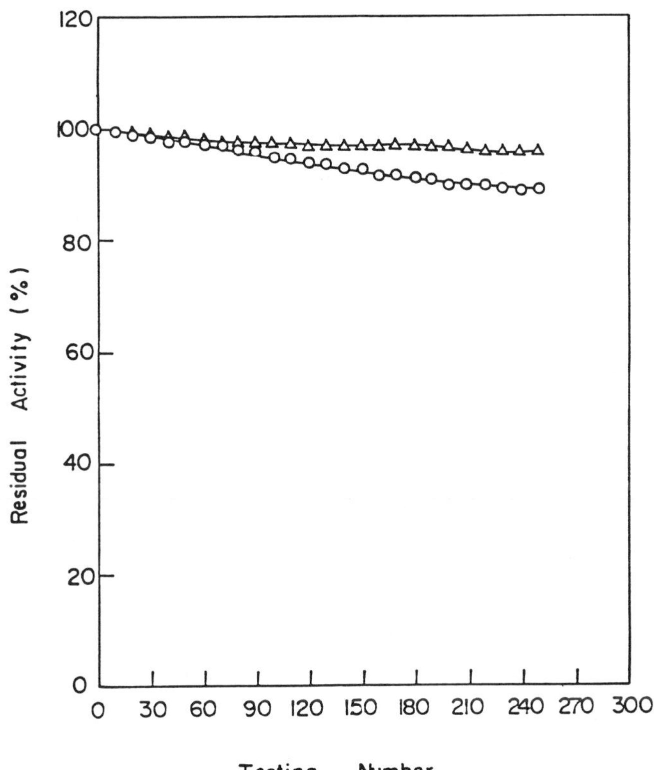

Figure 13. Effect of the repeated use on the residual activity of glucose oxidation at 30°C. (O) PAAc-PEO-GOD, and (Δ) PAAc-GOD.

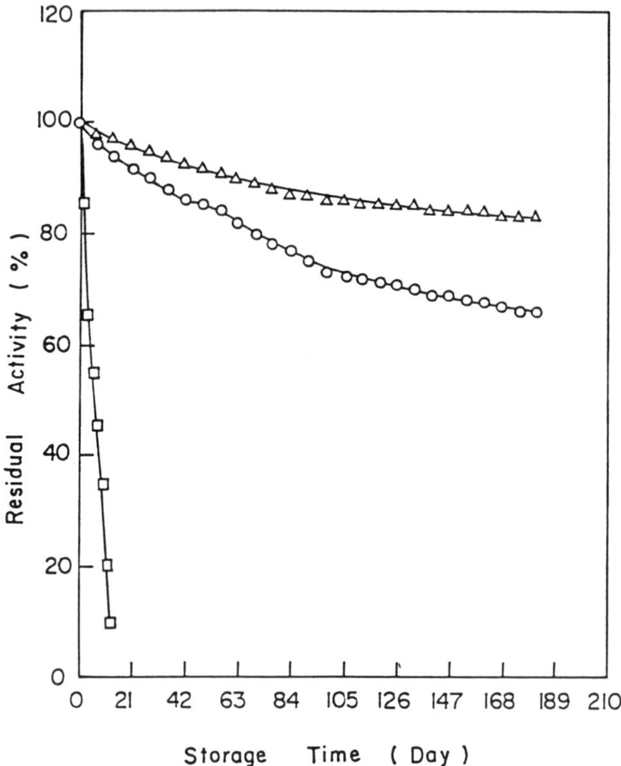

Figure 14. Effect of the storage in acetate buffer at pH 5.6and 4°C on the residual activity of glucose oxidation at 30°C. (□) native GOD, (O) PAAc-PEO-GOD, and (Δ) PAAc-GOD.

Acknowledgments

Financial support of this work by the National Science Council of the Republic of China (NCS-80-0405-E007-01) is gratefully acknowledged.

Literature Cited

1. Boenig, H.V., *Fundamentals of Plasma Chemistry and Technology,* **1988,** Technomic Publishing Co. Inc., Lancaster.
2. Yasuda, H., *Plasma Polymerization,* **1985,** Academic Press Inc., Tokyo.
3. Yasuda, H.K., *Polymerization and Plasma Interactions with Polymeric Materials,* **1990,** *J. Appl. Polym. Sci. Appl. Polym. Symp. , 46.*
4. Hoffman, A.S., *Ann. N.Y. Acad. Sci.,* **1987**, *146,* 96-101.
5. Hoffman, A.S.,*J. Appl. Polym. Sci. Appl. Polym. Symp.,* **1988,** *42,* 251-267.
6. Hoffman, A.S., *J. Appl. Polym. Sci. Appl. Polym. Symp.,* **1991,** *46,* 341-359.
7. Suzuki, M., Kishida, A., Iwata, H. and Ikada, Y., *Macromolecules,* **1986**, *19,* 1804-1808.
8. Fujimoto, K., Tadokoro, H., Minato, M., and Ikada, Y., *ACS Polym. Mater. Sci. Eng.,* **1990**, *62,* 736-740.
9. Fujimoto, K., Ueda, Y., Takebayashi, Y., and Ikada, Y., *ACS Polym. Mater. Sci. Eng.,* **1990**, *62,* 284-288.
10. Wang, C.C. and Hsiue, G.H., *J. Polym. Sci., Polym. Chem. Ed.,* **1993**, *31,* 1307-1314.
11. Hsiue, G.H. and Wang, C.C., *J. Appl. Polym. Sci.,* **1990,** *40,* 235-247.
12. Hsiue, G.H. and Wang, C.C., *Biotechnol. Bioeng.,* **1990**, *36,* 811-815.
13. Clark, D.T., *Adv. Polym. Sci.,* **1977**, *24,* 125-188.
14. Hayashi, T. and Ikada, Y., *Biotechnol. Bioeng.,* **1990**, *36,* 593-600.
15. Hayashi, T. and Ikada, Y., *ACS, Polym. Mater. Sci. Eng.,* **1990**, *62,* 512-516.
16. Ulbrich, R., Schellenberger, A. and Damerau, W., *Biotechnol. Bioeng.,* **1986**, *28,* 511-522.
17. Zaborsky, O., *Immobilized Enzymes,* **1973**, CRC Press Inc. Boca Raton, FL.
18. Henley, J.P. and Sasana, A., *Biotechnol. Bioeng.,* **1986**, *28,* 1277-1285.
19. Fortier, G.and Blanger, D., ibid., **1991**, *37,* 854-858.
20. Greenfield, P.F. and Laurence, R.L., *J. Food Sci,* **1975**, *40,* 906-918.

RECEIVED July 19, 1993

Chapter 23

Spatially Controlled On-Wafer and On-Chip Enzyme Immobilization Using Photochemical and Electrochemical Techniques

David J. Strike, Albert van den Berg, Nico F. de Rooij, and Milena Koudelka-Hep

Institute of Microtechnology, University of Neuchâtel, Rue A.-L. Breguet 2, CH–2000 Neuchâtel, Switzerland

Two methods for the spatially controlled deposition of proteins on microelectrodes are described. The first technique involves the entrapment of glucose oxidase in photopolymerized polyHEMA. The second uses electrochemically aided adsorption to deposit urease and to co-deposit glucose oxidase with bovine serum albumin. Both techniques were found to lead to active deposits and the properties and optimisation of the deposition procedures will be described. Further, to facilitate glucose measurement in complex media, the deposition of a thin film of polypyrrole following that of the proteins is described. The properties of this film with respect to two model interferents and complex yeast extract medium will be reported.

Increasingly, the world of biochemical sensors is meeting that of microelectronics. The size of the electrodes is shrinking as the techniques of semiconductor fabrication become more widely applied allowing better measurement *in vivo*, in smaller sample volumes, multisensor arrays, etc *(1)*. Although various transducer fabrication technologies seem to be well established the deposition of a biological component upon the miniature electrodes remains a problem. Several procedures to deposit spatially controlled enzyme containing layers have been reported in the literature e.g. screen printing, lift-off, photopolymerization and electrochemical deposition. Further to the problem of depositing the biological components it is frequently necessary to deposit protective anti-interference layers and diffusion barriers. For these too a global deposition method in which all electrodes are treated equally may not be optimal. The final choice of the deposition method will be mainly determined by the required sensor characteristics and pattern resolution.

In this paper the deposition of proteins, principally for a glucose biosensor, in two spatially well controlled ways, one photolithography and the other electrochemical will be described, as will be the electrochemical deposition of a thin anti-interference layer.

In our previous work, chemical co-crosslinking of glucose oxidase (GOx) and bovine serum albumin (BSA) by glutaraldehyde was used for membrane depositions either by casting *(2, 3)* or by spin coating followed by lift-off *(4)*. Although the latter

0097–6156/94/0556–0298$08.00/0

method allows good spatial and thickness control it is limited to rather thin (c. 1μm) membranes. Based on our previous studies of the deposition and patterning of poly-2-hydroxyethylmethacrylate (pHEMA) gel membranes of various thicknesses (10 - 100 μm) *(5)*, we have chosen to investigate the entrapment of glucose oxidase in this gel matrix. The choice of pHEMA was motivated by the good adhesion of these layers on planar transducers following chemical pretreatment such as a surface functionalization with methacrylic groups *(6)*. The photolithographic deposition involves the entrapment of the glucose oxidase (GOx) within a pHEMA gel layer that is photopolymerized and patterned on the wafer level. This simple enzyme immobilization method has been reported in the literature for pHEMA *(7, 8)* and several other gels e.g. polyacrylamide *(9)* and polyvinylalcohol *(10, 11)*, which have been deposited on potentiometric and amperometric transducers. Generally reasonably stable membranes can be obtained although care must be taken to avoid forming membranes having too large a diffusional barrier. In this paper we describe the entrapment of GOx in photopolymerized pHEMA and the properties of the resulting electrode.

Two principle techniques for electrochemical enzyme deposition have been reported, entrapment in an electrochemically grown polymer and electrochemically aided absorption. A wide range of electrochemically grown polymers have been used. The polymer can function as both an entrapment matrix and as an anti-interference layer *(1, 12-20)*, as a matrix for the immobilisation of the protein with an electron transfer mediator *(21-23)*, and as an electron transfer matrix alone *(24, 25)*. Electrochemically aided adsorption has received comparably less attention *(26-30)*. However, in our experience *(31)* the latter technique results in larger responses and is more appropriate to microelectrodes. Here we will present results on the electrochemically aided adsorption of GOx and BSA, and also of urease. Furthermore to reduce the interferences at the GOx/ BSA electrode we will describe the deposition of an anti-interference layer of polypyrrole, which is grown on the electrode after the deposition of the proteins.

Methods

Chemicals. Glucose oxidase used for the electrochemical deposition was type VII (100-200 $U.mg^{-1}$) obtained from Sigma, whilst that used for the photolithographic deposition was obtained from Calbiochem and had an activity of 250 $U.mg^{-1}$. The other materials were obtained as follows; dimethoxyphenylacetophenone (DMAP), polyvinylpyrrolidone K 90 (PVP), (trimethoxysilyl)propyl methacrylate (TMSM) and pyrrole were obtained from Aldrich; albumin bovine (BSA) was obtained from Calbiochem; K_2PtCl_4, hydroxyethyl methacrylate (HEMA), tetraethyleneglycol dimethacrylate (TEGDMA) and D(+)-glucose were obtained from Fluka; ethyleneglycol was obtained from Merck; Urease Type IX (66 μmolar $units.g^{-1}$) and urea were obtained from Sigma. All other products were of the highest grade available and were obtained from either Fluka or Merck. The glucose stock solution (1.0 M) was allowed at least 24 hr to mutarotate before use. The protein depositions and amperometric measurements were made in potassium phosphate buffered saline (PBS), pH 7.2, of composition 5.3 mM phosphate, 2.5 mM potassium chloride and 0.15 M sodium chloride. The conductometric measurements were made in 2 mM potassium phosphate buffer, pH 7.1.

Apparatus. The flow injection system was composed of a model MV-CA4 peristaltic pump (Ismatec, Switzerland), a rotary injection valve (Ig Instrumenten Gesellschaft) and a perspex flow through cell (on loan from Ciba-Geigy). For longer term measurements under continuous flow conditions two electronically controlled valves were used in conjunction with the pump so that the electrode was exposed to either streams of carrier or sample. In all experiments the electrodes were controlled using an EG&G 273A potentiostat/galvanostat or an IBM Voltammetric Analyzer EC/225. For the admittance measurements a Stanford Research Systems Model SR850 lock-in amplifier was used in conjunction with the EG&G 273A.

The thickness of dried electrodeposited protein films was measured using an Alpha-step 200 surface profiler from Tencor instruments and the thickness of the photolithographically deposited films was estimated using a microscope.

In the photolithographic depositions the films were exposed using a standard mask aligner.

For the protein deposition experiments a platinum wire was used as a counter electrode. For batch measurements an SCE reference electrode (Metrohm, Switzerland) was used together with the on-chip counter electrode, and for the Flow Injection Analysis (FIA) measurements all three on-chip electrodes were utilized.

Transducers (Microelectrodes). Two types of planar structures were used. Described elsewhere *(32)*, the first one is a three thin film electrode microcell (overall dimensions of 0.6 mm x 3 mm x 0. 38 mm) comprising Pt working (100 µm x 1000 µm) and counter electrodes and a Ag/AgCl reference electrode. The metallic layers were deposited by electron beam evaporation and patterned using lift-off. This planar cell was used for the deposition the pHEMA/ GOx membranes. The second structure used is a four microband thin film Pt electrode array (Figure. 1). This used polysilicon interconnections and a silicon nitride passivation layer to precisely determine the geometric area of the Pt electrodes. Four layouts of this array were realized having electrodes of length 1000 µm and widths of either 25 or 5 µm, separated by either by 5 or 25 µm. Following an initial electrochemical characterization, these arrays and the working electrode of the planar microcell were used for the electrochemical immobilization of GOx, BSA and urease.

Membrane Deposition.

Photolithographic. Different thicknesses of the enzyme containing pHEMA layer were deposited and patterned on wafer to dimensions of 600 µm x 1400 µm (width x length). Prior to the membrane deposition, the surface of the wafer was functionalized with methacrylic groups using a 10 % (trimethoxysilyl)propyl methacrylate (TMSM) toluene solution containing 0.5 % water at 60 °C. GOx (50 mg) was dissolved in water (100 µl) and added into 1 ml of the prepolymeric mixture (54 wt.% HEMA, 37.7 wt.% ethyleneglycol, 4.3 wt.% tetraethyleneglycol dimethacrylate, 1.5 wt.% dimethoxyphenylacetophenone and 2.5 wt.% polyvinylpyrrolidone K 90). An appropriate amount, depending on the required thickness, of the prepolymeric/GOx solution was deposited by micropipette on the modified wafer, covered with a Mylar sheet and exposed to UV light for 2 to 4 min, followed by development in ethanol. The sensors were preconditioned in PBS.

Electrochemical Aided Adsorption. The 1000 x 100 µm working electrode of the microcell was principally used for the GOx/ BSA depositions. Before any deposition was made the working electrode was pretreated by cyclic voltammetry in 0.5 M H_2SO_4 as follows: four sweeps at 50 $mV.s^{-1}$ were made , with the first two and the final one being made over the range -0.25 to 1.2 V, whilst for the third sweep the anodic limit was increased to 2.0 V. The final sweep was halted at 0.2V. The proteins were deposited out of PBS containing between 5 and 10% protein (wt/vol). The proteins were deposited simultaneously at, typically an applied current density of 5 $mA.cm^{-2}$ for a period of 2 min. The modified electrode was carefully washed in water for 5 s and then cross-linked with glutaraldehyde (2.5 % in PBS for 30 min or 25 % in PBS for 4 min) at room temperature. Finally the electrode was rinsed and stored in 10 mM potassium phosphate buffer at 4 °C until used.

The polypyrrole anti-interference layer was deposited after the deposition and cross-linking of the GOx/ BSA film. This layer was deposited in two stages. Firstly Pt was deposited on the modified working electrode by reduction of K_2PtCl_4 at -200 mV (vs SCE) out of 10 mM potassium acetate containing 50 mM KCl and 1 mM of the Pt salt, pH 5.0. The total charge passed during the deposition was 10 μC, equivalent to 10 $\mu g.cm^{-2}$ Pt. Secondly, a polypyrrole layer was deposited out of 0.1 M potassium phosphate buffer containing 0.1 M KCl and 0.15 M pyrrole at pH 7.0. The polymer was electrodeposited by 5 cyclic voltammetry sweeps over the range 0.1 to 1.0 V (vs SCE) at 25 mVs^{-1}, with the scan being held for 5 min at the anodic limit of the 4th scan. The total charge passed was 10μC.

Urease was simultaneously deposited on four 25 μm wide electrodes each separated by a 25 μm gap. Unlike the larger electrodes these were not pretreated prior to the deposition. The urease was deposited from phosphate buffered saline containing 50 $mg.cm^{-3}$ urease (c. 3,300 $U.cm^{-3}$) at an applied current density of 5 $mA.cm^{-2}$ over a 4 min period. Immediately after this the electrodes were soaked, without agitation, in water for 15 s. This step was repeated and then the electrodes were placed in a 25% glutaraldehyde solution in PBS for 4 min. A control electrode was prepared in the same way except that no current was passed.

Electrochemical Measurements. The amperometric glucose measurements were made at 650 mV (vs SCE). The admittance measurements were made between the outer pair of modified electrodes using a 10 mV (rms) 10 kHz ac potential, there was no applied dc potential. The measurements made with the polyHEMA membranes were at 37°C, the remainder were made at room temperature.

Results and Discussion

Photolithographic Deposition. A typical glucose calibration curve of the pHEMA/GOx sensor with a 20 μm thick membrane is shown in Figure. 2. The linear region is up to 12 mM glucose (r = 0.9986) and the apparent K_m calculated from the double-reciprocal plot (not shown) is 11.5 mM. The response time $t_{90\%}$ was of the order of 30s and the dry storage lifetime with intermittent testing of several weeks. The main problem encountered was obtaining a very fine enzyme dispersion in the prepolymeric mixture in order to limit the light scattering during the photopolymerization since this determines the pattern resolution. Although this parameter is not critical for the present deposition of quite large membranes (600 μm x 1400 μm), separated by 200 μm, it will become important for enzymatic membrane depositions on small structures. So far the pattern resolution is approximately 50 μm for membrane thicknesses of 10 μm.

Electrochemically Aided Adsorption. Figure. 3 shows the glucose response of an electrode modified with electrodeposited GOx/ BSA recorded under FIA conditions. The detection limit was 0.01 mM and the response was linear up to 0.6 mM (regression coefficient = 0.9999). Such an electrode was found to be stable both for long term storage *(33)* and during extended use in flow conditions, where measurement periods of several days have been obtained. When these electrodes were inspected visually a rough deposit localized on the working electrode was seen. The apparent activity of these films (ie. the size of their glucose response) as well as the precision of the deposition can be controlled for a given geometry. For a 1 mm by 100 μm wide band electrode negligible glucose response was observed until the deposition current density was above 0.5 $mAcm^{-2}$. The size of the response then increased rapidly with increasing current density until 5 $mAcm^{-2}$ after which the deposition was no longer confined to the electrode surface but spilt across the surrounding substrate, thus limiting the current density that can profitably be applied. During the deposition the applied voltage rises rapidly from c. 1.6 to c. 2.0 V. Although cyclic voltammetry indicates that solvent breakdown and gas evolution begins around 1.4 V the protein

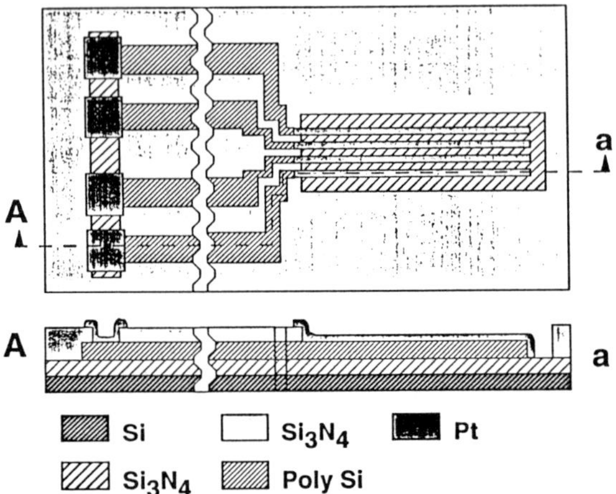

Figure. 1. Schematic of the 4 microband transducer.

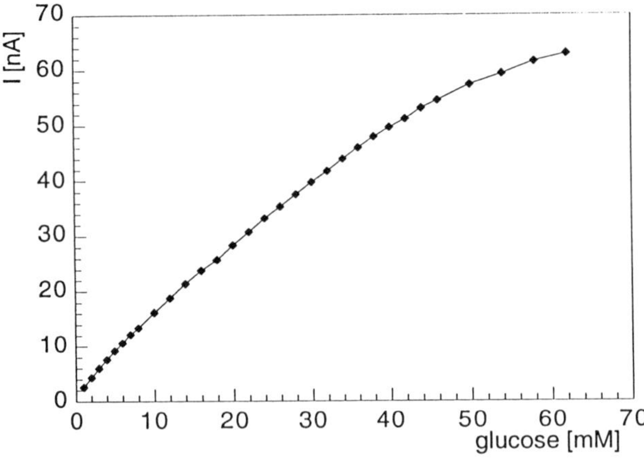

Figure. 2. Glucose response of a microcell transducer modified with a pHEMA/ GOD membrane. The measurements were recorded in stirred PBS at 37°C.

films do not appear to contain trapped bubbles. This suggests that the protein film may inhibit bubble formation. Although, at very high current densities or with constant potential deposition the deposited film contains many trapped bubbles. However, on the 25 and 5 μm electrodes a higher current density was required to obtain good deposits and these did not contain trapped bubbles.

Modification from solutions containing different ratios of GOx and BSA suggests a synergistic effect of the BSA, since the addition of 5% BSA to 5% GOx considerably increases the apparent activity of the deposited film, almost to the level of 10% GOx. It is well known that protein precipitation is concentration dependant and it is thought that BSA may act by increasing the total protein concentration. Although BSA may also allow more intermolecular bonds and reduce crowding of the enzyme *(34)*. The majority of our work has used depositions from 5% GOx/ 5% BSA since this seemed to offer the optimal compromise of current response against GOx consumption. The optimal deposition time was found to be 2 min, lengthening the time to 4 min resulted in a c. 30% increase in the glucose response, whilst reducing it caused a considerable decrease in the response. The dry thickness of a film deposited over a 2 min period from 5% GOx/ 5% BSA at 5 $mA.cm^{-2}$ was estimated to be 3-4 μm.

For this technique to be of more general use it must also be applicable to other enzymes. We have began by extending the procedure to urease, an enzyme whose entrapment in polypyrrole, as well as use in a conductometric biosensor for urea has already been reported *(35, 36)*. We have found that urease was readily electrodeposited and that the activity of the deposited enzyme was easily seen by admittance measurements (Figure. 4). The apparent K_m for the immobilised enzyme was calculated from the double-reciprical plot ($y = 0.17 + 0.50x$, regression coefficient = 0.9994, n = 9) to be 2.9 mM, which can be compared to the range 3-5.1 mM for the soluble enzyme *(36)*. A control electrode produced at the same time displayed negligible response to urea.

A problem with the, as deposited, GOx/ BSA films is that they are very prone to interference resulting from the rather high potential needed for the H_2O_2 detection. To attempt to circumvent this problem, whilst retaining the spatial control resulting from the use of electrodeposition we have investigated the electrodeposition of thin layers of polypyrrole, which have previously been reported to display anti-interference properties resulting from size exclusion effects *(19, 37)*. The polypyrrole was grown after the proteins had been deposited and cross-linked. Its growth appeared to be inhibited by the protein film. When cyclic voltammetry was used for the growth instead of the voltammogram characteristic of a nucleation and growth mechanism *(38)* an irreversible wave which diminished on subsequent sweeps was observed. This suggests the formation of a self-limiting film. Since this may stem from poisoning of the Pt surface by material from the protein deposition solution, a thin layer of Pt was electrodeposited on the surface prior to the polypyrrole deposition, so as to give a fresh surface of the polymer to grow. Although this step did not in fact aid the polypyrrole growth it is thought to increase electrode roughness and so was retained. The thickness of the polypyrrole film was estimated from the charge to be 20 nm. After polypyrrole deposition the electrode required a break-in period during which the glucose response climbed (Figure. 5), perhaps resulting from the chemical oxidation of the polymer by H_2O_2 *(39)*.

The effect of the polypyrrole film on the oxidation of 2 model interferences, acetaminophen and ascorbate was dramatic with the currents being reduced to <1% of their previous values. Furthermore the protective effect was reasonably durable since measurements of c. 250 injections, made over a period of some 17 hours showed only a c. 4 fold increase in the currents due to these compounds. Tests have also been made in diluted complex yeast extract medium (1:100 dilution). An unmodified electrode displayed considerable fouling when tested with this solution, the current falling to c. 30% of the initial value within 6 hours, accompanied by a large increase in the

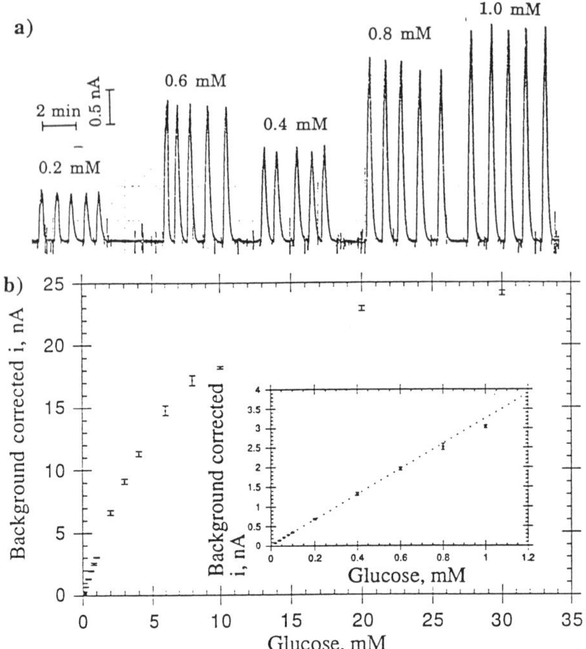

Figure. 3. Glucose response of a microcell transducer modified with electrochemically deposited GOD/ BSA recorded in FIA. The upper part, a), shows several injection peaks and the lower part, b), the whole response curve with an insert showing the linear region. The electrode was modified from a 5%/ 5% GOD/ BSA solution at $5 mAcm^{-2}$ for 2 min. The measurements were recorded at a flow rate of 1.0. $cm^3 min^{-1}$, with a 100µl injection volume.

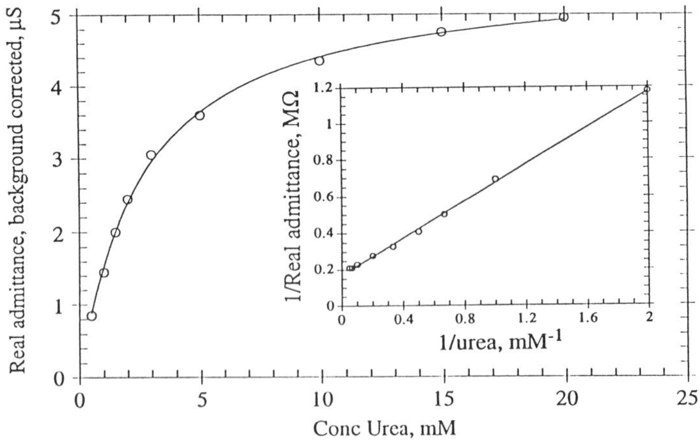

Figure. 4. Urea response of a urease modified electrode. The insert shows a double-reciprocal plot of the same data fitted to the line y=0.17+0.50x, (regression coefficient=0.9994, n=9).

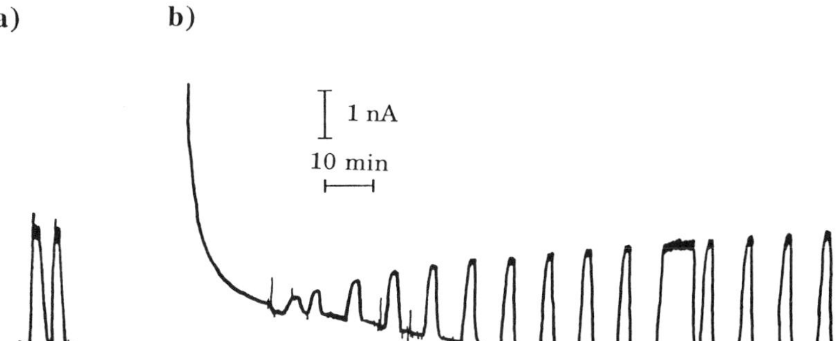

Figure. 5. Response of a microcell transducer modified with GOD/ BSA as described in Figure 3 to glucose. Section a) was recorded before, and b) after, the deposition of a polypyrrole anti-inteference layer, as described in the text. The measurements were made in PBS flow rate 0.5 $cm^3 min^{-1}$, with 1 cm^3 injections of 1.0 mM glucose.

response time. With the polymer film however, the electrode behaved much better, displaying a fall of only c. 20% during a 23 hour measurement period.

Conclusions

The above results demonstrate the use of electrochemical and photolithographic methods for the spatially controlled deposition of enzymatic and anti-interference layers in a controlled way upon the surface of miniature electrodes. Although the results concentrate upon glucose measurement, the example of urease indicates that the electrochemical method should be extendible to other enzymes and we are currently investigating this. We believe the measurements made using the electrodes with the thin anti-interference layer are particularly interesting and show that the modified electrodes can be used in complex matrixes such as complex yeast extract medium.

Entrapment of GOx in photopolymerized HEMA although limited at present by the pattern resolution, is another approach to the spatially control deposition of enzymatic layers. When performed at the on wafer level it allows strongly adhering membranes, displaying large glucose responses with linear ranges up to 12 mM to be obtained.

Acknowledgements

The authors would like to thank Ms. S. Pochon, Mr. G. Mondin and Mr J. van der Loo for their technical assistance. Ciba-Geigy for the loan of the FIA flow through cell and the kind donation of the complex yeast extract medium, and the Swiss National Science Foundation and the Committee for the Promotion of Applied Scientific Research for funding this work.

Literature Cited

1. Reynolds E.R., Geise R.J., Yacynych A.M., in *Biochemical and Chemical Sensors, Optimizing performance through polymeric materials*, Eds. Edelmean P.G., Wang J., American Chemical Society, Washington, DC, **1992,** 186-200.
2. Hintsche R., Möller B., Dransfeld I., Wollenberger U., Scheller F., Hoffmann B., *Sensors and Actuators*, **1991**, *B, 4*, 287-291.
3. Koudelka M., Gernet S., de Rooij N.F., *Senors and Actuators*, **1989**, *18*, 157-165.

4. Gernet S., Koudelka M., de Rooij N.F., *Sensors and Actuators*, **1989**, *17*, 537-540.
5. van den Berg A., Koudelka-Hep M., van der Schoot B.H., de Rooij N.F., Verney-Norberg E., Grisel A., *Analytica Chimica Acta*, **1992**, *269*, 75-82.
6. Sudhölter E.J.R., van der Wal P.D., Skowronska-Ptasinska M., van den Berg A., Bergveld P., Reinhoudt D.N., *Analytica Chimica Acta*, **1990**, *230*, 59-65.
7. Hinberg I., Kapoulas A., Korus R., O'Driscoll K., *Biotechnol. Bioeng.*, **1974**, *16*, 159-168.
8. Arica Y., Hasirci V.N., *Biomaterials*, **1987**, *8*, 489-495.
9. Guilbault G.G., Lubrano G.J., *Analytica Chimica Acta*, **1973**, *64*, 439-455.
10. Takatsu I., Moriizumi T., *Sensors and Actuators*, **1987**, *11*, 309-317.
11. Marty J-L., Mionetto N., Rouillon R., *Anal. Lett.*, **1992**, *25*, 1389-1398.
12. Foulds N.C., Lowe C.R., *J. Chem. Soc., Faraday Trans 1*, **1986**, *82*, 1259-1264.
13. Umana M., Waller J., *Anal. Chem.*, **1986**, *58*, 2979-2983.
14. Bartlett P.N., Whitaker R.G., *J. Electroanal. Chem.*, **1987**, *224*, 37-48.
15. Pandey P.C., *J. Chem. Soc., Faraday Trans.*, *1*, **1988**, *84(7)*, 2259-2265.
16. Sasso S.V., Pierce R.J., Walla R., Yacynych A.M., *Anal. Chem.*, **1990**, *62*, 1111-1117.
17. Malitesta C., Palmisano F., Torsi L., Zambonin P.G., *Anal. Chem.*, **1990**, *62*, 2735-2740.
18. Schalkhamer T., Mann-Buxbaum E., Urban G., Pittner F., *J. Chrom.*, **1990**, *810*, 355-366.
19. Centonze D., Guerrieri A., Malitesta C., Palmisano F., Zambonin P.G., *Fresenius J. Anal. Chem.*, **1992**, *342*, 729-733.
20. Bartlett P.N., Caruana D.J., *Analyst*, **1992**, *117*, 1287-1292.
21. Foulds N.C., Lowe C.R., *Anal. Chem.*, **1988**, *60*, 2473-2478.
22. Kajiya Y., Sugai H., Iwakura C., Yoneyama H., *Anal. Chem.*, **1991**, *63*, 49-54.
23. Bartlett P.N., Ali Z., Eastwick-Field V., *J. Chem. Soc., Faraday Trans.*, **1992**, *88*, 2677-2683.
24. Koopal C.G., Feiters M.C., Nolte R.J.M., de Ruiter B., Schasfoort R.B.M., *Biosensors and Bioelectronics*, **1992**, *7*, 461-471.
25. Cooper J.C., Hall E.A.H., *Biosensors and Bioelectronics*, **1992**, *7*, 473-485.
26. Suaud-Chagny M.F., Gonon F.G., *Anal. Chem.*, **1986**, *58*, 412.
27. Gunasingham H., Tan C.-B., *Electroanalysis*, **1989**, *1*, 223-227.
28. Ikariyama Y., Yamauchi S., Yukiashi T., Ushioda H., *J. Electrochem. Soc.*, **1989**, *136*, 702-706.
29. Johnson K.W., *Sensors and Actuators B*, **1991**, *5*, 85-89.
30. Wang J., Angnes L., *Anal. Chem.*, **1992**, *64*, 456-459.
31. Koudelka-Hep M., Strike D.J., de Rooij N.F., *Analytica Chimica Acta*, **1993**, in press.
32. Gernet S., Koudelka-Hep M., de Rooij N.F., *Sensors and Actuators*, **1989**, *18*, 59-70.
33. Strike D.J., de Rooij N.F., Koudelka-Hep M., *Sensors and Actuators*, **1993**, *13*, 61-64.
34. Yacynych A.M., *in Advances in Biosenors, A research annual, ed Turner A.P.F.*, **1992**, *2*, 1-52.
35. Watson L.D., Maynard P., Cullen D.C., Sethi R.S., Brettle J., Lowe C.R., *Biosensors* **1987/88**, *3*, 101-115.
36. Yon Hin B.F.Y., Sethi R.S., Lowe C.R., *Sensors and Actuators*, **1990**, *B*, *1*, 550-554.
37. Hämmerle M., Schuhmann W., Schmidt H.-L., *Sensors and Actuators B*, **1992**, *6*, 106-112.
38. Asavapiriyanont S., Chandler G.K., Gunawardena G.A., Pletcher D., *J. Electroanal. Chem.*, **1984**, *177*, 229-244.
39. Bélanger D., Nadreau J., Fortier G., *J. Electronal. Chem.*, **1989**, *274*, 143-155.

RECEIVED November 1, 1993

Chapter 24

Electrical Communication between Glucose Oxidase and Electrodes Based on Poly(vinylimidazole) Complex of Bis(2,2′-bipyridine)-*N,N*′-dichloroosmium

Ravi Rajagopalan, Timothy J. Ohara, and Adam Heller

Department of Chemical Engineering, The University of Texas at Austin, Austin, TX 78712–1062

Glucose oxidase (GOX) was electrically 'wired' to glassy carbon electrodes by redox hydrogels based on poly (N-vinyl imidazole) (PVI) complexed with $Os(bpy)_2Cl_2$ (PVI-Os). The hydrogels were formed by crosslinking PVI-Os and glucose oxidase with poly(ethylene glycol) diglycidyl ether (PEGDGE). Glucose was electrooxidized on the PVI-Os based 'wired' enzyme electrodes at 350 mV vs SCE. The dependence of the electrooxidation current on pH, ionic strength, film thickness, weight fraction of the enzyme in the redox hydrogel and oxygen partial pressure for electrodes rotating at 1000 rpm is described.

Polymeric redox mediators have been used for the transport of electrons between active sites of redox enzymes and electrodes (*1-11*). In complexes between high molecular weight redox polymers and redox enzymes, electrons are transferred efficiently from the substrate reduced redox active site of the enzyme to redox centers of the polymers (*7,12*)

Our research focuses on water soluble redox polymers that are crosslinked on electrode surfaces to form 3-dimensional redox hydrogel, to the polymer skeletons of which enzymes are covalently bound and in which segments of the polymer may also form complexes with the enzymes. Redox polymers based on poly (vinyl pyiridine) complexed with Os $(bpy)_2Cl^{+/2+}$ (bpy = 2,2' bipyridine) have been used to 'wire' i.e. to electrochemically connect enzymes to electrodes. By quarternizing part of the free pyridine rings on the polymer with bromoethylamine we formed earlier a water soluble, PEGDGE crosslinked redox polyamine, POsEA (*4,5*).

Electrooxidation of ascorbate, urate, acetaminophen and other readily electrooxidizable substrates often interferes with assays of analytes such as glucose. The problem of the resulting poor specificity for glucose can be alleviated by operating the electrodes at less oxidizing potentials, where the rate of electrooxidation of interferents can be slower. Alternatively, membranes can placed on electrodes to exclude anionic interferents or large interferent molecules. An example of such a membrane is Nafion (*13*). Recently we showed that interference by ascorbate, urate and acetaminophen can be altogether eliminated by their oxidation in a crosslinked overlayer of horseradish peroxidase by in situ generated or externally added H_2O_2 (*14*).

0097–6156/94/0556–0307$08.00/0

A goal of our research in glucose sensors is to develop a sensor that measures glucose levels accurately in whole blood samples and when implanted in the human body. Specific research objectives include improvement of the selectivity towards glucose, enhancement of the stability, enhanced current outputs, biocompatibility and insensitivity towards oxygen partial pressure. The last four issues are addressed elsewhere (*15,16*). The redox potentials of the PVI-Os hydrogels that are the subject of this paper are more reducing than the redox potential of POsEA, and their use may alleviate interference by electrooxidizable species.

Experimental Section

Chemicals. 2-Bromoethylamine hydrobromide (Aldrich), poly (ethylene glycol) diglycidyl ether (Polysciences, PEG400), K_2OsCl_6 (Johnson Matthey) and N-vinyl imidazole (VI) (Aldrich) were used as received. Glucose oxidase (D-glucose : oxygen reductase, EC 1.11.1.7) from Aspergillus niger was purchased from Sigma.

Synthesis of PVI. PVI was synthesized from N-vinyl imidazole (VI) by the method of Chapiro and Mankowski (*17*) A 7M aqueous solution of VI was purged with nitrogen for 30 min, and the bottle sealed. The solution was subjected to 2.5 MRad gamma radiation from a Co^{60} source. The polymer was precipitated from acetone, redissolved in methanol and reprecipitated from ethyl ether. The filtered precipitate was a pale yellow solid. Its molecular weight was determined by HPLC analysis using a Synchrom Catsec 300 column with 0.1% trifluoroacetic acid and 0.2 M NaCl as the eluent to be about 40kDa (M_n). The flow rate in the molecular weight determination was 0.4ml/min and poly (2-vinylpyridine) was used as the standard.

Redox Polymers PVI-Os, PVI-Os-NH_2. Osmium derivatized polymers were prepared by a method similar to that of Forster and Vos (*18*). Os $(bpy)_2Cl_2$ (0.456g, 0.8mmoles) and poly (N-vinyl imidazole) (0.380g, 4.0mmoles of imidazole groups) were refluxed in 100ml of ethanol under a nitrogen atmosphere for 7 days . The polymer was precipitated in ether, and dried at 70°C for 24 hr. This polymer was designated as PVI-Os. (Figure 1)

PVI-Os was quarternized with ethylamine functions by a method similar to that of Gregg and Heller (*5*). PVI-Os (500mg) and bromoethylamine hydrobromide (0.88g, 4.3mmoles) were reacted in an 18mL ethylene glycol : 30 mL DMF mixture at 60° C for 24 hr. After the solvent was evaporated, the polymer was dissolved in methanol and precipitated by adding it dropwise to a rapidly stirred solution of ethylether. It was then dissolved in water and stirred with 10 g of ion exchange (BioRad AG 1X-4, chloride form) beads for 12 h. This polymer is designated as PVI-Os-NH_2. (Figure 1)

Electrodes. The electrodes were glassy carbon discs, 3mm in diameter (V-10 grade vitreous carbon from Atomergic). Each glassy carbon rod was encased in a teflon cylinder with a deaerated slow setting epoxy (Armstrong) and the cylinder was fitted on an AFMSRX rotator (Pine Instruments).

All electrodes were prepared by polishing successively with four grades of alumina (20, 5, 1, 0.3 micron) with sonication and thorough washing between the polishing steps. Background scans were taken at 100mV/s in STD buffer to ensure that the voltammograms were featureless. The STD buffer is a solution of 0.15 M NaCl buffered with phosphate (33mM, pH 7.4) and was the buffer used in all electrochemical experiments. The electrodes were then washed and stored in a dessicator until use.

The electrodes were modified by syringing an appropriate volume of a 5 mg/mL solution of the polymer onto the electrode surface ($0.071 cm^2$). Then an aliquot of 5 mg/mL solution of glucose oxidase (10mM HEPES, pH 8.1) was added to the electrode, followed by an appropriate volume of a 1mg/mL solution of PEGDGE. The solutions were then mixed on the surface of the electrode and the resulting film allowed to cure for at least 24 hours under vaccum. Figure 2 shows the reactions of the crosslinker, PEGDGE with amines and with imidazole functions (19). The nitrogen of the imidazole group, the amine functions in the case of PVI-Os-NH_2 and amine functions on the lysine groups on the enzyme participate in the crosslinking.

Electrochemistry. The electrochemical experiments were performed with a Princeton Applied Research 173 potentiostat, a 175 PAR universal programmer equipped with a model 179 digital coulometer. Signals were recorded on an X-Y-Y' Kipp and Zonen recorder. The water jacketed cell used was thermostatted at 21.4^oC. The cell had an aqueous saturated calomel electrode as reference (SCE) and all potentials are reported with respect to this electrode. A platinum wire encased in a heat shrinkable sleeve with a frit was used as the counter electrode. Unless otherwise noted, the steady state current was monitored with the electrodes poised at 0.350 V, where the current no longer varied with potential. The electrodes were rotated at 1000 rpm.

Results and Discussion

Cyclic Voltametry of PVI-Os-NH_2. The cyclic voltammograms of electrodes coated with PEGDGE crosslinked PVI-Os-NH_2 films had the classical symmetrical shape characteristic of reversible oxidation and reduction of a surface bound species. Their half-wave potential was 190mV vs. SCE. This potential varied by about +/- 10mV with the amount of crosslinker in the film which ranged from 3wt% to 21wt%. At 12wt% PEGDGE the redox potential was 187mV and the peak separation was 10mV.

Glucose Response of PVI-Os and PVI-Os-NH_2 Glucose Electrodes. Figure 3 shows the variation of the current density with glucose concentration of PEGDGE crosslinked PVI-Os and PVI-Os-NH_2 glucose electrodes. The electrodes were prepared by immobilizing 7µg redox polymer and 3µg GOX with 0.6µg PEGDGE. Evidently, both PVI-Os or PVI-Os-NH_2 effectively 'wire' glucose oxidase.

Oxygen Competition. The steady state glucose response curves were measured in argon saturated and in oxygen saturated solutions for PVI-Os-NH_2. (30 wt% GOX, 6 wt% PEGDGE) (Figure 4a). Figure 4b shows the fraction of the retained current in the O_2 saturated solution with the electrode rotating at 1000 rpm, to emphasize the oxygen effect which would be very small in a stagnant solution, where the reaction of $FADH_2$ centers with O_2 would be severely mass transport limited because of the low solubility of O_2 in water.

At low and moderate glucose concentrations, the electrode current decreased substantially in O_2 saturated solutions because O_2 competes for the $FADH_2$ electrons with the Os^{III} centers (reactions (2) and (3)) (*5*) :

$$\text{GOX-FAD} + \text{glucose} \text{-------->} \text{GOX-FADH}_2 + \text{gluconolactone} \quad (1)$$
$$\text{GOX-FADH}_2 + 2\text{Os}^{III} \text{-------->} \text{GOX-FAD} + 2\text{Os}^{II} + 2\text{H}^+ \quad (2)$$
$$\text{GOX-FADH}_2 + \text{O}_2 \text{-------->} \text{GOX-FAD} + \text{H}_2\text{O}_2 \quad (3)$$

The loss in current under oxygen was higher at lower glucose concentrations. Apparently at high glucose concentrations the glucose flux maintains the enzyme, even in the presence of O_2, in the reduced form, which transfers electrons via the redox polymer to the electrode.

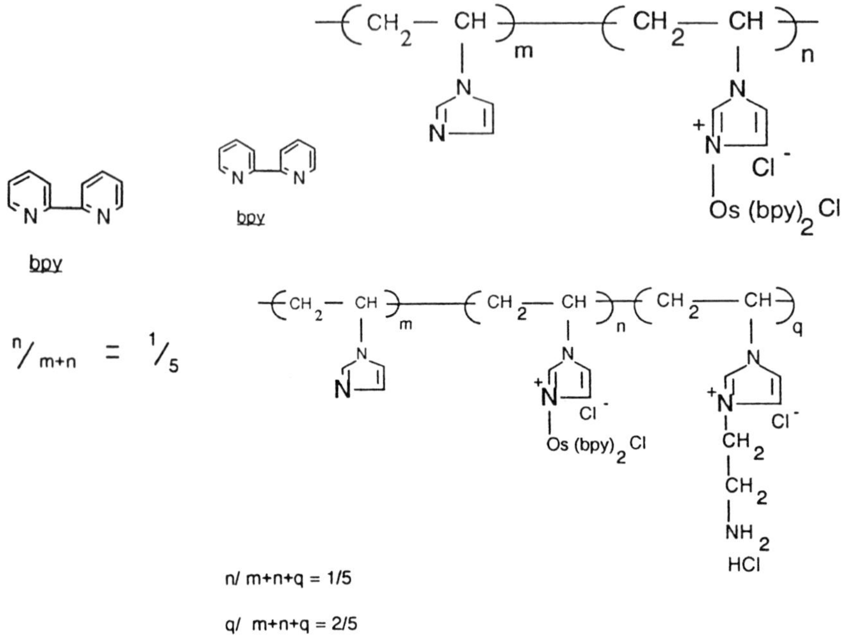

Figure 1. Structures of PVI-Os (top) and PVI-Os-NH_2 (bottom).

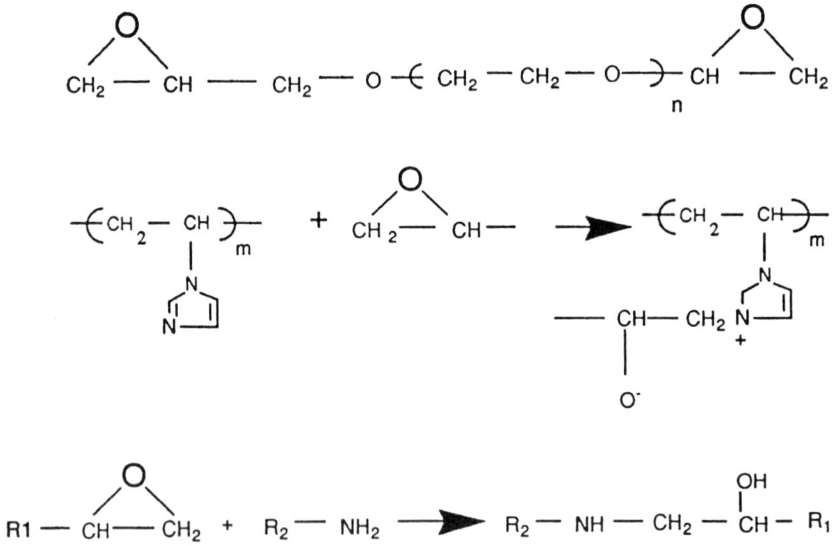

Figure 2. Structure of PEGDGE (top), its reaction with imidazole (center) and with an amine (bottom)

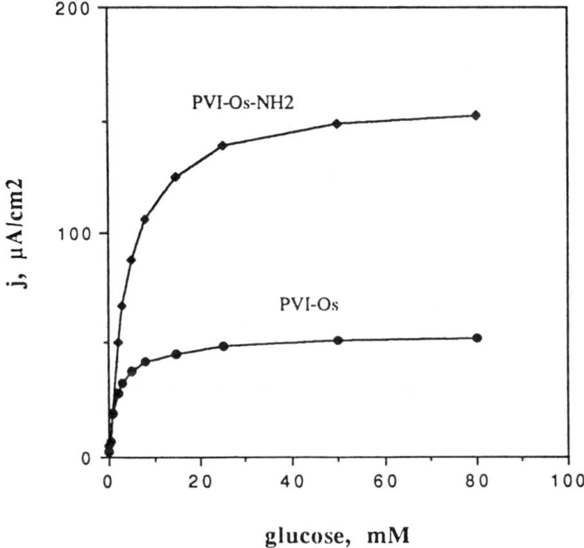

Figure 3. Steady state glucose electrooxidation current densities for PVI-Os and PVI-Os-NH_2. The films contain 30 wt % GOX and 6 wt.% PEGDGE. Argon atmosphere, pH 7.4, 1000 rpm, 0.35V (SCE). The total polymer + enzyme loading is 140 $\mu g/cm^2$.

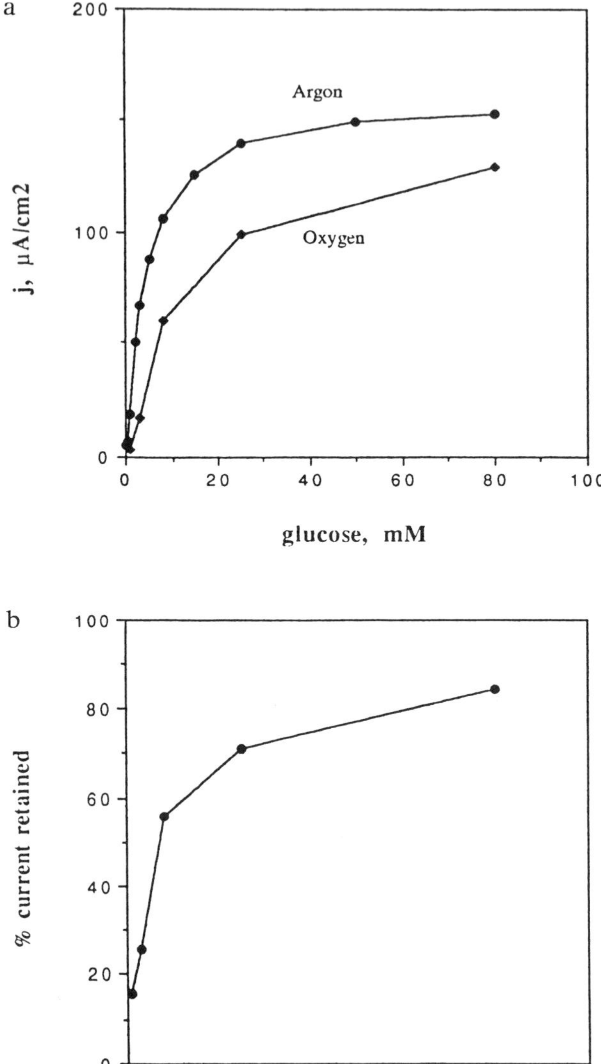

Figure 4. a) Steady state electrooxidation current densities for PVI-Os-NH_2 wired GOX electrodes. Electrode conditions as in Figure 3.
b) Glucose concentration dependence of the fraction of the current retained under O_2.

pH Dependence. Figure 5 shows the pH dependence of the currents of an electrode under argon at 100 mM glucose. (Henceforth, currents at 100mM glucose will be referred to as j_{max}. In Figure 5, the current is normalized with respect to the current at pH 9.0 where it reaches a maximum). The current varies only within +/- 10% through the 7.3 to 10.0 pH range. Below pH 7.3, the current decreases rapidly, and is practically nil at pH 3.4. POsEA based glucose electrodes show a similar dependence on pH (*6*).

The driving potential difference for electron transfer reaction from the $FADH_2$ centers of the enzymes to the redox polymer increases with pH, where the enzyme's redox potential becomes more reducing, while the redox polymer potential does not change with pH (*6*). The rate constant for electron transfer between the active center of GOX and ferrocene based diffusing mediators also decreases with decreasing pH (*20*) A second cause for the lesser current at low pH is the absence of electrostatic attraction leading to complex formation between PVI-Os-NH_2 and GOX. The pI of GOX is ca. 4.0. The polycationic polymer complexes the enzyme that is a polyanion at pH 7.0. At pH 3.0, the enzyme, like the redox polymer, is a polycation and the two repel each other.

Ionic Strength Dependence. Figure 6 shows the dependence of the current at 100mM glucose on the ionic strength. The currents shown are normalized with respect to the current at 150 mM NaCl. The current decreases as the ionic strength is increased, dropping to almost nil at 1M NaCl concentration.

The cause of this drop is probably due to the coiling of the redox polymer segments at high ionic strength. At low ionic strength the segments are stretched by repulsive forces between the cationic sites. When the positive charges are screened by counterions, i.e. anions, the segments assume an entropically favoured coiled configuration, i.e. they ball up. The balled up segments no longer fold along the surface contours of the enzyme and the redox centers of the enzyme and the polymer are too far apart for efficient electron transfer between the two (*1*)

Thickness Dependence. The increase in current density upon increasing the surface coverage by Os centers, i.e. polymer film thickness, is shown in Figure 7. The surface coverage by osmium centers was varied between 1.77 x 10^{-9} to 1.855 x 10^{-7} mol/cm^2 by varying the amount of total material applied to the electrode surface. The surface coverage was determined by integration of the voltammogram, i.e. of the current from the potential where electrooxidation starts, through the potential where it is complete. The films in these experiments contained 38wt % GOX and 10 wt % PEGDGE.

The current density increased approximately linearly with increasing thickness even for the highest thickness considered here, even though the rate of increase was less at high thickness. We conclude that the electroactive portion of the film, the reaction layer, extends through the entire film thickness. Again, the result is similar to that observed for POsEA (*5*). This suggests that the diffusion coefficient for electron transfer in the redox polymer film is large enough for electrons from polymer redox centers near active sites of enzymes to be transported efficiently to the electrode, even when the enzyme molecules are near the solution side of the electrodes.

Enzyme Loading Dependence.

Constant Polymer Loading. Glucose response curves were measured for a series of electrodes where the enzyme loading was varied from 20% to 65%. The polymer loading was kept constant at 70µg/cm^2 and 10 wt% of PEGDGE was used to crosslink the films (Figure 8). Except for one data point, at 65 wt% enzyme loading, the data points at different enzyme loadings show that j_{max} increases initially with increasing enzyme loading, peaks around 40 wt% GOX and then decreases upon further GOX loading. The trend is similar to that observed for POsEA (*5*).

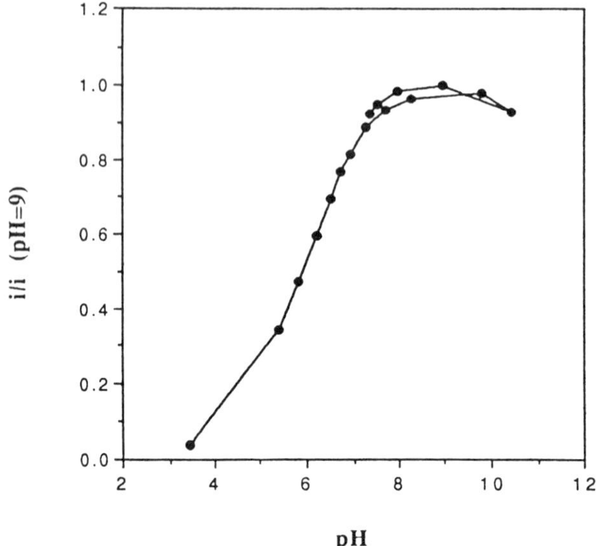

Figure 5. Dependence of the steady state current on pH, normalized with respect to the current at pH 9.0. Electrode conditions as in Figure 3.

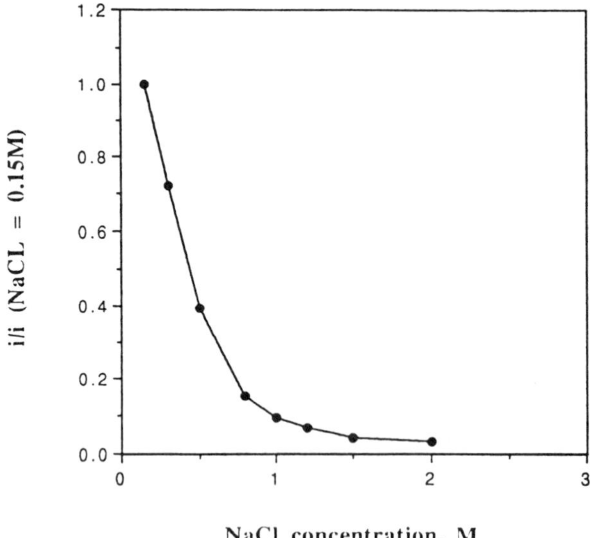

Figure 6. Dependence of the steady state current on ionic strength. The currents are normalized with respect to the current at 0.15M NaCl. Electrode conditions as in Figure 3.

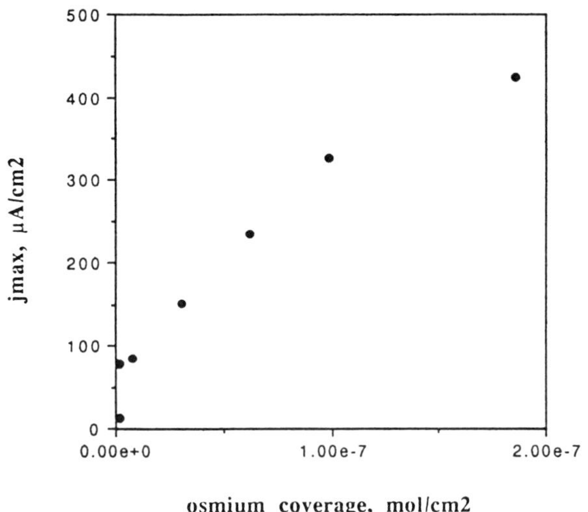

Figure 7. Dependence of the steady state current density on the thickness of the polymer films.Electrodes with 38wt. % GOX and 10 wt. % PEGDGE. [Glucose] = 100 mM; Electrode conditions as in Figure 3.

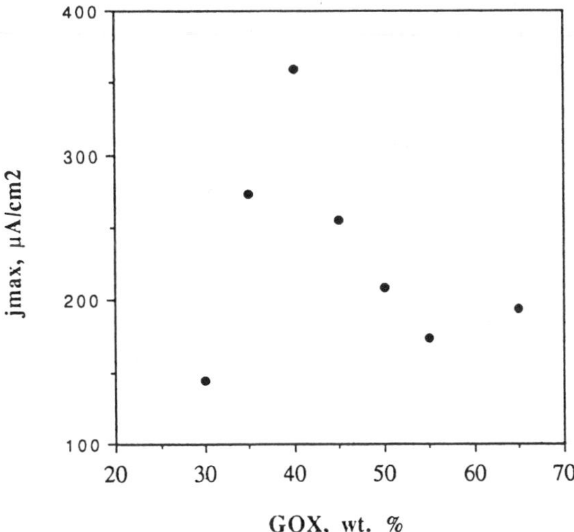

Figure 8. Dependence of the steady state current density on the GOX loading. The PVI-Os-NH_2 loading was held constant at 70 µg/cm^2. [Glucose] = 100 mM; Electrode conditions as in Figure 3.

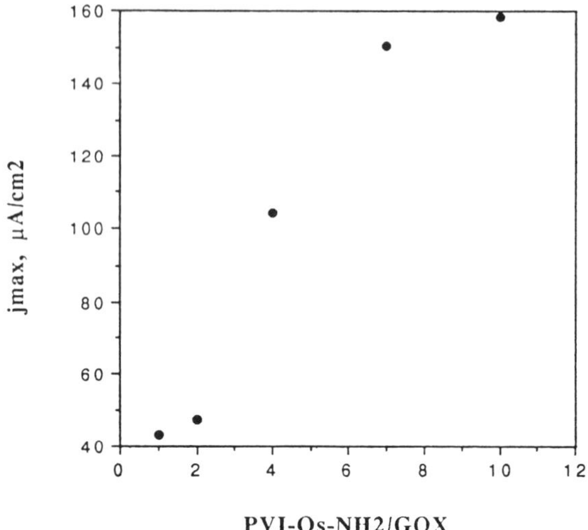

Figure 9. Dependence of the steady state current density on the redox polymer loading. The GOX loading was held constant at 14.2 $\mu g/cm^2$. [Glucose] = 100mM. Electrode conditions as in Figure 3.

IDA measurements show that in crosslinked POsEA based hydrogels with glucose oxidase, D_e decreases upon increasing the enzyme loading (*21*). It is conceivable that at enzyme loadings greater than 40%, the diffusion of electrons through the redox polymer becomes rate limiting. It is also possible that as the enzyme content is increased beyond 40wt% not all the enzyme is in electrical contact with the redox polymer, i.e. not all the enzyme is 'wired'. A third possibility is that the decline in current when an excess of enzyme is loaded into the polymer derives from loss of material that is rotating at 1000 rpm. The enzyme is not as readily crosslinked as the polymer by PEGDGE, and when the enzyme loading increases, more material may dissolve from the electrode surface.

Constant Enzyme Loading. Glucose response curves were measured for a series of electrodes where the enzyme loading was kept constant at 14.7$\mu g/cm^2$ and the polymer to enzyme weight ratio was increased from 1 to 10 (Figure 9). The purpose of this experiment was to determine the maximum current that can be obtained from a fixed amount of enzyme on the electrode since, at high polymer loading, we are presumably 'wiring' the enzyme with the maximum efficiency.

In this experiment j_{max} exhibited sigmoidal behaviour, increasing upon addition of polymer at low P/E ratios, and then becoming constant at P/E values near 10. If one assumes that the current is proportional to the rate of electron transfer between the enzyme and the polymer, i.e. that the electrodes are kinetically limited by the rate of electron transfer, then $j_{max} \propto C_{enz} * C_{Os}$. For a fixed amount of GOX on the electrode, increasing the amount of polymer increases the product, but once all the enzyme is electrically 'wired', the product can no longer increase. It should also be noted that for 14.7mg/cm^2 of GOX on the electrode surface, if all the enzyme were active and 'wired' by the redox polymer, the current output would approach 4mA/cm^2. We observe, however, a maximum current density of 200$\mu A/cm^2$, about 5% of the calculated valuc. Rcasons for the lower than theoretical current may include inactive enzyme, loss of enzyme from the electrode and the presence of non-'wired' enzyme molecules.

Conclusions

The PVI based complex of $[Os(bpy)_2Cl]^{+/2+}$ can be used as a molecular 'wire' to electrochemically connect glucose oxidase to electrodes. PVI-Os-NH_2 based glucose electrodes rotating at 1000 rpm exhibit characteristics similar to glucose electrodes based on POsEA. The glucose response of these electrodes is a function of pH, ionic strength, thickness, enzyme loading and O_2 partial pressure. The dependence of current on these factors mirrors that of POsEA, but glucose is electrooxidized at a potential 100mV more reducing than in electrodes based on POsEA.

Acknowledgments Support for the work by the Office of Naval Research, The Welch Foundation and the National Science Foundation is gratefully acknowledged. One of the authors (RR) thanks Dr. Yair Haruvy for help in synthesizing PVI, Lois Davidson for help in determination of the molecular weight of PVI, and in particular, Ioanis Katakis, for very helpful discussions and advice throughout the course of this work.

References

1) Degani, Y.; Heller, A. *J. Am. Chem. Soc.* 1989, 111, 2357-2358.
2) Pishko, M.V.; Katakis, I.; Lindquist, S.-L.; Ye,L.; Gregg, B.A.; Heller, A. *Angew. Chem., Int. Ed. Engl.* 1990, 29, 82-84.
3) Gregg, B.A.; Heller, A. *Anal. Chem.* 1990, 62, 258-263.
4) Gregg, B.A.; Heller, A. *J. Phys. Chem.* 1991, 95, 5970-5975.
5) Gregg, B.A.; Heller, A. *J. Phys. Chem.* 1991, 95, 5976-5980.
6) Katakis,I.; Heller, A. *Anal. Chem.* 1992, 64, 1008-1013.
7) Katakis, I.; Vreeke, M.; Ye, L.; Aoki, A.; Heller, A. in *Proceedings of the Mosbach Symposium* Petersson, B. Ed.; JAE Press, 1993.
8) Hale, P.D.; Inagaki, T.; Okamoto, Y.; Skotheim, T.A. *J. Am. Chem. Soc.* 1989, 111, 3482-3483.
9) Hale, P.D.; Boguslavski, L.I.; Inagaki, T.; Lee, H.S.; Skotheim,T.A. ; Karan, H.I.; Okamoto, Y. *Mol. Cryst. Liq. Cryst.* 1990, 190, 251-258.
10) Hale, P.D.; Boguslavski, L.I.; Inagaki, T.; Karan, H.I.; Lee, H.S.; Skotheim,T.A. ; Okamoto, Y. *Anal. Chem.* 1991, 63, 677-682.
11) Foulds, N.C.; Lowe, C.R. *Anal. Chem.* 1988, 60, 2473-2478.
12) Katakis, I.; Davidson, L.; Heller, A. *Unpublished results.*
13) Bindra, D.S.; Wilson, G.S. *Anal. Chem.* 1989, 61, 2566.
14) Maidan, R.; Heller, A. *Anal. Chem..* 1992, 64, 2889-2896.
15) Ye, L.; Hammerle, M.; Shumann, W; Schmidt, H.-L.; Duine, J.A.; Heller, A. *Anal. Chem.* 1992, 65, 238-241.
16) Ye, L.; Katakis, I.; Olsthoorn, A.J.J.; Shumann, W.; Schmidt, H.-L.; Duine, J.A.; Heller, A. *Submitted.*
17) Chapiro, A.; Mankowski, Z. *Eur. Polym. J.* 1988, 2424, 1019-1028.
18) Forster, R.J.; Vos, J.G. *Macromolecules* 1990, 23, 4372-4377.
19) Potter, W.G. *Epoxide Resins* The Plastics Institute, London, 1958, p 83
20) Bourdillon, C.; Demaille, C.; Moiroux, J.; Saveant, J.M. *J. Am. Chem. Soc.* 1993, 115, 2-10.
21) Aoki A. *Private comm..*

RECEIVED February 8, 1994

Author Index

Affiliation Index

Subject Index

C

D

Q

R

S

Production: Meg Marshall
Indexing: Deborah H. Steiner
Acquisition: Rhonda Bitterli
Cover design: Michele Telschow

Printed and bound by Maple Press, York, PA